O. I. Bogoyavlensky
Methods in the Qualitative Theory of Dynamical Systems
in Astrophysics and Gas Dynamics

Oleg I. Bogoyavlensky

Methods in the Qualitative Theory of Dynamical Systems in Astrophysics and Gas Dynamics

Translated from the Russian
by Dmitry Gokhman

With 40 Figures

Springer-Verlag
Berlin Heidelberg New York Tokyo

Oleg I. Bogoyavlensky
Steklov Mathematical Institute
Department of Mathematical Physics
ul. Vavilova 42, 117966 Moscow, USSR

Dmitry Gokhman
6425 Telegraph Avenue, Apt. 7
Oakland, CA 94609, USA

Title of the Russian original edition:
*Metody kachestvennoi teorii dinamicheskikh sistem
v astrofizike i gazovoi dinamike*
Publisher Nauka, Moscow 1980

This volume is part of the *Springer Series in Soviet Mathematics*
Advisers: L. D. Faddeev (Leningrad), R. V. Gamkrelidze (Moscow)

Mathematics Subject Classification (1980):
34C, 76L05, 76N, 83C, 85

ISBN-13: 978-3-642-64902-8   e-ISBN-13: 978-3-642-61661-7
DOI: 10.1007/978-3-642-61661-7

Library of Congress Cataloging in Publication Data
Bogoíavlenskiĭ, O. I. (Oleg Igorevich)
Methods of qualitative theory of dynamical systems in astrophysics and gas dynamics.
(Springer series in Soviet mathematics)
Translation of: Metody kachestvennoĭ teorii dinamicheskikh sistem v astrofizike i gazovoĭ dinamike.
Bibliography: p. Includes index.
1. Astrophysics. 2. Gas dynamics. 3. Cosmology.
I. Title. II. Series.
QB461.B5713 1985   523.01   84-26793

Typesetting: Asco Trade Typesetting Ltd., Hong Kong
2141/3140-543210

Preface

Homogeneous cosmological models, self-similar motion of self-gravitating gas and motion of gas with homogeneous deformation have important applications in the theory of evolution of the universe. In particular they can be applied to the theory of explosions of stars, formation of galaxies, pulsation of alternating stars etc. The equations of general relativity and Newtonian gas dynamics in the cases mentioned above are reduced to systems of a finite (but quite large) number of ordinary differential equations. In the last two decades these multi-dimensional dynamical systems were and still are being analyzed by means of traditional analytic and numerical methods. Important dynamical modes of some solutions were thus established. These include oscillatory modes of the space-time metric near a cosmological singularity, self-similar motion of self-gravitating gas with a shock wave and an expanding cavity inside (as in an explosion of a star), collapse of an ellipsoid of self-gravitating dust into a disc and others. However the multi-dimensional dynamical systems in question are so complex, that a complete analysis of all dynamical modes of the solutions by means of well-known traditional analytic methods does not seem feasible. Therefore the development of effective methods of qualitative analysis of multi-dimensional dynamical systems and their application to the problems of astrophysics and gas dynamics previously unsolved by traditional methods becomes especially urgent.

Here we present a detailed study of homogeneous cosmological models, self-similar motion of self-gravitating gas and motion of gas with homogeneous deformation based on the methods of qualitative theory of multi-dimensional dynamical systems.

Methods used here include maximally non-degenerate compactification, resolution of degenerate critical points and separatrix approximation of the trajectories of a dynamical system and are described in chapter I. They are a generalization and a modification in specific multi-dimensional problems of the classical methods of qualitative theory of two-dimensional dynamical systems created by Poincaré and Bendixon in the beginning of the twentieth century.

The method of maximally non-degenerate compactification of a dynamical system uses the resolution of degenerate critical points. It allows the study of the behavior of solutions in the limiting cases of the parameters (for example for large values of the energy or certain phase coordinates) as well as in neighborhoods of various singularities.

The method of separatrix approximation of the trajectories of a dynamical system allows a detailed study of complex non-linear modes of the dynamics of solutions. This method provides for more complete and precise results than the traditional method of fitting the approximate solutions obtained from consecu-

Chapter I
Methods of Qualitative Analysis of
Multi-Dimensional Dynamical Systems

The fundamentals of the qualitative theory of dynamical systems were laid down by the classical works of Poincaré [1] and Bendixon [2] completed in the beginning of the twentieth century in connection with the problems of celestial mechanics. Particularly in these works the qualitative theory of two-dimensional dynamical systems was created in its modern version. Subsequently beginning with the works of A.A. Andronov [3–6] it found a very important application in electrical engineering and the theory of oscillations. The two-dimensional qualitative theory has an important application in the problems of gas dynamics as well, mainly in the study of self-similar solutions [7]. Method of construction of solutions of dynamical systems in the form of convergent power series was developed in numerous papers starting with [1, 8]. Methods of finding the asymptotics of solutions in neighborhoods of degenerate critical points by means of several "shortened" systems and Newton's polygon were developed in [9]. Special classes of dynamical systems having the property of coarseness or structural stability were actively studied in the last decades in numerous mathematical works [10, 11].

In this chapter we describe methods of qualitative analysis of multidimensional dynamical systems [12–32]. These include the method of maximally non-degenerate compactification of a dynamical system, the method of resolution of degenerate critical points and the method of separatrix approximation of the trajectories of a dynamical system. The above methods are applied further in the book to the study of specific problems of astrophysics and gas dynamics.

1. Prerequisites from the Qualitative Theory of Two-Dimensional Dynamical Systems

The qualitative theory of autonomous dynamical systems in the plane is used to study the dynamics of trajectories of a system of two differential equations:

$$dx/dt = \dot{x} = P(x, y) \quad dy/dt = \dot{y} = Q(x, y). \tag{1.1}$$

In particular it is used to study the asymptotic behavior of the trajectories for $t \to \pm \infty$ and the subdivision of the (x, y)-plane into regions where the behavior of the system (1.1) is qualitatively homogeneous. The qualitative study of the dynamical system (1.1) is based on the study of its critical points (equilibrium or stationary points) (x_0, y_0), where $P(x_0, y_0) = Q(x_0, y_0) = 0$.

I. Classification of Non-Degenerate Critical Points [33–36]. The main characteristics of a critical point (x_0, y_0) are its eigenvalues, i.e. the eigenvalues of the system (1.1) at this point. They are defined as the roots of the following characteristic polynomial:

$$\det \begin{vmatrix} P_x(x_0, y_0) - \lambda & P_y(x_0, y_0) \\ Q_x(x_0, y_0) & Q_y(x_0, y_0) - \lambda \end{vmatrix} = \lambda^2 - \sigma\lambda + \Delta = 0. \tag{1.2}$$

Here the subscripts signify differentiation with respect to the given variable. From now on we assume that all functions are infinitely differentiable. The eigenvalues λ_1, λ_2 of (x_0, y_0) are invariant under changes of coordinates x, y, which are regular at (x_0, y_0). A critical point (x_0, y_0) is called *non-degenerate*[1] if $\mathrm{Re}\,\lambda_1 \neq 0$ and $\mathrm{Re}\,\lambda_2 \neq 0$.

Trajectories of the system (1.1) can approach a critical point (x_0, y_0) for $t \to +\infty$ or $t \to -\infty$ along certain directions $k = (k_x, k_y)$. These are the eigenvectors of the characteristic matrix of the system (1.1) evaluated at (x_0, y_0) (see (1.2)). If $\lambda_1 \neq \lambda_2$, then there are two eigenvectors with slopes $k = k_y/k_x$. They can be expressed in terms of the eigenvalues by

$$k_1 = (\lambda_1 - P_x(x_0, y_0))/P_y(x_0, y_0)$$
$$k_2 = (\lambda_2 - P_x(x_0, y_0))/P_y(x_0, y_0). \tag{1.3}$$

The behavior of the trajectories of the system (1.1) in some small neighborhood of (x_0, y_0) is qualitatively equivalent to the behavior of the trajectories of its linear part:

$$\dot{x} = P_x(x_0, y_0) \cdot (x - x_0) + P_y(x_0, y_0) \cdot (y - y_0)$$
$$\dot{y} = Q_x(x_0, y_0) \cdot (x - x_0) + Q_y(x_0, y_0) \cdot (y - y_0). \tag{1.4}$$

The linear system (1.4) is easily integrated explicitly:

$$x - x_0 = \mathrm{Re}(C_1 e^{\lambda_1 t} + C_2 e^{\lambda_2 t})$$
$$y - y_0 = \mathrm{Re}(C_1 e^{\lambda_1 t} + C_2 e^{\lambda_2 t}). \tag{1.5}$$

These formulas allow us to obtain a graphic representation of the qualitative behavior of the system (1.1) in a neighborhood of (x_0, y_0) (a phase portrait).

By means of a linear change of coordinates, the system (1.1) in a neighborhood of a non-degenerate critical point (x_0, y_0) is reduced to one of the following canonical forms (the choice of the form depends on λ_1 and λ_2, (x_0, y_0) goes to $(0,0)$, and the functions $\varphi(u, v)$ and $\psi(u, v)$ are second order in u and v):

---

[1] *Translator's note: i.e. hyperbolic*

Case 1. If λ_1 and λ_2 are distinct real numbers, then we have

$$\dot{u} = \lambda_1 u + \varphi(u, v) \quad \dot{v} = \lambda_2 v + \psi(u, v). \tag{1.6}$$

Case 2. If $\lambda_1 = \lambda_2$, then we have

$$\dot{u} = \lambda u + \varphi(u, v) \quad \dot{v} = \lambda v + \mu u + \psi(u, v), \tag{1.7}$$

where $\lambda = \lambda_1 = \lambda_2$ and μ is a real parameter, which is either 0 or 1.

Case 3. If $\lambda_1 = \bar{\lambda}_2$ (i.e. $\lambda_1 = \alpha + i\beta$, $\lambda_2 = \alpha - i\beta$, $\alpha \neq 0$, $\beta \neq 0$), then we have

$$\dot{u} = \alpha u - \beta v + \varphi(u, v) \quad v = \alpha u + \beta v + \psi(u, v). \tag{1.8}$$

Let us classify the non-degenerate critical points using these canonical forms (the corresponding phase portraits of the dynamical system are given in Fig. 1 in the coordinates x, y).

Case 1. *A node* $(\lambda_1$ and λ_2 are real and have the same sign).

a) *A non-degenerate node* $(\lambda_1 \neq \lambda_2)$. A node is called attracting (stable) if λ_1, $\lambda_2 < 0$. In this case all trajectories in a neighborhood of (x_0, y_0) approach (x_0, y_0) for $t \to \infty$. The eigenvector corresponding to the smallest (by absolute value) eigenvalue is tangent to all but two trajectories, whereas the other eigenvector is tangent to the two exceptional trajectories (see Fig. 1 (a)). A node is called repelling (unstable) if $\lambda_1, \lambda_2 > 0$. In this case the phase portrait is obtained from Fig. 1 (a) by reversing time t.

b) *A degenerate node* $(\lambda_1 = \lambda_2 = \lambda$, $\mu \neq 0$ in the canonical form (1.7)). For $\lambda < 0$ (attracting node) all trajectories in a neighborhood of (x_0, y_0) approach (x_0, y_0) for $t \to \infty$ along the only eigenvector (see Fig. 1 (b)). For $\lambda > 0$ the node is called repelling.

c) *A degenerate node* $(\lambda_1 = \lambda_2 = \lambda$, $\mu = 0)$. For $\lambda < 0$ (attracting node) all trajectories in a neighborhood of (x_0, y_0) approach (x_0, y_0) for $t \to \infty$ along arbitrary directions (see Fig. 1 (c)) in such a way that to each direction there corresponds only one trajectory. The case when $\lambda > 0$ (repelling node) is described analogously by reversing time.

2. *A saddle* $(\lambda_1, \lambda_2$ are real and have opposite signs: $\lambda_1 < 0$, $\lambda_2 > 0)$. The phase-portrait of the dynamical system in a neighborhood of a saddle is shown in Fig. 1 (d). A saddle is unstable even if time is reversed. Each of the four exceptional trajectories approaching the saddle for $t \to +\infty$ or $t \to -\infty$ (see Fig. 1 (d)) is called a *separatrix of the saddle*.

3. *A focus* $(\lambda_1 = \bar{\lambda}_2)$. If $\mathrm{Re}\,\lambda_i = \alpha < 0$, then the focus is called attracting: all trajectories in a neighborhood of this focus (x_0, y_0) are spirals winding down to (x_0, y_0) for $t \to \infty$, while going around (x_0, y_0) infinitely many times. The direction of rotation of the trajectories is determined by the sign of $\beta = \mathrm{Im}\,\lambda_1$ in the canonical form (1.8). The corresponding phase portraits are shown in Fig. 1 (e) $(\beta > 0)$ and (f) $(\beta < 0)$. If $\mathrm{Re}\,\lambda_i = \alpha > 0$, then the focus is repelling and is described analogously by reversing time.

If the eigenvalues of (x_0, y_0) are purely imaginary $(\lambda_1 = i\beta, \lambda_2 = -i\beta)$, such a point is called a *center* and the canonical form in a neighborhood of (x_0, y_0) is still

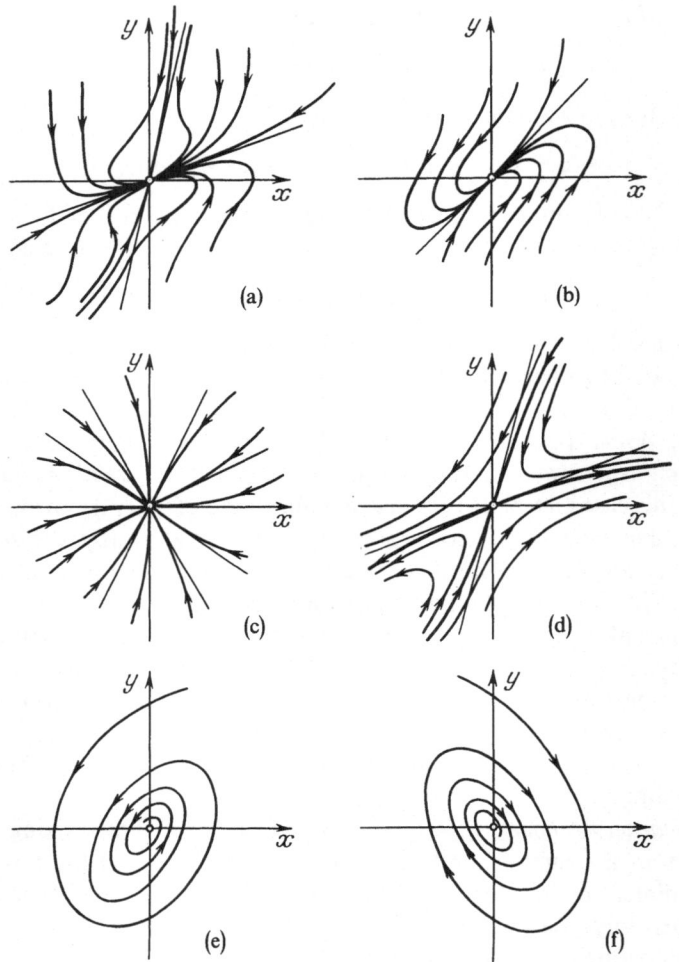

Fig. 1. Phase portraits of the dynamical system in neighborhoods of critical points: (a) a non-degenerate node, (b) a degenerate node $(\mu \neq 0)$, (c) a degenerate node $(\mu = 0)$, (d) a saddle, (e) a focus $(\beta > 0)$, (f) a focus $(\beta < 0)$

$(1.8) (\alpha = 0, \beta \neq 0)$. However the qualitative behavior of the dynamical system in a neighborhood of a center is largely dependent on the nature of functions $\varphi(u, v)$ and $\psi(u, v)$ (we can only assert that for some direction of time all trajectories initially close to (x_0, y_0) remain in a small neighborhood of (x_0, y_0) and rotate around it infinitely many times).

II. Analysis of Degenerate Critical Points. A simple classification of critical points with eigenvalues $\lambda_1 = 0$ and $\lambda_2 \neq 0$ is described in [34]. As in Bendixon [2], a polar system of coordinates is used to study degenerate critical points with eigenvalues $\lambda_1 = \lambda_2 = 0$. Suppose a degenerate critical point lies at the origin $(0,0)$. Suppose that the right hand sides of the dynamical system (1.1) are expanded starting with terms of power $m > 1$

$$P(x, y) = P_m(x, y) + \varphi(x, y)$$
$$Q(x, y) = Q_m(x, y) + \psi(x, y),$$
(1.9)

where $P_m(x, y)$ and $Q_m(x, y)$ are homogeneous polynomials of degree m and functions $\varphi(x, y)$ and $\psi(x, y)$ are power series containing terms of higher order. After a change of coordinates from cartesian to polar

$$r = (x^2 + y^2)^{1/2}$$
$$\varphi = \arctan(y/x)$$

and a time substitution

$$d\tau/dt = r^{m-1}$$

the dynamical system (1.1) takes the following form:

$$\dot{r} = r[\cos \varphi P_m(\cos \varphi, \sin \varphi) + \sin \varphi Q_m(\cos \varphi, \sin \varphi) + r\Phi_1]$$
$$\dot{\varphi} = \cos \varphi Q_m(\cos \varphi, \sin \varphi) - \sin \varphi P_m(\cos \varphi, \sin \varphi) + r\Phi_2,$$
(1.10)

where Φ_1 and Φ_2 are analytic functions of r and φ. Note that this change of coordinates is bijective everywhere except the origin $(0,0)$. In polar coordinates the origin corresponds to an entire circle

$$S^1: r = 0, 0 \leqslant \varphi \leqslant 2\pi.$$

Apparently the dynamical system (1.10) can be smoothly continued to the circle S^1, which is an integral trajectory of this system. If in the initial coordinates (x, y) a trajectory of the system (1.1) approached the critical point $(0,0)$ at an angle φ_* with the x-axis, then in polar coordinates the corresponding trajectory of the system (1.10) approaches the critical point

$$r = 0, \quad \varphi = \varphi_*$$

lying on the glued-in circle S^1 replacing the degenerate critical point $(0,0)$ (see Fig. 2). Thus, all possible directions of approach of the trajectories of the system (1.1) to the critical point $(0,0)$ are determined by the roots of

$$R(\varphi_*) = \cos \varphi_* Q_m(\cos \varphi_*, \sin \varphi_*) - \sin \varphi_* P_m(\cos \varphi_*, \sin \varphi_*) = 0. \quad (1.11)$$

These directions correspond to the critical points of the system (1.10) for $r = 0$. Suppose that all critical points of this system which lie on the circle S^1 ($r = 0$, $0 \leqslant \varphi \leqslant 2\pi$) are non-degenerate. Then based on the classification outlined in part I a phase portrait of the system (1.10) can be constructed in a neighborhood of the circle S^1 thus giving a full description of the behavior of the trajectories of the system (1.1) in a neighborhood of the critical point $(0,0)$. If however some critical points ($r = 0$, $\varphi = \varphi_*$) are degenerate, then to analyze them we can use a change

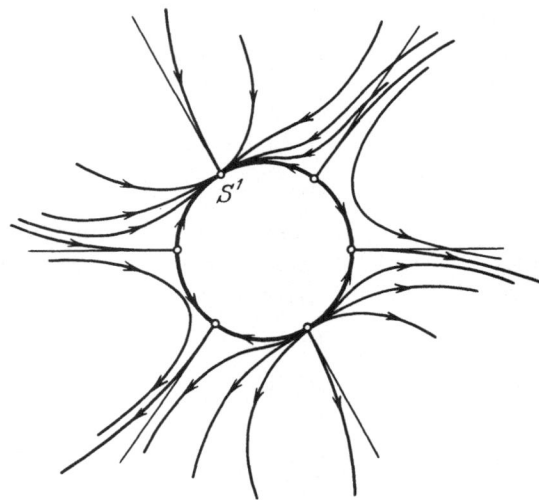

Fig. 2. Possible behavior of the trajectories of the dynamical system in a neighborhood of the circle S^1 glued-in to replace a degenerate critical point

to polar coordinates centered at a given critical point $(r = 0, \varphi = \varphi_*)$ and so on. If there are no critical points of the system (1.10) on the circle S^1, i.e. $R(\varphi) \neq 0$ for all φ, then the circle S^1 is a non-critical closed trajectory of the system (1.10). In this case all trajectories in a neighborhood of S^1 (i.e. all trajectories of the system (1.1) in a neighborhood of the degenerate critical point $(0,0)$) execute an infinite number of turns around it for some direction of time.

Note that the transformation of the dynamical system (1.1) into polar coordinates described above is equivalent to a transformation into two systems of coordinates

$$V_1: x, u = y/x \quad V_2: y, v = x/y. \tag{1.12}$$

Here the critical point $(0,0)$ corresponds to a circle covered by two straight lines

$$L_1: x = 0, \ -\infty < u < +\infty \quad \text{and} \quad L_2: y = 0, \ -\infty < v < +\infty.$$

This remark is especially important for the study (by a similar method) of degenerate critical points of multi-dimensional dynamical systems (see I.2).

III. Analysis of the Dynamical System at Infinity. In order to construct a full phase portrait of the dynamical system (1.1) in the (x, y) plane it is necessary to know the behavior of the trajectories of this system at infinity. To analyze this question a transformation of the dynamical system (1.1) to the Poincarè sphere is used. Figure 3 shows a mapping of the (x, y) plane onto the lower hemisphere of S^2 by means of rays passing through the center of the sphere (a point of the plane and its image on the sphere lie on the same ray). Here the points at infinity of the

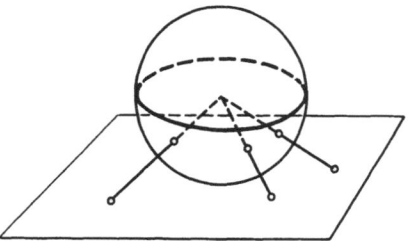

Fig. 3. Mapping of the plane to the lower hemisphere of S^2

plane correspond to the equator of the sphere. Thus the entire (x, y) plane completed at infinity is equivalent to a closed disc.

To obtain such a transformation of the plane in analytic form the following inversion mapping can be used:

$$(x, y) \rightarrow (\rho, \varphi), \quad \text{where } \rho = r^{-1} = (x^2 + y^2)^{-1/2} \quad \text{and} \quad \varphi = \arctan(y/x).$$

In this case the points at infinity of the plane correspond to the points of the circle S^1: $\rho = 0, 0 \leqslant \varphi \leqslant 2\pi$. Analysis of the system (1.1) after a change to the (ρ, φ) coordinates in a neighborhood of the circle S^1 at infinity is fully analogous to the analysis described in part II. Usually however to study a dynamical system (1.1) with polynomial right hand sides $(P(x, y)$ and $Q(x, y)$ are polynomials of degree $n)$ a transformation of the system (1.1) into projective coordinates

$$z = 1/x, u = y/x \quad \text{and} \quad w = 1/y, v = x/y$$

is used (the two systems of coordinates (z, u) and (w, v) with $u \neq 0$ and $v \neq 0$ are equivalent). In projective coordinates the points at infinity of the (x, y) plane correspond to the circle S^1 covered by two straight lines:

$$z = 0, -\infty < u < +\infty \quad \text{and} \quad w = 0, -\infty < v < +\infty.$$

After a change to projective coordinates (z, u) and a time substitution

$$d\tau_1/dt = x^{n-1}$$

the dynamical system (1.1) takes the following form:

$$\dot{z} = -zP^*(z, u)$$
$$\dot{u} = Q^*(z, u) - uP^*(z, u), \tag{1.13}$$

where the functions

$$P^*(z, u) = z^n P(1/z, u/z) \quad \text{and} \quad Q^*(z, u) = z^n Q(1/z, u/z)$$

are polynomials in z and u of degree no higher than n. Obviously the straight line $z = 0$ is an integral trajectory of the system (1.13). Analysis of the system (1.13) in a neighborhood of the straight line $z = 0$ (corresponding to the points at infinity of the (x, y) plane) is fully analogous to the analysis described in part II.

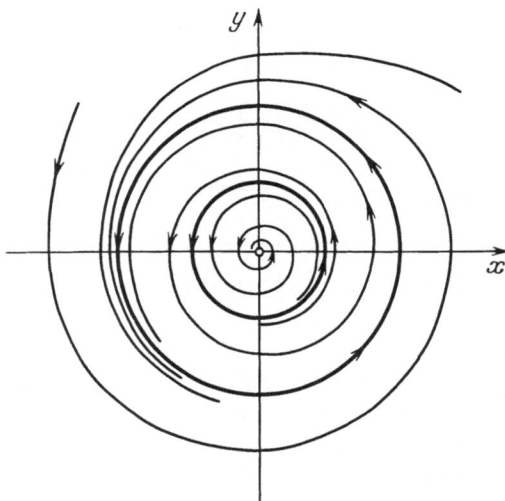

Fig. 4. The behavior of the trajectories of a dynamical system in neighborhoods of limit cycles

Note that the projective coordinates and the coordinates (1.12) have an advantage over (ρ, φ) coordinates and polar (r, φ) coordinates. The advantage is that when the system (1.1) is transformed into these coordinates, the right hand sides remain polynomial $($in (z, u) or (x, u) respectively$)$.

IV. Criteria for Existence and Absence of Limit Cycles of a Dynamical System. A closed non-critical trajectory of the system (1.1) is called a cycle. A closed trajectory divides the (x, y) plane into two parts: internal and external. An isolated closed trajectory is called a limit cycle. All trajectories of the dynamical system (1.1) in a neighborhood of an isolated cycle X tend to X for some direction of time. A limit cycle X is called stable if all trajectories of the system (1.1) in a neighborhood of it tend to X for $t \to +\infty$. It is called unstable if such limiting process occurs for $t \to -\infty$. For example the dynamical system

$$\dot{x} = -y - x(x^2 + y^2 - 1)(x^2 + y^2 - 4)$$
$$\dot{y} = x - y(x^2 + y^2 - 1)(x^2 + y^2 - 4)$$

has two limit cycles:

$$x^2 + y^2 = 1 \quad \text{and} \quad x^2 + y^2 = 4$$

(see Fig. 4). Note that the inner cycle is unstable and the outer cycle is stable. A limit cycle X is called semi-stable if from one side of the plane (e.g. the internal part relative to X) the trajectories tend to it for $t \to +\infty$ and from the other side for $t \to -\infty$.

A proof of existence or absence of limit cycles is the most difficult problem of the two-dimensional qualitative theory, lacking a universal method of solution. However for dynamical systems of a special kind several methods are known to be quite effective in many concrete cases. Suppose for example that the trajectories of the dynamical system (1.1) cross some closed curve $\mathscr{L}$ and enter the region U bounded by this curve. Further assume that the region U contains the sole critical point of the system (1.1), which happens to be repelling. In this case the region U contains at least one stable limit cycle. More precisely the number of stable cycles in the region U exceeds the number of unstable cycles by one (if all cycles are isolated).

To obtain a proof of the existence of limit cycles, methods of bifurcation theory of dynamical systems [6, 34] can be applied. For example a stable cycle is generated from a focus if a change of parameters of the problem makes an attracting focus become a repelling one.[2] The method of small parameters is widely used [33, 34, 37–41]. Among other things it provides sufficient conditions for the existence of limit cycles of dynamical systems, which are close to Hamiltonian systems having only closed trajectories.

To obtain a proof of the absence of closed trajectories (cycles) of a dynamical system of type (1.1) the Dulac-Bendixon criterion can be effectively used in many cases. If in some region G there exists such a smooth function $F(x, y)$ that

$$\partial FP/\partial x + \partial FQ/\partial y > 0, \tag{1.14}$$

then the region G contains no closed trajectories of the system (1.1) contractible to a point in this region. Indeed if such a cycle l bounding a contractible region s existed then we would have (by the Gauss-Ostrogradsky theorem)

$$\oint_l (FP\,dy - FQ\,dx) = \int_s (\partial FP/\partial x + \partial FQ/\partial y)\,dx\,dy > 0.$$

However the integrand on the left is identically 0 on the trajectory l of the system (1.1).

V. Analysis of a Concrete Example. To illustrate the subject let us use the methods described above to analyze the following dynamical system:

$$\begin{aligned}
\dot{x} &= -y^3(x^2 - 1)(2 + xy) \\
\dot{y} &= x^3(y^2 - 1)(2 - xy).
\end{aligned} \tag{1.15}$$

The system (1.5) has nine critical points in the finite part of the (x, y) plane:

---

[2] *Translator's note.* This assumes that the bifurcation is supercritical; a certain higher order coefficient must have the correct sign. Generally this is called a Hopf bifurcation. See for example Hassard, Kazarinoff and Wan "Theory And Applications Of Hopf Bifurcation" London Math. Soc. Lect. Note Series, No. 41, Cambridge University Press, 1981.

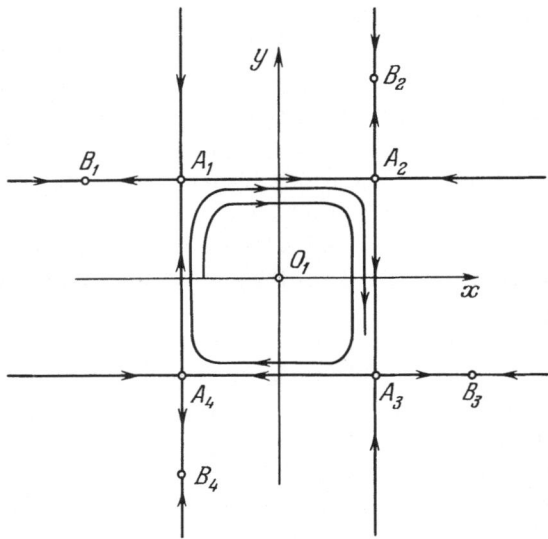

Fig. 5. Example of a dynamical system, where the separatrices of four unstable critical points A_i (saddles) from a cycle

$$A_1\,(-1,1) \qquad A_2\,(1,1) \qquad A_3\,(1,-1)$$
$$A_4\,(-1,-1) \qquad B_1\,(-2,1) \qquad B_2\,(1,2)$$
$$B_3\,(2,-1) \qquad B_4\,(-1,-2) \qquad O_1\,(0,0)$$

(see Fig. 5). Each critical point A_i has eigenvalues $\lambda_1 = -6$ and $\lambda_2 = 2$, i.e. it is unstable (a saddle). Each critical point B_i has eigenvalues $\lambda_1 = -64$ and $\lambda_2 = -3$, i.e. it is an attracting node. Critical point $O_1\,(0,0)$ is degenerate $(\lambda_1 = \lambda_2 = 0)$.

The separatrices of the unstable critical points A_i can be integrated explicitly. Along each separatrix one of the coordinates (x,y) is constant and is either 1 or -1 (see Fig. 5). There is a sequence of four separatrices $(A_1 A_2, A_2 A_3, A_3 A_4, A_4 A_1)$ each going from one critical point A_i into another thus forming a closed diagram. A trajectory of the system (1.15), which is inside the square $A_1 A_2 A_3 A_4$ and starts sufficiently close to its boundary, will travel for an arbitrarily long time along the sequence of separatrices of unstable critical points A_i. This proves the existence of oscillations[3] in the dynamical system (1.15) for $|x| < 1$, $|y| < 1$. The above discussion is the simplest illustration of the method of separatrix approximation of the trajectories of a dynamical system (see I.4), which can be effectively used in the analysis of some multi-dimensional dynamical systems.

To study the degenerate critical point $O_1\,(0,0)$, the system (1.15) is trans-

---

[3] *Translator's note.* Note that unlike common English usage, "oscillations" do not necessarily correspond to periodic orbits.

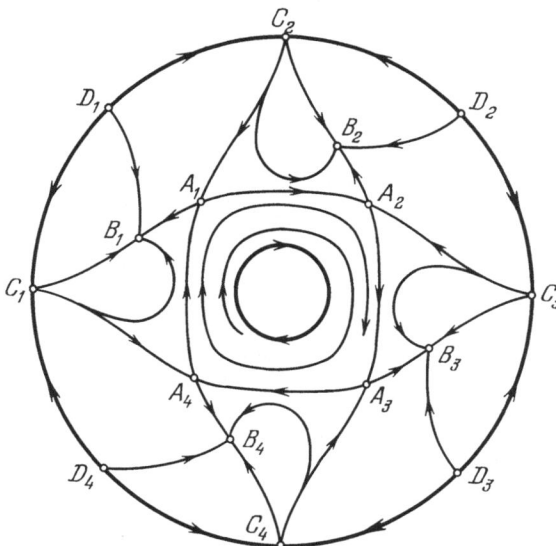

Fig. 6. A complete phase portrait of the dynamical system (1.15)

formed into polar coordinates (ρ, φ) (see part II). Instead of the point O_1 $(0,0)$ a circle S^1: $r = 0$, $0 \leqslant \varphi \leqslant 2\pi$ is glued in (see Fig. 6). After a time substitution $d\tau/dt = r^2$ in polar coordinates the system (1.15) takes the following form:

$$\dot{r} = r\left(-\tfrac{1}{2}\sin 4\varphi + \tfrac{1}{4}r^2 \sin^2 2\varphi - \tfrac{1}{8}r^4 \sin^4 2\varphi\right)$$
$$\dot{\varphi} = -2 + \sin^2 2\varphi + \tfrac{1}{2}r^2 \sin 2\varphi(\sin 2\varphi + \cos 2\varphi) - \tfrac{1}{16}r^4 \sin^2 2\varphi \sin 4\varphi. \tag{1.16}$$

On the circle S^1 $(r = 0)$ the system (1.16) has no critical points. The circle S^1 is a closed integral trajectory of this system. Therefore for $r \ll 1$ all trajectories of the system (1.15–1.16) execute an infinite number of turns around the critical point O_1 $(0,0)$.

Let us show that all trajectories of the dynamical system (1.15) inside the square $A_1 A_2 A_3 A_4$ approach the critical point O_1 for some direction of time and for the other direction wind onto the cycle of separatrices $A_1 A_2$, $A_2 A_3$, $A_3 A_4$, $A_4 A_1$. It is sufficient to prove that inside the square $A_1 A_2 A_3 A_4$ ($|x| < 1$, $|y| < 1$) the system (1.15) has no closed trajectories. Let us use the Dulac-Bendixon criterion (see part IV). The function

$$F = (x^2 - 1)^{-1}(y^2 - 1)^{-1}$$

is obviously regular inside the square. The fact that

$$\frac{\partial FP}{\partial x} + \frac{\partial FQ}{\partial y} = \frac{x^4}{1 - x^2} + \frac{y^4}{1 - y^2} > 0$$

for $|x| < 1$, $|y| < 1$ proves the absence of closed trajectories.

To analyze the behavior of the trajectories of the system (1.15) at infinity the system is transformed into projective coordinates

$$z = 1/x, u = y/x \quad \text{and} \quad w = 1/y, v = x/y$$

(see part III). During such a change of coordinates the (x, y) plane is completed by a circle at infinity covered by two straight lines $z = 0$ and $w = 0$. In the (z, u) coordinates after a time substitution we get the following dynamical system:

$$\dot{z} = zu^3(1 - z^2)(u + 2z^2)$$
$$\dot{u} = (u^2 - z^2)(-u + 2z^2) + u^4(1 - z^2)(u + 2z^2). \tag{1.17}$$

On the attached boundary $(z = 0)$ the system (1.17) has three critical points in the region $z \geqslant 0$ $(x \geqslant 0)$:

$$D_2 \; (u = 1) \quad D_3 \; (u = -1) \quad C_3 \; (u = 0)$$

(see Fig. 6). There are three analogous points D_1, D_4, C_1 in the region $z \leqslant 0$ $(x \leqslant 0)$. At the critical points D_2 and D_3 the eigenvalues are $\lambda_1 = 1$ and $\lambda_2 = 2$, i.e. these points are repelling nodes. The critical point C_3 is degenerate and further analysis shows that it is unstable. In projective coordinates (w, v) the system (1.15) has analogous critical points C_2 and C_4.

A complete phase portrait of the dynamical system (1.15) after resolving the degenerate critical point O_1 $(0, 0)$ and completing the (x, y) plane by a boundary at infinity is shown in Fig. 6.

2. Analysis of Degenerate Critical Points of a Dynamical System

I. A general system of ordinary differential equations defined in phase space with variables $x_1, \ldots, x_n$ has the form

$$\frac{dx_i}{dt} = \dot{x}_i = f_i(x_1, \ldots, x_n) \quad i = 1, \ldots, n. \tag{2.1}$$

From now on it is always assumed that the system (2.1) is autonomous, i.e. the right hand sides f_i are time independent. Any non-autonomous system can be reduced to the form (2.1) by introducing a new variable $x_{n+1} = t$: $dx_{n+1}/dt = 1$. If the initial system of differential equations contains time derivatives of higher order $(\dot{y}, \ddot{y}, \ldots, y^{(k)})$, then such a system is reduced to the form (2.1) by introducing new variables

$$z_1 = \dot{y}, \ldots, z_{k-1} = y^{(k-1)}.$$

Other methods are often used to change a system to the form (2.1). For example in many problems the time evolution is determined by some principle of least action

$$\delta \int_{t_0}^{t_1} L(q_i, \dot{q}_i) \, dt = 0 \tag{2.2}$$

Therefore the initial equations are of second order and have the Lagrangian form

$$\frac{d}{dt}\frac{\partial L}{\partial \dot{q}_i} = \frac{\partial L}{\partial q_i}, \quad i = 1, \ldots, n \tag{2.3}$$

To change a Lagrangian system (2.3) to the form (2.1) it is most convenient to use the Legendre transformation: $p_i = L/\dot{p}_i$. In $2n$-dimensional phase space $(q_1, \ldots, q_n, p_1, \ldots, p_n)$ the system (2.3) takes the Hamiltonian form

$$\dot{p}_i = -\frac{\partial H}{\partial q_i}, \quad \dot{q}_i = \frac{\partial H}{\partial p_i}, \quad H = \sum_{i=1}^{n} q_i \frac{\partial L}{\partial \dot{q}_i} - L \tag{2.4}$$

and consequently becomes a special case of (2.1).

The dimension of phase spaces $(x_1, \ldots, x_n)$ of dynamical systems arising in problems of astrophysics and gas dynamics can be quite large. For example the dynamical systems describing the most general homogeneous cosmological models in the general theory of relativity are defined in 12-dimensional phase space. Here the coordinates x_i consist of six components of the metric g_{ij} ($i, j = 1, 2, 3$) and six corresponding momenta p_{ij}.

Dynamical systems describing self-similar solutions in gas dynamics (with Newtonian gravitation taken into account) are defined in the 4-dimensional space of components of velocity, mass, density and pressure of the gas. The motion of a gas with uniform deformation is described by a Hamiltonian system in 18-dimensional phase space. For systems of hydrodynamical type the dimensionality of the phase space can be arbitrarily large.

The variable t in a system of the type (2.1) can have various physical meanings. Depending on the problem it can represent time as well as some combination of time and space coordinates. In this chapter we shall forget the intrinsic meaning of the variable t and for simplicity's sake shall call it time.

In many dynamical systems of the type (2.1) arising in problems of astrophysics and gas dynamics the right hand sides f_i are rational functions of their arguments $x_1, \ldots, x_n$. In such cases the functions f_i can be brought to a common denominator and written in the form $f_i = F_i/F$, where F_i and F are polynomials in $x_1, \ldots, x_n$. Often it is convenient to perform a time change $t \to \tau$:

$$\frac{d\tau}{dt} = \frac{1}{F(x_1, \ldots, x_n)} \tag{2.5}$$

and to write the system in the form

$$\frac{dx_i}{d\tau} = F_i(x_1, \ldots, x_n), \tag{2.6}$$

where all F_i are polynomials. The transformation (2.5) removes the singularities of the system (2.1), where $F = 0$. Usually time changes of the form $d\tau/dt = 1/|F|$ are used. These do not change the direction of time. However in some cases, such as the study of self-similar solutions, it is necessary to use the non-monotonic time change (2.5) as well.

II. A point $(x_1^0, \ldots, x_n^0)$ is called a critical point of the dynamical system (2.6) if $F_i(x_1^0, \ldots, x_n^0) = 0$ for all i. In this case the system (2.6) has the following simple solution: $x_i(t) \equiv x_i^0$. Critical points of the system (2.6) often correspond to important non-stationary solutions of a studied problem. Such is the case, because as a rule the functions x_i are some combinations of physical parameters, which separately may change with t. The problem of finding all critical points of the system (2.6) and the corresponding solutions is thus reduced to the problem of solving a system of algebraic equations $F_i(x_1, \ldots, x_n) = 0$.

Consider the characteristic matrix $A_i^j = \partial F_i/\partial x_j$ of the system (2.6) at the critical point $(x_1^0, \ldots, x_n^0)$. The characteristic roots of this matrix, i.e. the roots λ_1, $\ldots, \lambda_n$ of the equation

$$\det\left(\frac{\partial F_i}{\partial x_j}(x_i^0) - \lambda \delta_i^j\right) = 0, \tag{2.7}$$

are called the eigenvalues of the system (2.6) at the critical point $(x_1^0, \ldots, x_n^0)$. A critical point is called *non-degenerate*[4], if all the eigenvalues λ_i have non-zero real parts: $\text{Re}\,\lambda_i \neq 0$.

Consider the linear part of the system (2.6)

$$\frac{dx_i}{d\tau} = \sum_{j=1}^{n} \frac{\partial F_i}{\partial x_j}(x_1^0, \ldots, x_n^0)(x_j - x_j^0). \tag{2.8}$$

Recall Hartman's theorem (see [42]): if a critical point $(x_1^0, \ldots, x_n^0)$ is non-degenerate, then there exists a continuous bijective transformation of a neighborhood of this point, which changes the trajectories of the systems (2.6) to the trajectories of the system (2.8). Thus the qualitative behavior of the system (2.6) in a neighborhood of a non-degenerate critical point is equivalent to the behavior of its linear part.

Let $(\xi_i^1), \ldots, (\xi_i^n)$ be the eigenvectors of the matrix $A_i^j = \partial F_i/\partial x_j(x_1^0, \ldots, x_n^0)$:

$$\sum_{j=1}^{n} A_i^j \xi_j^k = \lambda_k \xi_i^k.$$

If the eigenvalues are complex $\lambda_k = \alpha_k + i\beta_k$, then so are the coordinates of the eigenvectors (ξ_i^k). Usually the matrix A_i^j has n distinct eigenvectors (i.e. the Jordan canonical form of the matrix A_i^j is diagonal). In this case all solutions of the system (2.8) have the form

$$x_i(t) - x_i^0 = \text{Re} \sum_{k=1}^{n} C_k \xi_i^k e^{\lambda_k t} \tag{2.9}$$

Therefore the qualitative behavior of the trajectories of the system (2.8) is determined by the eigenvalues λ_k.

A non-degenerate critical point $(x_1^0, \ldots, x_n^0)$ is called *attracting*, if for all i,

---

[4] *Translator's note:* i.e. hyperbolic

Re $\lambda_i < 0$. In this case all trajectories in a neighborhood of the critical point approach the point for $\tau \to +\infty$. A non-degenerate critical point is called *repelling*, if for all i, Re $\lambda_i > 0$. In this case all trajectories in a neighborhood of $(x_1^0, \ldots, x_n^0)$ approach the critical point for $\tau \to -\infty$.

In the study of arbitrary critical points of the system (2.6) the following theorem is continually used: if at some critical point $(x_1^0, \ldots, x_n^0)$ the system (2.6) has d eigenvalues $\lambda_1, \ldots, \lambda_d$ with negative real parts $\alpha_1 \leqq \cdots \leqq \alpha_d < 0$ (eigenvalues are counted with their multiplicities), then there exists a (locally) invariant d-dimensional stable manifold W_s^d, on which all trajectories of the system (2.6) approach the critical point $(x_1^0, \ldots, x_n^0)$ for $t \to \infty$.[5] For each such solution we have

$$\lim_{t \to \infty} t^{-1} \ln\left(\left(\sum_{j=1}^{n} (x_j(t) - x_j^0)^2\right)^{1/2}\right) = \alpha_j \tag{2.10}$$

for some i. If an eigenvalue λ_d with the maximum negative real part α_d is unique, then the corresponding eigenvector ξ^d is tangent to almost all trajectories on the invariant manifold W_s^d and the asymptotics are given by (2.10) with $i = d$. Analogously if at some critical point $(x_1^0, \ldots, x_n^0)$ the system (2.6) has k eigenvalues with positive real parts, then there exists a k-dimensional invariant unstable manifold W_u^k on which all trajectories leave this point.

Trajectories approaching some critical point for $t \to \pm\infty$ are called *separatrices*. The preceding theorem gives in particular the asymptotics of separatrices in neighborhoods of critical points.

A non-degenerate critical point is called a *saddle* (unstable), if at this critical point the dynamical system (2.6) has d eigenvalues with negative real parts and $n - d$ eigenvalues with positive real parts. According to the above there exist two invariant manifolds W_s^d and W_u^{n-d} passing through a saddle and filled with separatrices approaching or leaving this critical point. These manifolds are called respectively the *stable* and the *unstable* manifolds of the system (2.6) passing through $(x_1^0, \ldots, x_n^0)$. All other trajectories not lying on the invariant manifolds W_s^d and W_u^{n-d} do not approach the saddle in question.

In dynamical systems. which possess a certain symmetry, critical points often fill out continuous sets (manifolds M^k of some dimension k). At such critical points the dynamical system (2.6) must have k zero eigenvalues. A critical set M^k is called *non-degenerate*, if at almost all critical points of this set there are $n - k$ eigenvalues of the system (2.6) with non-zero real parts. All the above definitions (of attracting nodes, repelling nodes and saddles) can be extended to non-degenerate critical sets in an obvious way.

---

[5] A manifold M is called an *invariant manifold* of a dynamical system, if each trajectory passing through some non-critical on M lies entirely in this manifold (for all $-\infty < t < +\infty$). An invariant manifold M can only be crossed transversally by trajectories approaching critical points of the dynamical system on this manifold.

III. The behavior of the trajectories of the dynamical system (2.6) in neighborhoods of degenerate critical points (for example where all eigenvalues $\lambda_i = 0$) is usually significantly more complex than the behavior of the trajectories in neighborhoods of non-degenerate critical points. For example the number of separatrices and entire invariant manifolds approaching a degenerate critical point at various angles can be arbitrarily large.

To study degenerate critical points we used the method of resolution of singularities with the aid of a special transformation of the phase space (blowing up of the phase space at a critical point), which attaches an entire invariant manifold in place of the critical point.[6] To be definite suppose that the critical point has coordinates $(0,\ldots,0)$. Let us introduce n systems of coordinates (local charts) $V_1, \ldots, V_n$ in the phase space X $(x_1,\ldots,x_n)$. In a local chart define coordinates

$$y_1^{(i)} = x_1/x_i, \ldots, y_j^{(i)} = x_j/x_i, \ldots, x_i, \ldots, y_n^{(i)} = x_n/x_i. \qquad (2.11)$$

Trajectories approaching the critical point $(0,\ldots,0)$ along some tangent vectors $(\alpha_1,\ldots,1_i,\ldots,\alpha_n)$ approach the points $\left(y_j^{(i)} = \alpha_j, x_i = 0 \right)$ in the local charts V_i. Thus the critical point $(0,\ldots,0)$ corresponds to an entire hyperplane L_i: $x_i = 0$ in the coordinates V_i (2.11). Note that each straight line passing through $(0,\ldots,0)$ contains two rays approaching $(0,\ldots,0)$ from opposite directions. In the coordinates (2.11) the end points of two such rays are identified. Such identification is not always convenient. In order to avoid it, two systems of coordinates V_i: V_i^+ and V_i^- are introduced in correspondence with the sign of the x_i coordinate. The whole collection of $2n$ systems of coordinates $V_i^\pm$ covers the transformed phase space X_1, which is in a one-to-one correspondence with the initial phase space X except for the critical point $(0,\ldots,0)$. On the manifold X_1 the critical point corresponds to an $(n-1)$-dimensional sphere S^{n-1} covered by the hyperplanes $L_i^\pm$ $(x_i = \pm 0)$ in the local charts $V_i^\pm$. If only the local charts V_i are used, then on the $(n-1)$-dimensional sphere S^{n-1} attached in place of $(0,\ldots,0)$ the antipodal points would be identified, i.e. the critical point $(0,\ldots,0)$ would be replaced by an attached real projective space RP^{n-1}. Such transformation is known as a σ-process in algebraic geometry.

After a transformation of the system (2.6) into the local coordinates V_i (2.11) we get the following system

$$\dot{y}_j = \frac{1}{x_i}\left(F_j\left(y_1 x_i,\ldots,x_i,\ldots,y_n x_i\right) - y_j F_i\left(y_1 x_i,\ldots,x_i,\ldots,y_n x_i\right)\right)$$

$$\dot{x}_i = F_i\left(y_1 x_i,\ldots,x_i,\ldots,y_n x_i\right), \quad \text{where } j = 1,\ldots,n \text{ and } j \neq i. \qquad (2.12)$$

---

[6] The method of resolution of singularities in the analysis of specific multi-dimensional dynamical systems was also used in [43, 44] (1974) independently from the joint paper [12] by the author and S.P. Novikov (1973). The works [43, 44] deal mainly with the analysis of motion of four masses on a straight line attracted according to Newton's laws.

Suppose that at the degenerate critical point $(0,\ldots,0)$ the expansion of the right hand sides of the system (2.6) (polynomials F_i) starts with terms of degree $k > 1$. For further constructions it is sufficient that the functions F_i be represented by convergent power series in a neighborhood of $(0,\ldots,0)$. Take out a common factor x_i^{k-1} from the right hand sides of the system (2.6). After a time substitution

$$\frac{d\tau_1}{d\tau} = |x_i^{k-1}| \tag{2.13}$$

the system (2.12) takes the form

$$\frac{dy_j}{d\tau_1} = P_j(y_1,\ldots,x_i,\ldots,y_n) - y_j P_i(y_1,\ldots,x_i,\ldots,y_n)$$

$$\frac{dx_i}{d\tau_1} = x_i P_i(y_1,\ldots,x_i,\ldots,y_n), \quad \text{where } P_j = \text{sign}(x_i^{k-1})F_j/x_i^k. \tag{2.14}$$

Thus the right hand sides of the transformed system (2.14) are still polynomials. The hyperplane $L_i^{\pm}$ $(x_i = \pm 0)$ is an invariant manifold of the system (2.14). Therefore the entire $(n-1)$-dimensional sphere S^{n-1} attached in place of the critical point $(0,\ldots,0)$ as a result of the above transformations is an invariant manifold of the dynamical system (2.6) after a transformation to the manifold X_1 and appropriate time substitutions of the type (2.13).

In general the dynamical system (2.14) has a number of new critical points in the invariant hyperplane $x_i = 0$. If among these critical points there are non-degenerate critical points, then their analysis provides important information about the behavior of the trajectories of the initial system (2.6) in a neighborhood of the degenerate critical point $(0,\ldots,0)$. For example the separatrices of the non-degenerate critical points, which do not lie in their entirety in the plane $x_i = 0$ determine the trajectories of the system (2.6) approaching $(0,\ldots,0)$. If the system (2.14) has degenerate critical points in the hyperplane $x_i = 0$, then at these critical points the above transformations of the type (2.11) should be repeated (and so on).

Consider an important special case when the polynomials F_i are homogeneous of degree k (i.e. the system (2.6) is invariant relative to the group of rigid motions). Then the equation (2.14) for the coordinate x_i is separated, i.e. the order of the system (2.6) is reduced by one after a transformation into the coordinates (2.11). In this case the dynamics of the system (2.14) are completely determined by the dynamics of the system (2.14) on the sphere S^{n-1} attached in place of the degenerate critical point $(0,\ldots,0)$ as part of the transformation.

Note that on the invariant plane $x_i = 0$ the dynamical system is often substantially simpler than the intial dynamical system (for example due to the reduction of order). Therefore an analysis of the system (2.14) for $x_i = 0$ usually provides much more information than a simple analysis of non-degenerate critical points.

In cases when the dynamical system (2.6) possesses a certain symmetry (for example can be written in tensor form as in the theory of homogeneous cosmological models) in order to resolve the degenerate singularities it is convenient to use spherical coordinates:

$$y_i = x_i/G, \quad G = (x_1^2, \ldots, x_n^2)^{1/2}. \tag{2.15}$$

After a transformation into the coordinates (2.15) the system of n equations (2.6) becomes a system of $n + 1$ equations considered on the level surface

$$y_1^2 + y_2^2 + \cdots + y_n^2 = 1. \tag{2.16}$$

In the coordinates (2.15) the phase space of the system (2.6) is the product of the $(n - 1)$-dimensional sphere S^{n-1} and the half-line $G \geq 0$. The critical point $(0, \ldots, 0)$ corresponds to the invariant manifold S^{n-1} determined by the conditions (2.16) and $G = 0$. A change to spherical coordinates (2.15) is completely equivalent to the introduction of local charts $V_i^{\pm}$ of the type (2.11). In various cases one of these two methods of resolution of degenerate critical points is the most convenient (requires less calculations).

Resolution of degenerate critical sets is realized by means of transformations analogous to (2.11) (or (2.15)) along these sets. Suppose that a set of degenerate critical points M^k of dimension k is defined locally by equations $x_1 = \cdots = x_k = 0$. Then to resolve the critical set M^k the transformations (2.11) should be used with $k + 1 \leq i$ and $j \leq n$. As a result of these transformations an $(n - k - 1)$-dimensional sphere S^{n-k-1} is attached in place of each point of the set M^k.

3. Maximally Non-Degenerate Compactification of a Dynamical System

Dynamical systems arising in physical problems are usually defined in some region S_1 of the phase space determined by a number of physical conditions of the type

$$\Phi_j(x_1, \ldots, x_n) > 0 \quad j = 1, \ldots, k. \tag{3.1}$$

Such conditions may define regions where the energy or some coordinates have the same sign etc. To study the dynamical system it pays to add on the boundary of the region S_1. The boundary of S_1 consists of several components Γ_j (lying in some region of finite values of the coordinates) on which the corresponding conditions become equalities:

$$\Gamma_j: \ \Phi_j(x_1, \ldots, x_n) = 0. \tag{3.2}$$

If the trajectories of the dynamical system never leave the region S_1 for any direction of time, then all components of the boundary Γ_j become invariant manifolds of the dynamical system after an appropriate time change.

In many cases it is important to know the behavior of the trajectories of the dynamical system (2.1) for large values of the coordinates x_i. For example in

Hamiltonian systems of the type (2.4) it is often necessary to study the asymptotic behavior of the trajectories for large energies, i.e. for $H \gg 1$. In other problems (for example in the problem of a powerful explosion in the atmosphere) only isolated trajectories of the system, which approach infinity for some direction of time, are of any interest. In order to study the behavior of the trajectories for large values of the coordinates x_i it is necessary to supplement the phase space by the boundary at infinity. Such completion is achieved by introducing $2n$ charts $U_i^{\pm}$ of projective coordinates:

$$y_1^{(i)} = x_1/x_i, \ldots, y_j^{(i)} = x_j/x_i, \ldots, z_i = 1/x_i, \ldots, y_n^{(i)} = x_n/x_i. \qquad (3.3)$$

In a local chart $U_i^{\pm}$ we have $\text{sign}(x_i) = \pm 1$. In the coordinates (3.3) the points infinitely far away correspond to the points of the hyperplane $L_i^{\pm}$: $z_i = \pm 0$. Trajectories going to infinity along some asymptote $(\alpha_1, \ldots, \alpha_n)$ with $\alpha_i \neq 0$ approach the point $y_j^{(i)} = \alpha_j/\alpha_i$, $z_i = \pm 0$ lying on the hyperplane $L_i^{\pm}$.

Transformations of coordinates of the type (3.3) in phase space in the region of infinitely large values of the coordinates x_i result in the attachment of the boundary. This boundary is an $(n-1)$-dimensional sphere S^{n-1} covered by the hyperplanes $L_i^{\pm}$.

After a transformation into the projective coordinates (3.3), the dynamical system (2.1) takes the form:

$$\dot{y}_j = (1/x_i)(F_j(y_1 x_i, x_i, y_n x_i) - y_j F_i(y_1 x_i, x_i, y_n x_i))$$
$$\dot{z}_i = (-1/x_i) z_i F_i(y_1 x_i, x_i, y_n x_i). \qquad (3.4)$$

Suppose that the highest degree of the polynomials F_i is equal to m.[7] Take out a common factor of x_i^{m-1} from the right hand sides of the system (3.4). In the remaining expressions the variable x_i appears only in negative powers, i.e. these expressions contain the variable $z_i = 1/x_i$ in positive powers. Therefore after a time change

$$d\tau_2/d\tau = |x_i|^{m-1}$$

the system (3.4) becomes

$$\dot{y}_j = P_j(y_1, \ldots, z_i, \ldots, y_n) - y_j P_i(y_1, \ldots, z_i, \ldots, y_n)$$
$$\dot{z}_i = -z_i P_i(y_1, \ldots, z_i, \ldots, y_n), \quad \text{where } i \neq j = 1, \ldots, n \qquad (3.6)$$

and $P_j = \text{sign}(x_i^{m-1}) F_j/x_i^m$. The right hand sides of the resultant system (3.6) are still polynomials of their arguments.

The hyperplane L_i ($z_i = 0$) is obviously an invariant manifold of the dynamical system (3.6). Consequently the boundary sphere S^{n-1} attached to the phase space in the region of infinitely large values of the coordinates x_i is also an

---

[7] Projective completion of the phase space can also be effectively applied to the analysis of dynamical systems, in which the right hand sides F_i are not polynomials, but have some properties of homogeneity.

invariant manifold of the dynamical system in the supplemented phase space.

The system (3.6) can have a number of non-degenerate critical points in the invariant hyperplane L_i. Separatrices approaching these critical points and not lying entirely in the boundary sphere S^{n-1} determine the trajectories of the dynamical system (2.1) approaching infinity along some linear asymptotes. If the system (3.6) has degenerate critical points on the attached boundary S^{n-1} ($z_i = 0$), then these singularities should be resolved by means of the transformations described in section 2.

In cases when the system (2.1) possesses a certain symmetry, in order to supplement the phase space at infinity it is convenient to use (as in Sect. 2) spherical coordinates

$$y_i = x_i \mathfrak{G} \qquad \mathfrak{G} = (x_1^2 + \cdots + x_n^2)^{-1/2}. \tag{3.7}$$

After a transformation into the coordinates (3.7) the dynamical system (2.1) is considered on the level surface (2.16): $y_1^2 + \cdots + y_n^2 = 1$. The coordinates (3.7) cover the manifold, which is the product of the $(n-1)$-dimensional unit sphere S^{n-1} (2.16) and the half-line $\mathfrak{G} = 0$.

In some dynamical systems the phase variables are naturally divided into several groups. For example in the Hamiltonian case (2.4) there are n coordinates q_i and n moments p_i. In such cases it is sometimes convenient to proceed with completion (compactification) of the phase space separately for each group of the coordinates. In each group the transformation can still be taken in the form (3.3) or (3.7). Note that the whole boundary attached to the phase space at infinity consists of several components, each of which corresponds to the completion of the phase space for some one group of variables. The components of the boundary intersect along the corners of the boundary, which are also invariant manifolds of the dynamical system in the supplemented phase space.

As a result of the above transformations (completion of the phase space by a boundary at infinity and resolution of degenerate critical points of the dynamical system) the physical region S_1 becomes a compact manifold S with boundary Γ. This manifold S is basically a polyhedron, whose boundary consists of several components $\Gamma_1, \ldots, \Gamma_k$ which intersect pairwise along the various corners of the boundary $\Gamma_i \cap \Gamma_j$. The manifold S is covered by some system of local charts, in which along with the initial coordinates x_i new coordinates of the types (2.11), (2.15), (3.3) and (3.7) are introduced. The dynamical system is defined on the manifold S as a result of the transformations of the system (2.6) into all these local charts and appropriate (monotone) time changes of the type (2.13) and (3.5). The dynamical systems thus obtained at the intersections of pairs of local charts are completely equivalent and differ solely by monotone time changes.

Components of the boundary Γ_i, which supplement the phase space at infinity and are attached to the phase space as a result of the resolution of degenerate critical points, are invariant manifolds of the dynamical system transformed to the compact manifold S. Components of the boundary Γ_j of the type (3.2), which supplement the physical region S_1 for some finite values of the coordinates x_i may not be invariant manifolds. This happens if the trajectories of the dynamical system intersect the above components of the boundary only in one direction, namely inside the manifold S.

The transformations described in Sect. 2 and 3 comprise the main part of the contents of the following general method of analysis of multi-dimensional dynamical systems. The method of maximally non-degenerate compactification of a dynamical system defined in some physical region S_1 consists of the construction of such a compact manifold S with boundary Γ (consisting of several components) and such a dynamical system on S that the constructed dynamical system can be continued smoothly to the boundary Γ, has maximally non-degenerate critical points and is equivalent inside the manifold S to the initial dynamical system in the physical region S_1. The construction of the compact manifold S uses the methods prescribed above. These include the completion of the phase space by a boundary at infinity and the resolution of degenerate critical points of the dynamical system.

4. Separatrix Approximation Method for the Trajectories of a Dynamical System

After constructing a compact manifold S with the properties outlined in Sect. 3 we begin the analysis of non-degenerate critical points and critical sets of the dynamical system defined on the manifold S and their separatrices. Critical points of the dynamical system lying inside the manifold S correspond to the simplest exact solutions of the problem. Non-degenerate critical points lying inside the manifold S and on various components of the boundary Γ away from the corners of the boundary have separatrices approaching them from the physical region. These separatrices correspond to some special solutions of the problem. Asymptotics for $\tau \to +\infty$ (or $\tau \to -\infty$) and stability of such solutions are completely determined by the properties of the critical points.

Unstable critical points of the dynamical system lying on the corners of the boundary $\Gamma_i \cap \Gamma_j$ have separatrices, which lie entirely in the components of the boundary Γ_i and Γ_j (invariant manifolds of the dynamical system). They can have no separatrices approaching them from the physical region S_1. Each such separatrix by itself does not correspond to any exact solution of the dynamical system. This happens because it lies on an attached component of the boundary, which was absent from the initial phase space. However in many cases the separatrices lying on various components of the boundary Γ_i go from one unstable critical point to another, from that one to a third and so on thus forming long sequences of separatrices. These sequences approximate complex modes of the dynamics of true physical trajectories of the dynamical system.

Indeed suppose that all separatrices leaving some unstable critical point X_i go to some other unstable critical point X_j and so on thus forming a sequence of steps:

$$\cdots \to X_i \to X_j \to X_k \to X_l \to \cdots. \qquad (4.1)$$

Then by the theorem of continuous dependence of a solution on the initial data, the trajectories of the dynamical system, which start sufficiently close to one of these separatrices or critical points will move along some sequence of separatrices (4.1) for an arbitrary length of time.

The situation outlined here occurs if the unstable manifold $W_u(X_i)$ of a critical point X_i (cf. Sect. 2) is contained in the stable manifold $W_s(X_j)$ of X_j and so on. In general the intersection of any two invariant manifolds $W_u(X_k)$ and $W_s(X_l)$ is some new invariant manifold W. On W all trajectories leave X_k for $t \to -\infty$ and approach X_l for $t \to +\infty$. Trajectories lying in the intersection of two invariant manifolds $W_u(X_k)$ and $W_s(X_l)$ are called *heteroclinic*. Trajectories lying in the intersection of the stable and the unstable manifolds $W_s(X_i)$ and $W_u(X_i)$ of the same critical point X_i are the *homoclinic* trajectories of X_i. Stable manifolds $W_s(X_k)$ and $W_s(X_l)$ are pairwise disjoint and the same goes for the unstable manifolds as follows from their respective definitions.

Stable and unstable invariant manifolds exist not only for critical points of the dynamical system but also for non-critical closed trajectories (cycles) Z_i. Homoclinic and heteroclinic trajectories arising in the intersections of such invariant manifolds $W_u(Z_i)$ and $W_s(Z_j)$ have very complex dynamics in the phase space. For $t \to -\infty$ and $t \to +\infty$ these solutions oscillate near the manifolds $W_u(Z_i)$ and $W_s(Z_j)$ but never intersect them. Homoclinic solutions were found for the first time by H. Poincaré in the three body problem [1]. According to Poincaré in order to study homoclinic solutions we consider hypersurfaces L in the phase space, which intersect the trajectories of the dynamical system transversally. In a neighborhood of a cycle Z_i intersecting a hypersurface $\mathscr{L}$ at a point Y_i we define a diffeomorphism f of $\mathscr{L}$ to itself called the return map. Each point x on $\mathscr{L}$ is the initial point of some trajectory $l(x)$. The return map f takes x to the next point of intersection of $l(x)$ and $\mathscr{L}$. Closed trajectories (cycles Z_i) correspond to the fixed points Y_i of f. These also have stable and unstable invariant manifolds, which are the intersections of invariant manifolds $W_u(Z_i)$ and $W_s(Z_j)$ with the hypersurface $\mathscr{L}$. Homoclinic and heteroclinic points can be analogously defined for the fixed points Y_i of the diffeomorphism f. Homoclinic points and the properties of the trajectories related to them were subject to detailed and numerous studies starting with [1, 8]. A survey of recent results can be found in [11].

Dynamical systems considered in this book and arising from the problems of astrophysics and gas dynamics are special in the sense that the majority of critical points of these dynamical systems lie on the various components of the boundaries Γ_i of their respective compact manifolds S. On the components of the boundary Γ_i the dynamical systems are greatly simplified and as a consequence the separatrices of critical points X_i can be completely analyzed in many cases. In fact sometimes they can be integrated explicitly. The proof of the fact that some separatrix goes from one point to another often depends on the existense of a special monotone function U $(dU/d\tau > 0)$ of the dynamical system on the given components of the boundary. In order to study the motion of separatrices on the two-dimensional components of the boundary it is often useful to apply the Dulac-Bendixon criterion for the absense of closed trajectories (cycles) (cf. Sect. 1, part IV).

It is convenient to systematize all the separatrix steps between critical points and critical sets, which occur in a given dynamical system as part of a separatrix diagram. A separatrix diagram can be either in the form of a $n \times n$ matrix, where n is the number of critical points and critical sets, or in the form of a graph in the plane (e.g. (4.1)). If the sequence of separatrix steps (4.1) contains a finite number of separatrices, begins at some repelling critical point X_1 and ends at some attracting critical point X_n, then the separatrix approximation (4.1) provides a complete (for all t) qualitative description of the dynamics of nearby trajectories of the dynamical system.

In some dynamical systems there are infinite sequences of separatrix steps (4.1) containing cycles, returns to a neighborhood of the initial critical point and so on. In such cases the dynamical system possesses complex non-linear oscillatory modes, which can be approximated by sequences of separatrices (4.1) so that there exist physical trajectories going through an arbitrarily large number of separatrix steps. If critical points form continuous critical sets, then the separatrix steps define mappings of one critical set to another. The initial point of a separatrix gets mapped to its final point. The mappings of critical sets so obtained are discrete combinatorial models of complex non-linear oscillatory modes of the dynamical system under study. In some cases (such as unlimited repetition) the mappings of critical sets possess ergodic properties and therefore set up conditions for stochastization (by some parameters) of the trajectories of the dynamical system going along the sequences of separatrices between these critical sets.

Another mechanism of stochastization of the trajectories is connected with the possibility of multi-valued separatrix steps in the separatrix diagram (4.1). This happens if there is a number of separatrices which emerge from some critical point X_k and after unlimited continuation approach distinct points $X_1, \ldots, X_l$. This also provides conditions for the stochastic behavior of the trajectories and is connected to the instability of the trajectories depending on the initial data in neighborhoods of the dynamical system.

Note that in some cases it is possible to integrate the separatrices of critical points, which lie inside the manifold S. Approximation of the trajectories of the dynamical system by such separatrix steps allows one to study the behavior of the trajectories not only in a neighborhood of the boundary Γ, but also inside the manifold S.

Thus the separatrix approximation method for the trajectories of a dynamical system based on the method of maximally non-degenerate compactification of a dynamical system in principle allows a detailed analysis of complex non-linear oscillatory modes of the dynamics of the trajectories possessing a number of stochastic properties. Also it provides a picture of the behavior of the trajectories of a dynamical system as a whole. The effectiveness of this chapter's methods of qualitative analysis of multi-dimensional dynamical systems is demonstrated in later chapters of this book.

Chapter II
Qualitative Theory of Homogeneous Cosmological
Models Without the Motion of Matter

Homogeneous cosmological models form one of the most important classes of solutions for the equations of the general theory of relativity. The study of homogeneous cosmological models, which began with the classical works of A.A. Friedmann [45, 46], was the starting point of modern relativistic cosmology and subsequently became a deciding influence on its development. The complete class of homogeneous solutions for the equations of the general theory of relativity was first determined in [47]. Several works [48–52] were devoted to the study of singularities in general solutions of Einstein's equations, culminating in the discovery of the general oscillatory mode of behavior of the metric near cosmological singularities [49, 50]. General theorems proved in [53–56] are of great importance to the study of singularities in the solutions of Einstein's equations. In the analysis of homogeneous cosmological models the most important physical problems are the determination of concrete characteristics of the singularity of the solution for the contraction of space, the study of isotropization of solutions for the expansion of space, the determination of the most general modes of the dynamics of homogeneous cosmological models for the contraction and the expansion of space, the study of concrete properties of the oscillatory mode of behavior of the metric near a singularity and so on. Many works [57–76] are devoted to the study of these questions and a detailed bibliography of these is provided in [74, 75]. All works mentioned here use traditional methods of theoretical physics to study the systems of ordinary differential equations, to which Einstein's equations are reduced for homogeneous cosmological models. These methods include approximation of the trajectories of a complex system of equations by solutions of simpler systems with subsequent patching of solutions, the method of small parameters and numerical methods.

The classical methods of two-dimensional qualitative theory of Poincaré and Bendixon were first used in [77, 78] in the study of the simplest homogeneous cosmological models, for which Einstein's system of equations is reduced to a plane. Later, similar methods were used in [79–81]. However more complex homogenous cosmological models are described by dynamical systems in spaces of higher dimension. In fact for the most general models this dimension is equal to 12. The methods of qualitative theory of multi-dimensional dynamical systems were first used in a joint paper by S.P. Novikov and the author [12] to study a six-dimensional dynamical system describing an important homogeneous cosmological model of type IX without the motion of matter. The results of this paper were published in detailed form in [15, 21]. Further research on homogeneous cosmological models [13, 16, 18, 19, 24, 82, 83] has shown that the qualitative theory of multi-dimensional dynamical systems is the most adequate mathematical apparatus for the analysis of homogeneous cosmological models, which by means of rigorous methods provides a full picture of their dynamics. This chapter and chapter III are devoted to a detailed analysis of all homogeneous cosmological models with a hydrodynamical stress-energy tensor and an electromagnetic field by means of the methods of qualitative theory of multi-dimensional dynamical systems.

1. Equations of the General Theory of Relativity

Classical textbooks [84, 85], which present in detail the basic notions and various special problems of the general theory of relativity, are widely known. Therefore in this and the next sections only the most general and necessary information for later sections is provided.

The main objects of study in the general theory of relativity are the various properties of four-dimensional space-time manifolds M^4 and the motion of matter in these manifolds. Geometric properties of space time are determined by the metric

$$ds^2 = g_{ij} dx^i dx^j, \tag{1.1}$$

where there is implied summation over repeated indices and the symmetric matrix $g_{ij}(x_1, x_2, x_3, x_4)$ has signature $(+ - - -)$. At each point there exists such a system of coordinates (called the local Lorentz frame of reference), in which the metric (1.1) can be reduced to

$$ds^2 = dt^2 - dx^2 - dy^2 - dz^2 \tag{1.2}$$

up to terms of second order.

The type of matter occupying the space-time and its properties are determined by the stress-energy tensor of matter T_{ij}. In this book we consider two types of tensors T_{ij}, namely the hydrodynamical stress-energy tensor

$$T_{ij} = (p + \varepsilon)u_i u_j - pg_{ij} \tag{1.3}$$

and the stress-energy tensor of the electromagnetic field

$$T_{ij} = (1/4\pi)(F_{ik}F_{jl}g^{kl} + \tfrac{1}{4}g_{ij}F_{lm}F^{lm}). \tag{1.4}$$

In the first case ε is the energy density of matter, p is the pressure and u^i is the 4-vector of velocity of matter, which by definition satisfies $g_{ij}u^i u^j = 1$. In the second case (1.4) the skew-symmetric tensor F_{ij} is the electromagnetic tensor. By definition

$$F_{ij} = \partial A_j/\partial x_i - \partial A_i/\partial x_j, \tag{1.5}$$

where A^i is the vector potential of the electromagnetic field.

The equations of the general theory of relativity (Einstein's equations) connect the geometric properties of the space-time manifold M^4 with the properties of matter which occupies it. They have the following form:

$$R_{ij} - \tfrac{1}{2}g_{ij}R = (8\pi k/c)T_{ij}. \tag{1.6}$$

The Ricci tensor R_{ij} is a contraction of the Riemann tensor R_{ikj}^l ($R_{ij} = R_{ilj}^l$). It completely determines the properties of curvature of space-time (see [84–86]). The scalar curvature R is the complete contraction of the Riemann tensor: $R = R_{ilj}^l g^{ij} = R_{ij}g^{ij}$. From now on we will use a system of units in which the speed of light c equals 1 and the gravitational constant k is $1/8\pi$.

The equations of relativity (1.6) (with boundary conditions) completely determine the evolution of the metric g_{ij}. These equations also determine the motion of matter, because (1.6) and the Bianchi identity imply that

$$T_{i;j}^j = 0 \tag{1.7}$$

(here we use the standard notation for the covariant derivative [84, 86]). Equations (1.7) for the hydrodynamical stress-energy tensor (1.3) are the equations hydrodynamics (in invariant four-dimensional notation). Equations (1.7) for the stress-energy tensor of the electromagnetic field (1.4) are a generalization to curved space of the second pair of Maxwell's equations.

Einstein's equations (1.6) follow from Hilbert's principle of least action

$$\delta \int (R\sqrt{|g|} + \Lambda\sqrt{|g|})\,d\Omega = 0, \tag{1.8}$$

where $R\sqrt{|g|}$ and $\Lambda\sqrt{|g|}$ are actions respectively for the gravitational field and the metric $|g| = |\det(g_{ij})|$. Note that in the variational principle (1.8) the variations of the metric δg_{ij} are assumed to be finite along the space-time manifold M^4. In the case of the electromagnetic field the action is

$$\Lambda\sqrt{|g|} = -(1/16\pi)F_{ik}F^{ik}.$$

For the hydrodynamical stress-energy tensor (1.3) we can formally assume that

$$\Lambda\sqrt{|g|} = -(p + \varepsilon)u_i u_j g^{ij}\sqrt{|g|} - (p - \varepsilon)\sqrt{|g|},$$

where the connection $u_i u_j g^{ij} = 1$ is used only after varying the action along the components of the metric.

Various general properties of solutions of Einstein's equations can be seen best in a number of special highly symmetric exact solutions.

2. Classical Solutions to the Equations of the General Theory of Relativity

One of the basic methods of obtaining exact solutions of the equations of the general theory of relativity consists of the analysis of those solutions, which are invariant relative to some groups of transformations of the coordinates. These solutions possess deep symmetries and are the simplest and at the same time natural models of many astrophysical and cosmological phenomena. Classification of various types of invariant solutions based on group theory appears in [87].

If a solution is invariant relative to some group of transformations G, then in particular, the group G is a group of isometries of the metric (1.1) on a given manifold M^4. The flat Minkowski space possesses the highest invariance. Its metric (1.2) has a 10-dimensional group of isometries called the Poincaré group. This group is generated by the 6-dimensional group of Lorentz transformations $O(1,3)$ (the group of linear transformations which preserve the quadratic form (metric) (1.2)) and by the 4-dimensional commutative group R^4 of arbitrary translations of flat space-time.

I. Friedmann Solutions. Open, closed and flat Friedmann solutions have 6-dimensional groups of isometries. These solutions were proposed by A.A. Friedmann in 1922 as non-stationary models of a homogeneous and isotropic Universe, whose metric is the metric of constant curvature. The matter in Friedmann solutions is at rest, i.e. the 4-velocity u^i of matter has components $(1, 0, 0, 0)$. The metric of the closed Friedmann model has the form

$$ds^2 = dt^2 - a^2(t)(d\chi^2 + \sin^2 \chi(d\theta^2 + \sin^2 \theta \, d\varphi^2)). \tag{2.1}$$

Spatial sections $t = t_0$ of this model are 3-dimensional spheres S^3. The metric (2.1) restricted to such a sphere has constant positive curvature $\lambda = a(t_0)^{-2}$ (i.e. the sphere S^3 has radius $a(t_0)$). For any dependence of $a(t)$ on t the metric (2.1) is invariant relative to the 6-dimensional group $O(4)$ of rotations of the standard 3-dimensional sphere. If the equation of state of the matter is

$$p = k\varepsilon, \quad \text{where } 0 \leq k < 1,$$

then Einstein's equations for the metric (2.1) can be reduced to two equations:

$$\varepsilon a^4 = 3(da/d\eta)^2 + a^2, \quad \varepsilon a^{3(1+k)} = \text{const}, \tag{2.2}$$

where the variable η is determined by the relation

$$dt = a \, d\eta.$$

In the case of dust-like matter $(p = k = 0)$ the equations (2.2) are solved in the following form:

$$a = a_0(1 - \cos \eta), \quad t = a_0(\eta - \sin \eta). \tag{2.3}$$

Thus as η changes from 0 to π, the radius of the universe $a(t)$ in the closed Friedmann model grows monotonically from $a(0) = 0$ (which signifies a singularity of the solution, since $\varepsilon = Ca^{-3} \to \infty$) to the state of maximal expansion $a(\pi a_0) = 2a_0$. Then as η changes from π to 2π, the radius of the universe decreases monotonically to zero: $a(2\pi a_0) = 0$. Closed Friedmann solutions for arbitrary $k \neq 0$ have analogous properties.

The metric of the open Friedmann model has the form

$$ds^2 = dt^2 - a^2(t)(d\chi^2 + \sin h^2 \chi(d\theta^2 + \sin^2 \theta \, d\varphi^2)). \tag{2.4}$$

Here the spatial sections $t = t_0$ are manifolds of constant negative curvature $\lambda = -a(t_0)^{-2}$ (3-dimensional Lobachevsky spaces), which have a 6-dimensional group of isometries namely the Lorentz group $O(1, 3)$. Einstein's equations for the metric (2.4) are reduced to two equations:

$$\varepsilon a^4 = 3(da/d\eta)^2 - a^2, \quad \varepsilon a^{3(1+k)} = \text{const}, \tag{2.5}$$

which in the case of dust-like matter can be solved in the following form:

$$a = a_0(\cos h\eta - 1), \quad t = a_0(\sin h\eta - \eta). \tag{2.6}$$

According to (2.6) the open Friedmann solution (2.4) has a singularity for $t = 0$, since $a(0) = 0$. As t increases from 0 to ∞ the function $a(t)$ monotonically increases to infinity.

The metric of Friedmann solutions with a flat three-dimensional space has the form

$$ds^2 = dt^2 - a^2(t)(dx_1^2 + dx_2^2 + dx_3^2). \tag{2.7}$$

These solutions (2.7) are obviously invariant relative to the 6-dimensional group of isometries of the flat 3-dimensional space (shifts and orthogonal rotations). Einstein's equations for the metric (2.7) are reduced to two equations

$$\varepsilon a = 3(da/d\eta)^2, \quad \varepsilon a^{3(1+k)} = \text{const}, \tag{2.8}$$

which can be easily solved for all k:

$$a = a_0 t^{2/(3(1+k))}. \tag{2.9}$$

Consequently for a flat Friedmann solution the function $a(t)$ grows monotonically from the singularity $a(0) = 0$ to infinity as t grows from 0 to ∞, just as for an open Friedmann solution.

All three Friedmann solutions have a common property. It is the existence of a physical singularity $a(0) = 0$, where the energy density ε approaches infinity. Note that at the same time $|R| \to \infty$, because $R = -T_{ij}g^{ij} = (3k - 1)\varepsilon$. The existence of some physical singularity in the general solution of Einstein's equations is guaranteed by a number of theorems, proved in [54–57]. Particular characteristics of the singularity may vary significantly. For example the simplest generalization of the flat Friedmann solution has a completely different (unisotropic) singularity.

II. Kasner Solutions. Metrics of the form

$$ds^2 = dt^2 - q_1(t)\,dx_1^2 - q_2(t)\,dx_2^2 - q_3(t)\,dx_3^2 \tag{2.10}$$

are generalizations of the flat Friedmann solution, which is characterized by $q_1(t) \equiv q_2(t) \equiv q_3(t)$. The group of isometries of the general metric (2.10) consists of arbitrary translations of the three-dimensional space x_1, x_2, x_3, i.e. it is the three-dimensional commutative group R^3. Einstein's equations for the metric (2.10) are reduced to the following system of equations:

$$\left((q_1 q_2 q_3)^{1/2}(\ln q_i)'\right)^{\cdot} = (\varepsilon - p)(q_1 q_2 q_3)^{1/2},$$

$$2(\ln(q_1 q_2 q_3))^{\cdot\cdot} + \sum_{i=1}^{3} \left((\ln q_i)'\right)^2 = -2(\varepsilon + 3p),$$

(2.11)

where the dots mean differentiation with respect to time.

In the case of dust-like matter $(p = 0)$ these equations (2.11) can be easily solved [88, 59] and have the following two types of solutions:

$$\varepsilon = 0: \quad q_i(t) = C_i t^{2p_i},$$

(2.12)

$$\varepsilon \neq 0: \quad q_i(t) = C_i t^{2p_i}(t + t_0)^{(4/8)-2p_i}, \ \varepsilon = 4/3t(t + t_0).$$

(2.13)

The constants p_1, p_2, p_3 satisfy two conditions

$$p_1 + p_2 + p_3 = 1, \quad p_1^2 + p_2^2 + p_3^2 = 1$$

(2.14)

and are called the Kasner exponents.

Kasner solutions [88] (2.12) define metrics of the type (2.10) in empty space. For $t = 0$ these solutions have a physical singularity (which cannot be removed by any change of coordinates) for all sets of Kasner exponents, except $(1, 0, 0)$. The exceptional solution

$$ds^2 = dt^2 - C_1 t^2 \, dx_1^2 - C_2 \, dx_2^2 - C_3 \, dx_3^2$$

(2.15)

becomes a Minkowski solution (1.2) after a change to the coordinates:

$$\tau = t \cosh x_1 \quad \zeta = t \sinh x_2.$$

This solution (2.15) is the simplest example, showing that the presence of a singularity for a given type of metric does not necessarily mean the existence of a physical singularity of the solution, because in several cases a singularity of the metric can be removed by an appropriate change of coordinates.

Solutions (2.13) in the extended space, obtained for the first time in [59], contain (as a special case when $C_1 = C_2 = C_3, t = 0$) the flat Friedmann solution (2.7–9). In general $(t_0 > 0)$ these solutions (2.13) have an unisotropic singularity for $t \to 0$ of the same type as the Kasner solutions (2.12) in vacuum. Thus for the contraction of space $(\det(g_{ij}) \to 0, t \to 0)$ the flat Friedmann solution (2.7–9) is unstable in the class of perturbations of the type (2.10). For the expansion of space $(t \to \infty)$ all such solutions (2.13) have the same asymptotics as the Friedmann solution. Consequently for the expansion of space we observe the isotropization of initially unisotropic solutions (2.13).

Metrics of the type (2.10) define the simplest homogeneous cosmological model called the model of type I. In the considered case two important physical questions, namely the analysis of the behavior of the metric in a neighborhood of a cosmological singularity $(\det \| g_{ij} \|) = 0$ and the analysis of the isotropization of the metric during the expansion of space, have very simple solutions. For the more complex homogeneous cosmological models, Einstein's equations can no longer be integrated explicitly. A full analysis of all modes of behavior of the metric for homogenous cosmological models is conducted in the following sections by means of the methods of qualitative theory of dynamical systems.

III. Schwarzschild Solutions. This classical solution describes the gravitational field due to a spherically symmetric body of mass $m = r_g/2$ in vacuum. In this case the space-time metric is necessarily static and has the following form:

$$ds^2 = \left(1 - (r_g/r)\right) dt^2 - dr^2/(1 - r_g/r) - r^2(d\theta^2 + \sin^2\theta\, d\varphi^2). \quad (2.16)$$

For $r \to \infty$ the Schwarzschild solution approaches the flat Minkowski solution and has a physical singularity at $r = 0$ as well a non-physical removable singularity at the gravitational radius $r = r_g$.

The complete spherically symmetric solution in vacuum without non-physical singularities was first obtained by Kruskal [89] by means of the following transformation of the solution (2.16) into the coordinates τ and ζ:

$$
\begin{aligned}
r > r_g: \ \tau &= 2r_g \exp(r/2r_g)(r - r_g)^{1/2} \sinh(t/2r_g) \\
\zeta &= 2r_g \exp(r/2r_g)(r - r_g)^{1/2} \cosh(t/2r_g), \\
r < r_g: \ \tau &= 2r_g \exp(r/2r_g)(r_g - r)^{1/2} \cosh(t/2r_g) \\
\zeta &= 2r_g \exp(r/2r_g)(r_g - r)^{1/2} \sinh(t/2r_g).
\end{aligned}
\quad (2.17)
$$

In the new coordinates τ, ζ the metric takes the following form:

$$ds^2 = (1/r)\exp(-r/r_g)(d\tau^2 - d\zeta^2) - r^2(d\theta^2 + \sin^2\theta\, d\varphi^2). \quad (2.18)$$

Note that (2.17) imply that the function $r(\tau, \zeta)$ is determined by the relation

$$\tau^2 - \zeta^2 = 4r_g^2 \exp(r/r_g)(r_g - r). \quad (2.19)$$

This implies that the lines $r = $ const in the τ, ζ plane are hyperbolas. The solution (2.18) is defined in the region of the τ, ζ plane enclosed by the two branches of the hyperbola $r(\tau, \zeta) = 0$, in which the solution (2.18) has a non-removable physical singularity. The line $r = r_g$ consists of two straight lines $\tau = \pm\zeta$, on which the metric (2.18) is completely regular. These straight lines $\tau = \pm\zeta$ divide the plane into four regions: two regions T_+ $(\tau > 0)$ and T_- $(\tau < 0)$ inside the gravitational radius $(r < r_g, |\tau| > |\zeta|)$ and two regions R_1 and R_2 outside the gravitational radius $(r > r_g, |\tau| < |\zeta|)$. By (2.17) the regions T_+ and R_1 correspond to the two regions $r < r_g$ and $r > r_g$ of the Schwarzschild solution (2.16).

In the complete Kruskal solution (2.18) the two regions R_1 and R_2 cannot communicate by any physical signals. In the region T_+ all time-like and light lines approach the physical singularity in a finite amount of their time. Space-like sections $\tau = \tau_0$, $-\infty < \zeta < +\infty$, for $\tau_0^2 < 4r_g$ are products of a two-dimensional sphere S^2 and a straight line R^1. Note that the minimal radius of S^2 is determined by (2.19) with $\zeta = 0$. Later on in chapter IV we will construct self-similar solutions of Einstein's equations in the extended space with the equation of state of matter $p = k\varepsilon$, which possess a number of analogous properties.

3. General Properties of Homogeneous Cosmological Models

I. Homogeneous cosmological models are defined to be space-time manifolds M^4 together with a metric ds^2 satisfying Einstein's equations and invariant under some three-dimensional Lie group G of transformations of the manifold M^4 such that the action has three-dimensional orbits (an orbit is a set obtained from a given point by the action of all transformations of the group G). In cosmology the most important models are the space-homogeneous models for which the orbits of the Lie group G are space-like (restrictions of the metric ds^2 to the orbits of the group G are negative definite). From now on we shall consider only space-homogeneous models. Let X_0 be a vector field, which is orthogonal to the orbits of G. Since the metric is assumed to be space-homogeneous, the vector field X_0 is time-like. Therefore we can normalize the length of the vector field X_0 ($\langle X_0, X_0 \rangle = 1$) and consider the integral curves of X_0 to be time lines so that on each orbit of the group G we have $t = $ const. It can be shown (using the existence of a monotone function) that if the metric ds^2 satisfies Einstein's equations, then each integral curve of the field X_0 intersects any orbit of the group G only once. Therefore the space-time manifold M^4 can be written as the product of the time axis R^1 and some orbit G_1: $M^4 = R^1 \times G_1$. Note that the orbit G_1 is itself a three-dimensional Lie group, which has the same Lie algebra as the Lie group G.

Suppose that X_1, X_2, X_3 are vector fields, which generate the action of G. These vector fields are tangent to the orbits of G and satisfy the commutation relations determined by the Lie algebra $\mathfrak{G}$ of the Lie group G:

$$[X_i, X_j] = C_{ij}^k X_k. \tag{3.1}$$

By construction these vector fields X_i commute with X_0 ($[X_0, X_i] = 0$). In the basis of vector fields X_0, X_1, X_2, X_3 the metric ds^2 has the form

$$\begin{pmatrix} 1 & 0 & 0 & 0 \\ 0 & & & \\ 0 & & -g_{ij}(t) & \\ 0 & & & \end{pmatrix}, \quad i, j = 1, 2, 3, \tag{3.2}$$

where the matrix $g_{ij}(t)$ is positive definite. In homogeneous cosmological models all parameters of the matter, which fills the space, have to be invariant relative to the transformations of the Lie group G. In the case when the matter has a hydrodynamical stress-energy tensor (1.3), the components of velocity u^i in the basis of invariant vector fields X_0, X_1, X_2, X_3 as well as the energy density ε and pressure p depend only on time t. Thus the local properties of homogeneous cosmological models are completely determined by their corresponding three-dimensional Lie algebras $\mathfrak{G}$.

What follows is a simple classification of three-dimensional Lie algebras. It was first obtained by Bianchi in 1897 [90]. The structure constants C_{ij}^k (3.1) are anti-symmetric in the indices i and j, so they can be written in the form

Table 1. The Bianchi classification of
three dimensional Lie algebras

Bianchi type	a	n_1	n_2	n_3
I	0	0	0	0
II	0	1	0	0
VI_0	0	1	−1	0
VII_0	0	1	1	0
VIII	0	1	1	−1
IX	0	1	1	1
III	1	1	−1	0
IV	1	1	0	0
V	1	0	0	0
VI $(a \neq 1)$	a	1	−1	0
VII	a	1	1	0

$$C_{jj}^k = \varepsilon_{ijl} n^{kl} + \delta_j^k a_i - \delta_i^k a_j, \tag{3.3}$$

where ε_{ijl} is a completely anti-symmetric three-dimensional tensor, n^{kl} is a three-dimensional tensor of rank two and a_i is a three-dimensional vector. The Jacobi identity

$$[[X_i, X_j], X_k] + [[X_j, X_k], X_i] + [[X_k, X_i], X_j] = 0$$

implies that $n^{kl} a_1 = 0$, i.e. if the vector a_1 is non-zero, then it is an eigenvector of the matrix n^{kl} with the eigenvalue zero. In some new basis X_1, X_2, X_3 the symmetric matrix n^{kl} assumes a diagonal form. Let n_1, n_2, n_3 be the eigenvalues of n^{kl}. Since $n^{kl} a_1 = 0$ we can assume that in the new basis the vector a_1 has coordinates $(0, 0, a)$. Note that either $n_3 = 0$ or $a = 0$. In the new basis X_1, X_2, X_3 the commutation relations (3.1)–(3.3) take the form

$$[X_1, X_2] = n_3 X_3,$$

$$[X_2, X_3] = n_1 X_1 - a X_2, \tag{3.4}$$

$$[X_3, X_1] = n_2 X_2 + a X_1.$$

By means of linear deformations of the vectors X_1, X_2, X_3 these commutation relations (3.4) and the corresponding three-dimensional Lie algebras can be reduced to one of nine types listed in Table 1 (the parameter a is any positive number). Types VI and VII form one-parameter families of non-isomorphic Lie algebras.

Table 1 contains the following known three-dimensional Lie algebras: type I is the commutative Lie algebra $\mathbb{R}^3$, type VIII is the Lie algebra of the group $SO(2, 1)$ or $SL(2, R)$ and type IX is the Lie algebra of the group $SO(3)$. According to the standard algebraic classification of Lie algebras (see [91]), Lie algebras of type II are nilpotent, Lie algebras of types III–VII are solvable and Lie algebras of types VIII and IX are the so-called simple Lie algebras.

The Lie algebra $\mathfrak{G}$ of a Lie group G has another Lie group G_2 associated with it, namely the group of all inner automorphisms of the Lie algebra $\mathfrak{G}$ or in other words the group of all linear transformations of the Lie algebra $\mathfrak{G}$, which preserve the commutation relations for the vectors X_i. For the commutative Lie algebra (type I) the group of inner automorphisms G_2 is the complete nine-dimensional group of all linear transformations of $\mathbb{R}^3$. For the simple Lie algebras of types VIII and IX the groups of inner automorphisms G_2 coincide with the three-dimensional Lie groups SO(2, 1) and SO(3) corresponding to these algebras. For the other three-dimensional Lie algebras the groups of inner automorphisms G_2 have dimensions four and higher.

The three-dimensional Lie algebras and the associated homogeneous cosmological models are naturally divided into two classes [62] corresponding to the two parts of table 1: class A $(a = 0)$ and class B $(a \neq 0, n_3 = 0)$. The homogeneous cosmological models in each of these two classes possess a number of important common properties.

II. For homogeneous cosmological models all components of the Riemann curvature tensor R^i_{jkl} of the metric (3.2) can be expressed in the basis of vector fields X_0, X_1, X_2, X_3 in terms of the components of the metric $g_{ij}(t)$, its first and second time derivatives and the structure constants C^k_{ij} of the corresponding Lie algebra. All necessary computations are carried out according to standard formulas determining the curvature of the metric in an anholonomic basis (see [84, 85]). The components of the Ricci curvature tensor have the following form:

$$R^0_0 = -\frac{1}{2}\varkappa^\alpha_\alpha - \frac{1}{4}\varkappa^\beta_\alpha\varkappa^\alpha_\beta, \quad \alpha, \beta, i, j, k, l = 1, 2, 3,$$

$$R^0_i = -\frac{1}{2}\varkappa^j_k(C^k_{ji} - \delta^k_i C^l_{lj}), \tag{3.5}$$

$$R^j_i = -P^j_i - \frac{1}{2\sqrt{|g|}}\frac{d}{dt}(\sqrt{|g|}\,\varkappa^j_i),$$

where $\varkappa^j_k = g_{ki}g^{ij}$, $|g| = |\det(g_{ij})|$ and the components of the three-dimensional Ricci tensor P_{ij} are given by the following formulas:

$$P_{ij} = -\Gamma^k_{il}\Gamma^l_{jk} - C^l_{lk}\Gamma^k_{ij},$$

$$\Gamma^k_{ij} = \frac{1}{2}(C^k_{ij} + C^m_{il}g_{mj}g^{kl} + C^m_{jl}g_{mi}g^{kl}). \tag{3.6}$$

These formulas (3.5, 3.6) show that Einstein's system of equations $R^j_i - \frac{1}{2}\delta^j_i R = T^j_i$ for homogeneous cosmological models with a hydrodynamical stress-energy tensor (1.3) is a system of six "tensor" equations $(i, j) \neq 0$ containing the second time derivatives of the metric $g_{ij}(t)$ and four equations $R^0_0 - \frac{1}{2}R = T^0_0$ and $R^0_i = T^0_i$ containing only the first derivatives of the components of the metric

$g_{ij}(t)$. These last four equations allow us to express the energy density ε and the three independent components of velocity u^i in terms of the components of the metric $g_{ij}(t)$ and their first time derivatives. If these expressions are substituted into Einstein's six tensor equations, then we obtain a closed system of differential equations of second order for the components of the metric $g_{ij}(t)$.

Let us show that for $R_i^0 = 0$ the metric of homogeneous cosmological models of class A is diagonalizable simultaneously for all t and in the case of a homogeneous cosmological model of class B it can be reduced to the form with only one element off the diagonal. The proof is based on an important property of Einstein's system of equations for homogeneous cosmological models, namely invariance under the action of the group G_2 of inner automorphisms of the associated Lie algebra. Indeed if a transformation $X_i \rightarrow X_i' = A_i^j X_j$ does not change the commutation relations satisfied by the vectors X_i (i.e. does not change the coefficients C_{ik}^j), then Einstein's system of equations has the same form in the basis of vector fields X_0, X_i' as in the basis X_0, X_i. Consequently the corresponding transformation of the metric

$$g_{ij} \rightarrow g_{ij}' = A_i^k g_{kl} A_j^l \tag{3.7}$$

preserves the set of solutions of this system.

In the case of a homogeneous cosmological model of type I the group G_2 of inner automorphisms of the commutative Lie algebra $\mathbb{R}^3$ is just the group of all linear transformations of $\mathbb{R}^3$. Therefore by means of some inner automorphism we can simultaneously change the metric $g_{ij}(t)$ and the matrix of its first derivatives with respect to time into diagonal form. A solution with diagonal initial data remains diagonal for all time, which means that in a homogeneous cosmological model of type I an arbitrary solution can be written in diagonal form.

For any homogeneous cosmological model the metric $g_{ij}(t)$ can be diagonalized for a given time $t = t_0$ by means of an inner automorphism of the associated Lie algebra. Note that for homogeneous cosmological models of class A Einstein's equations $R_i^0 = 0$ (see 3.5) have the following form at $t = t_0$:

$$(-2)R_1^0 = g_{23}(n_2 g^{33} - n_3 g^{22}) = 0,$$
$$(-2)R_2^0 = g_{13}(n_3 g^{11} - n_1 g^{33}) = 0, \tag{3.8}$$
$$(-2)R_3^0 = g_{21}(n_1 g^{22} - n_2 g^{11}) = 0.$$

For a homogeneous model of type II $(n_1 = 1, n_2 = n_3 = 0)$ an appropriate inner automorphism of the associated Lie algebra simultaneously diagonalizes the metric $g_{ij}(t_0)$ and ensures that $\dot{g}_{23}(t_0) = 0$. In this case it follows from (3.8) that $\dot{g}_{13}(t_0) = \dot{g}_{21}(t_0) = 0$. For other homogeneous models of class A it follows from (3.8) that whenever $g^{ii} \neq g^{jj}$ we have $\dot{g}_{ij} = 0$. If for example $g^{11} = g^{22}$ and $n_1 = n_2 = 1$, then an orthogonal rotation of the vectors X_1 and X_2 preserves the commutators and the diagonality of the metric $g_{ij}(t_0)$ thus giving $\dot{g}_{12}(t_0) = 0$.

Therefore for every homogeneous cosmological model of class A, at an initial time t_0 we can simultaneously diagonalize the two matrices $g_{ij}(t_0)$ and $\dot{g}_{ij}(t_0)$ by means of a linear transformatioń of the vectors X_1, X_2, X_3 preserving the commutation relations. Einstein's equations also show that the diagonality of the initial data $g_{ij}(t_0)$ and $\dot{g}_{ij}(t_0)$ implies the diagonality of the solution for all time. This gives the desired simultaneous reduction of the solution to the diagonal form for all homogeneous cosmological models of class A.

For homogeneous cosmological models of class B if the metric $g_{ij}(t_0)$ is diagonalized, Einstein's equations $R_i^0 = 0$ (3.5) have the following form at $t = t_0$:

$$(-2)R_1^0 = (n_2\dot{g}_{23} + 3a\dot{g}_{13})g^{33} = 0,$$

$$(-2)R_2^0 = (3a\dot{g}_{23} - n_1 g_{13})g^{33} = 0, \tag{3.9}$$

$$(-2)R_3^0 = -a(\varkappa_1^1 + \varkappa_2^2 - 2\varkappa_3^3) + n_1\varkappa_1^2 - n_2\varkappa_2^1 = 0.$$

If $9a^2 + n_1 n_2 \neq 0$, then from the first two equations (3.9) it follows that $\dot{g}_{13}(t_0)$ $= \dot{g}_{23}(t_0) = 0$. For the solutions of Einstein's equations for homogeneous models of class B with such initial data we have that $g_{13}(t)$ and $g_{23}(t)$ are identically 0. This provides the possibility of reducing the solution to a form with only one non-zero non-diagonal coefficient $g_{12}(t)$. The exceptional case $9a^2 + n_1 n_2 = 0$ occurs only in homogeneous models of type VI with $a = 1/3$. In this case it is necessary to consider the solutions of Einstein's equations in the full space of three-dimensional matrices (see Chapt. III). In the case of a homogeneous cosmo-logical model of type V $(a = 1, n_i = 0)$ an appropriate inner automorphism of the associated Lie algebra simultaneously diagonalizes the metric $g_{ij}(t_0)$ and gives $\dot{g}_{12}(t_0) = 0$. This and the equations (3.9) imply that the solutions of Einstein's equations in a homogeneous model of type V can be simultaneously diagonalized for all time.

For the homogeneous cosmological models of class A and homogeneous models of type V let us use as coordinates in the space of diagonal metrics $g_{ij}(t)$ their eigenvalues $q_i(t) = g_{ii}(t)$. For other homogeneous cosmological models of class B let us diagonalize the metric $g_{ij}(t)$ (which has one element $g_{12}(t) \neq 0$ off the diagonal) by means of a time dependent linear transformation $\mathscr{P}(t)$, which pre-serves the commutation relations in the associated Lie algebra. Such transfor-mations have the following form:

$$
\begin{array}{cc}
\text{type VII} & \text{type III, VI} \\
\mathscr{P}(t) = \begin{pmatrix} \cos\varphi & \sin\varphi & 0 \\ -\sin\varphi & \cos\varphi & 0 \\ 0 & 0 & 1 \end{pmatrix}, & \mathscr{P}(t) = \begin{pmatrix} \operatorname{ch}\varphi & \operatorname{sh}\varphi & 0 \\ \operatorname{sh}\varphi & \operatorname{ch}\varphi & 0 \\ 0 & 0 & 1 \end{pmatrix},
\end{array}
$$

$$
\begin{array}{c}
\text{type IV} \\
\mathscr{P}(t) = \begin{pmatrix} 1 & \varphi & 0 \\ 0 & 1 & 0 \\ 0 & 0 & 1 \end{pmatrix}.
\end{array}
\tag{3.10}
$$

By definition we have

$$\mathscr{P}^t \circ g \circ \mathscr{P} = q, \quad g = (\mathscr{P}^t)^{-1} \circ q \circ \mathscr{P}^{-1}, \tag{3.11}$$

where g is the matrix of the metric (g_{ij}) and q is a diagonal matrix with entries q_1, q_2, q_3.

Note that for homogeneous cosmological models of class A the diagonal metrics $g_{ii}(t)$ satisfy Einstein's equations $R_i^0 = 0$ identically. For homogeneous models of class B the metrics with one off-diagonal coefficient $g_{12}(t)$ identically satisfy the equations $R_1^0 = R_2^0 = 0$, whereas $R_3^0 = 0$ (see 3.9) is an extra condition, which will be considered later on.

III. The tensor components of Einstein's equations can be written in the convenient Lagrangian form. To do this we will use the derivation of Einstein's equations from the principle of least action (1.8) based on the following well-known identity (below the indices of tensors take the following values: $i, j, k, \alpha, \beta = 1, 2, 3; l, n, m = 0, 1, 2, 3$):

$$\delta(R\sqrt{|g|}) = (R_{mn} - \tfrac{1}{2}g_{nm}R)\sqrt{|g|}\,\delta g^{nm} + \sqrt{|g|}\,g^{nm}\delta R_{nm}. \tag{3.12}$$

Here and below $|g| = |\det(g_{ij})|$. The last summand in (3.12) is a divergence (see [84]):

$$\sqrt{|g|}\,g^{nm}\delta R_{nm} = \partial(\sqrt{|g|}\,w^l)/\partial x^l. \tag{3.13}$$

Therefore, in the case of finite variations of the metric δg_{ij} the integral of the above expression over the entire four-dimensional space-time is identically zero. For homogeneous cosmological models the variations of the metric are not finite by the very definition of homogeneity. Therefore, in the derivation of the Lagrangian form of Einstein's equations for homogeneous cosmological models the last summand in (3.12) must be taken into account (the homogeneous cosmological model of type IX is an exception, because the compactness of the group SO(3) implies that the homogeneous variations of the metric are finite). Using the formulas (3.5) we see that

$$\sqrt{|g|}\,g^{nm}\delta R_{nm} = -\sqrt{|g(t)|}\,g^{ik}(t)\delta P_{ik} + \frac{d}{dt}(\sqrt{|g|}\,w^0), \tag{3.14}$$

where $w^0 = -\tfrac{1}{2}(g^{ik}\delta g_{ik} + \delta(\ln|g|))$. The first summand in (3.14) can be written in the following form:

$$-\sqrt{|g(t)|}\,g^{ik}(t)\delta P_{ik} = h_i'\delta q_i + h_\varphi'\delta_\varphi. \tag{3.15}$$

Here and further on we use the representation of the metric in the form (3.11): $g = (\mathscr{P}^t)^{-1}q\mathscr{P}^{-1}$; the coordinate φ for the models of class B is defined by the formulas (3.10); for the models of class A (and the model of type V), the coordinate $\varphi \equiv 0$, $\mathscr{P}(\varphi) \equiv 1$. The coefficients h_i and h_φ will be computed later on.

Let us perform the necessary transformations of the remaining summands in the equation (3.12). Einstein's equations imply that in the absence of motion of the matter (i.e. the equation of state of matter is $p = k\varepsilon$, see (1.3)) we obtain

$$(R_{nm} - \tfrac{1}{2}g_{nm}R)\sqrt{|g|}\,\delta g^{nm} = -T_n^l g^{nm}\sqrt{|g|}\,\delta g_{ml}$$

$$= T_{nm}\sqrt{|g|}\,\delta g^{nm}$$

$$= -\varepsilon\sqrt{|g|}(g_{00}^{-1}\delta g_{00} - kg^{ij}\delta g_{ji}) \tag{3.16}$$

$$= -\varepsilon\sqrt{|g|}(g_{00}^{-1}\delta g_{00} - kq_i^{-1}\delta q_i).$$

Here we use the equation $g^{il}\delta g_{il} = |g|^{-1}\delta|g| = q_i^{-1}\delta q_l$. In the synchronous system of coordinates (3.2) the formulas (3.5) imply that

$$R\sqrt{|g|} = (R_0^0 + R_i^i)\sqrt{|g|}$$

$$= \frac{1}{4}\sqrt{|g|}(\varkappa_\alpha^\alpha\varkappa_\beta^\beta - \varkappa_\alpha^\beta\varkappa_\beta^\alpha) - P_\alpha^\alpha\sqrt{|g|} - \frac{d}{dt}(\varkappa_\alpha^\alpha\sqrt{|g|}) \tag{3.17}$$

$$= L_1 - \frac{d}{dt}L_2,$$

where we introduce the following notation:

$$L_2 = \varkappa_\alpha^\alpha\sqrt{|g|}, \quad L_1 = T' - V_G' \quad \text{и}$$

$$T' = \tfrac{1}{4}(q_1 q_2 q_3)^{1/2}(\varkappa_\alpha^\alpha\varkappa_\beta^\beta - \varkappa_\alpha^\beta\varkappa_\beta^\alpha), \quad V_G' = P_\alpha^\alpha(q_1 q_2 q_3)^{1/2}. \tag{3.18}$$

The function L_1 contains only the first time derivatives of the metric $g_{ij}(t)$ and further plays the role of the Lagrangian.

Consider a system of coordinates which differs from the synchronous system (see (3.2)) by a time change

$$\frac{d\tau}{dt} = |g(t)|^{-k/2} = (q_1 q_2 q_3)^{-k/2}. \tag{3.19}$$

In this system of coordinates we have $g_{00} = (q_1 q_2 q_3)^k$. Therefore, the expression (3.16) is identically zero. The expression (3.14) in the new system of coordinates takes on the form

$$\sqrt{|g|}g^{nm}\delta R_{nm} = -(q_1 q_2 q_3)^{(1+k)/2}g^{ik}(t)\delta P_{ik} \tag{3.20}$$

$$+ \frac{d}{d\tau}(w^0(q_1 q_2 q_3)^{(1+k)/2}).$$

Because of invariance of the scalar curvature R in the new system of coordinates we have

$$R\sqrt{|g|} = L_0 - \frac{d}{d\tau}L_2, \quad L_0 = L_1(q_1 q_2 q_3)^{k/2}. \tag{3.21}$$

Later on it turns out to be convenient to use the Lagrangian $L = \tfrac{1}{2}L_0$. In the new system of coordinates let us substitute the newly found expressions (3.16), (3.20) and (3.21) into the equation (3.12) and then integrate with respect to time τ from τ_1 to τ_2 (we assume that at the endpoints of the interval of integration we have $\delta g_{ij}(\tau_k) = 0$). This results in the following equation:

$$\int \left(\frac{\partial L}{\partial q_i} - \frac{d}{d\tau} \frac{\partial L}{\partial \dot{q}_i} \right) \delta q_i \, d\tau + \int \left(\frac{\partial L}{\partial \varphi} - \frac{d}{d\tau} \frac{\partial L}{\partial \varphi} \right) \delta \varphi \, d\tau$$

$$= \int h_i \delta q_i \, d\tau + \int h_\varphi \delta \varphi \, d\tau,$$

(3.22)

where $h_i = \frac{1}{2} h_i' (q_1 q_2 q_3)^{k/2}$. Since the variations δq_i, $\delta \varphi$ are arbitrary, the above equation (3.22) implies that Einstein's equations for homogeneous cosmological models without motion of matter are equivalent to the following system of equations

$$\frac{\partial L}{\partial q_i} - \frac{d}{d\tau} \frac{\partial L}{\partial \dot{q}_i} = h_i, \qquad \frac{\partial L}{\partial \varphi} - \frac{d}{d\tau} \frac{\partial L}{\partial \dot{\varphi}} = h_\varphi.$$

(3.23)

Note that this derivation allows us to eliminate the energy density ε from Einstein's equations and obtain a closed system of differential equations (3.23), which determine the changes of the metric $g_{ij}(t)$. The energy density ε can be expressed in terms of the metric g_{ij} and its first derivatives $\dot{g}_{ij}$ by means of Einstein's equation $R_0^0 - \frac{1}{2} R = T_0^0$. Indeed the formulas (3.5) imply that

$$2\varepsilon \sqrt{|g|} = 2(R_0^0 - \frac{1}{2}R) \sqrt{|g|} = (R_0^0 - R_\alpha^\alpha) \sqrt{|g|}$$

$$= \frac{1}{4} (\varkappa_\alpha^\alpha \varkappa_\beta^\beta - \varkappa_\alpha^\beta \varkappa_\beta^\alpha) \sqrt{|g|} + P_\alpha^\alpha \sqrt{|g|} = T' + V_G'.$$

(3.24)

The function $H' = T' + V_G'$ (see (3.18)) is the energy for the Lagrangian L_1 (3.17). The energy corresponding to the Lagrangian $L = \frac{1}{2} L_0$ has the form (see (3.21)):

$$H = T + V_G = \frac{1}{2}(q_1 q_2 q_3)^{k/2}(T' + V_G') = \varepsilon(q_1 q_2 q_3)^{(1+k)/2}.$$

(3.25)

Later we will show that this function is the first integral of the system (3.23).

IV. Let us move on to finding an explicit form for the function L and the coefficients h_i and h_φ. According to (3.25) and (3.18) the kinetic energy T has the form:

$$T = \frac{1}{8}(q_1 q_2 q_3)^{(1+k)/2} \left[\left(\text{Tr} \left(\frac{dg}{dt} \circ g^{-1} \right) \right)^2 - \text{Tr} \left(\frac{dg}{dt} \circ g^{-1} \circ \frac{dg}{dt} \circ g^{-1} \right) \right].$$

(3.26)

Now we use the matrix g in the form (3.11) and perform a few simple transformations, which use standard properties of the trace of a matrix and the explicit form of the matrices $\mathscr{P}(t)$ (for homogeneous models of class A we have $\mathscr{P}(t) = 1$, $\varphi = 0$), to obtain the final expression for T:

$$T = \frac{1}{4(q_1 q_2 q_3)^{(1+k)/2}} (\dot{q}_1 \dot{q}_2 q_3 + q_1 \dot{q}_2 \dot{q}_3 + \dot{q}_1 q_2 \dot{q}_3 - \dot{\varphi}^2 q_3 (n_1 q_1 - n_2 q_2)^2).$$

(3.27)

Here the dot represents differentiation with respect to time τ, defined according to the substitution (3.19). Note that the kinetic energy T considered as a quadratic form in the velocities $\dot{q}_i$ is indefinite. This is one of the reasons why the dynamical systems describing homogeneous cosmological models in the general theory of relativity substantially differ from Hamiltonian systems studied in classical mechanics.

Potential energy $V_G = (q_1 q_2 q_3)^{(1+k)/2} P_\alpha^\alpha$ is a scalar, so it is sufficient to compute the value of V_G for a diagonal metric q. According to (3.6) and (3.4) the Ricci tensor for a diagonal metric has the following components:

$$\bar{P}_{11} = -2a^2 q_1 q_3^{-1} + \tfrac{1}{2}(n_1^2 q_1^2 - (n_2 q_2 - n_3 q_3)^2)q_2^{-1} q_3^{-1},$$

$$\bar{P}_{22} = -2a^2 q_2 q_3^{-1} + \tfrac{1}{2}(n_2^2 q_2^2 - (n_3 q_3 - n_1 q_1)^2)q_3^{-1} q_1^{-1},$$

$$\bar{P}_{33} = -2a^2 + \tfrac{1}{2}(n_3^2 q_3^2 - (n_1 q_1 - n_2 q_2)^2)q_1^{-1} q_2^{-1}, \qquad (3.28)$$

$$\bar{P}_{12} = \bar{P}_{21} = a(n_1 q_1 - n_2 q_2)q_3^{-1}, \quad \bar{P}_{13} = \bar{P}_{31} = \bar{P}_{23} = \bar{P}_{32} = 0.$$

From this for the homogeneous models of class A $(a = 0)$ we obtain

$$V_G = \frac{1}{4(q_1 q_2 q_3)^{(1-k)/2}}\left(2\sum_{i<j}^{3} n_i n_j q_i q_j - \sum_{i=1}^{3} n_i^2 q_i^2\right). \qquad (3.29)$$

For the homogeneous models of class B $(a \neq 0, n_3 = 0)$ from (3.28) we find that

$$V_G = \frac{1}{4(q_1 q_2 q_3)^{(1-k)/2}}\left(-12a^2 q_1 q_2 - (n_1 q_1 - n_2 q_2)^2\right). \qquad (3.30)$$

Thus, the formulas (3.27), (3.29) and (3.30) so obtained determine the explicit form of the function $L = T - V_G$ for all homogeneous cosmological models without the motion of matter.

In order to compute the coefficients $h_i = \tfrac{1}{2}h_i'(q_1 q_2 q_3)^{k/2}$ given by the relation (3.15) we will use matrix notation

$$g^{ik}\delta P_{ik} = \mathrm{Tr}(g^{-1} \circ \delta P) \qquad (3.31)$$

and the following expressions:

$$g^{-1} = \mathscr{P}q^{-1}\mathscr{P}^t, \quad P = (\mathscr{P}^t)^{-1}\bar{P}\mathscr{P}^{-1},$$

$$\delta P = \delta((\mathscr{P}^t)^{-1})\bar{P}\mathscr{P}^{-1} + (\mathscr{P}^t)^{-1}\delta\bar{P}\mathscr{P}^{-1} + (\mathscr{P}^t)^{-1}\bar{P}\delta(\mathscr{P}^{-1}). \qquad (3.32)$$

Here the Ricci tensor P for the metric g (which satisfies the relations (3.11)) is expressed in terms of the Ricci tensor $\bar{P}$ computed for the diagonal metric q (the components of $\bar{P}$ are given by (3.28)). Using standard properties of the trace of a product of matrices we obtain

$$\mathrm{Tr}(g^{-1}\delta P) = \mathrm{Tr}(q^{-1}\delta\bar{P}) + 2\mathrm{Tr}(q^{-1}\bar{P}\delta(\mathscr{P}^{-1})\mathscr{P}). \qquad (3.33)$$

According to the formulas (3.28) and (3.10), the matrices $q^{-1}\bar{P}$ and $\delta(\mathscr{P}^{-1})\mathscr{P}$ have the following form:

$$q^{-1}\bar{P} = \begin{pmatrix} q_1^{-1}\bar{P}_{11} & q_1^{-1}\bar{P}_{12} & 0 \\ q_2^{-1}\bar{P}_{12} & q_2^{-1}\bar{P}_{22} & 0 \\ 0 & 0 & q_3^{-1}\bar{P}_{33} \end{pmatrix}, \quad \delta(\mathscr{P}^{-1})\mathscr{P} = \delta\varphi \begin{pmatrix} 0 & -n_1 & 0 \\ n_2 & 0 & 0 \\ 0 & 0 & 0 \end{pmatrix}. \quad (3.34)$$

From this we find that

$$2\mathrm{Tr}(q^{-1}\bar{P}\delta(\mathscr{P}^{-1})\mathscr{P}) = 2\bar{P}_{12}(n_2 q_1^{-1} - n_1 q_2^{-1})\delta\varphi. \quad (3.35)$$

Computing the first summand in (3.33) and using the formulas (3.28) gives

$$\mathrm{Tr}(q^{-1}\delta\bar{P}) = 4a^2 q_3^{-2}\delta q_3 - 2a^2 q_1^{-1} q_3^{-1}\delta q_1 - 2a^2 q_2^{-1} q_3^{-1}\delta q_2. \quad (3.36)$$

The coefficients h_i and h_φ appear in the expansion of the quantity

$$-\tfrac{1}{2}(q_1 q_2 q_3)^{(1+k)/2} g^{ij}\delta P_{ij}.$$

Therefore, according to the expressions (3.33), (3.35) and (3.36), we obtain the final formulas

$$h_1 = a^2 q_2/Q, \quad h_2 = a^2 q_1/Q, \quad h_3 = -2a^2 q_1 q_2/q_3 Q,$$
$$h_\varphi = a(n_1 q_1 - n_2 q_2)^2/Q, \quad Q = (q_1 q_2 q_3)^{(1-k)/2}. \quad (3.37)$$

Thus for all homogeneous models of class A $(a = 0)$, $h_i = h_\varphi = 0$. Consequently Einstein's equations for models of class A are equivalent to a Lagrangian system of the type (3.23) and the corresponding energy $H = T + V_G$ (see (3.25)) is conserved: $H = \text{const}$.

For homogeneous models of class B we have $h_i \neq 0$, so the system of equations (3.23) is not Lagrangian. This system is equivalent to the system of Einstein's equations (in the absence of motion of the matter) if another additional condition $R_3^0 = 0$ is satisfied (see (3.9)). After a simple calculation this condition takes on the following form:

$$\frac{\dot{q}_1}{q_1} + \frac{\dot{q}_2}{q_2} - 2\frac{\dot{a}_3}{q_3} + \frac{(n_1 q_1 - n_2 q_2)}{aq_1 q_2}\dot{\varphi} = 0. \quad (3.38)$$

V. In order to study the behavior of the trajectories of the system (3.23) let us transform this system into a dynamical system defined in the phase-space with coordinates p_i, q_i, p_φ, φ. The momenta p_i and p_φ are given by the Legendre transformation:

$$p_i = \frac{\partial L}{\partial \dot{q}_i} = \frac{1}{4(q_1 q_2 q_3)^{(1+k)/2}}(\dot{q}_j q_k + q_j \dot{q}_k), \quad i, j, k = 1, 2, 3,$$

$$p_\varphi = \frac{\partial L}{\partial \dot{\varphi}} = -\frac{1}{2(q_1 q_2 q_3)^{(1+k)/2}}\dot{\varphi}q_3(n_1 q_1 - n_2 q_2)^2. \quad (3.39)$$

The inverse transformation has the form

$$\frac{\dot{q}_i}{q_i} = \frac{2}{(q_1 q_2 q_3)^{(1-k)/2}}(p_j q_j + p_k q_k - p_i q_i). \quad (3.40)$$

The Lagrangian system (3.23) $(h_i = h_\varphi = 0)$, which determines the evolution of the metric for homogeneous cosmological models of class A $(a = 0)$, becomes a Hamiltonian system under the Legendre transformation (3.39):

$$\dot{p}_i = -\frac{\partial H}{\partial q_i}, \quad \dot{q}_i = \frac{\partial H}{\partial p_i}. \tag{3.41}$$

The Hamiltonian H looks like this:

$$H = \dot{q}_i \frac{\partial L}{\partial q_i} = T + V_G$$

$$= \frac{1}{(q_1 q_2 q_3)^{(1-k)/2}} \left[2 \sum_{i<j}^{3} p_i p_j q_i q_j - \sum_{i=1}^{3} p_i^2 q_i^2 \tag{3.42} \right.$$

$$\left. + \frac{1}{4}\left(2 \sum_{i<j}^{3} n_i n_j q_i q_j - \sum_{i=1}^{3} n_i^2 q_i^2 \right) \right].$$

In the case of homogeneous cosmological models of class B the system of equations (3.23) under the Legendre transformation (3.39) turns into the following dynamical system $(i = 1, 2, 3)$:

$$\dot{p}_i = -\frac{\partial H}{\partial q_i} - h_i, \quad \dot{q}_i = \frac{\partial H}{\partial p_i},$$

$$\tag{3.43}$$

$$\dot{p}_\varphi = -\frac{\partial H}{\partial \varphi} - h_\varphi, \quad \dot{\phi} = \frac{\partial H}{\partial p_\varphi}.$$

Here the coefficients h_i and h_φ are given by the formulas (3.37) and the Hamiltonian H is given by

$$H = \frac{1}{(q_1 q_2 q_3)^{(1-k)/2}} \left[2 \sum_{i<j}^{3} p_i p_j q_i q_j - \sum_{i=1}^{3} p_i^2 q_i^2 \right.$$

$$\left. - \frac{p_\varphi^2 q_1 q_2}{(n_1 q_1 - n_2 q_2)^2} - \frac{1}{4}\left(12 a^2 q_1 q_2 + (n_1 q_1 - n_2 q_2)^2\right) \right]. \tag{3.44}$$

The dynamical system (3.43) is considered on the level of constraint $R_3^0 = 0$ (3.38). After a substitution of the expressions (3.40) this condition becomes

$$2p_3 q_3 - p_1 q_1 - p_2 q_2 - p_\varphi/2a = 0. \tag{3.45}$$

In the case of a homogeneous model of type V $(n_i = 0, a = 1)$ without the motion of matter we have $\varphi = p\varphi \equiv 0$, so the system of equations (3.43)–(3.45) is greatly simplified.

The dynamical systems (3.41) and (3.43)–(3.45) have two common properties: conservation of the Hamiltonian H and the existence of a monotone function[1]

---

[1] *Translator's note.* This is commonly known as a Liapunov function.

$$F = \frac{d}{dt}|g|^{1/6}, \tag{3.46}$$

where $|g| = \det \|g_{ij}\|$. The existence of the monotone function (3.46) is a general property of Einstein's equations in the synchronous system of coordinates (in the absence of homogeneity partial derivatives with respect to t are used everywhere). We shall give a proof of this statement, which was in fact obtained in [84]. The derivative of the function F with respect to t has the following form:

$$
\begin{aligned}
\frac{dF}{dt} &= \frac{d^2 |g|^{1/6}}{dt^2} \\
&= \frac{|g|^{1/6}}{6}\left(\left(\frac{|g|^{\cdot}}{|g|}\right)^{\cdot} + \frac{1}{6}\left(\frac{|g|^{\cdot}}{|g|}\right)^2\right) \\
&= \frac{|g|^{1/6}}{6}\left(\dot{\varkappa}^{\alpha}_{\alpha} + \frac{1}{6}(\varkappa^{\alpha}_{\alpha})^2\right) \leqslant \frac{|g|^{1/6}}{3}\left(\frac{1}{2}\dot{\varkappa}^{\alpha}_{\alpha} + \frac{1}{4}\varkappa^{\alpha}_{\beta}\varkappa^{\beta}_{\alpha}\right) \\
&= \frac{|g|^{1/6}}{3}(-R^0_0).
\end{aligned}
\tag{3.47}
$$

Here we use the algebraic inequality $(\varkappa^{\alpha}_{\alpha})^2 \leqslant 3\varkappa^{\alpha}_{\beta}\varkappa^{\beta}_{\alpha}$ and the explicit formula (3.5) for the Ricci tensor in the synchronous system of coordinates. Einstein's equations imply that

$$R^0_0 = T^0_0 - \frac{1}{2}T = \frac{1}{2}(\varepsilon + 3p) + \frac{(p+\varepsilon)v^2}{1-v^2} \geqslant 0 \tag{3.48}$$

where v^2 is the square of the three-dimensional velocity of matter. Therefore from (3.47) and (3.48) we get

$$\frac{dF}{dt} = \frac{d^2|g|^{1/6}}{dt^2} \leqslant \frac{|g|^{1/6}}{3}(-R^0_0) \leqslant 0, \tag{3.49}$$

This and Einstein's equations imply the monotonicity of the function F. Note that the inequality (3.49) shows that in the synchronous system of coordinates the function $|g|^{1/6}$ is concave up. Therefore $|g|$ has no local minima with respect to t other than $|g| = 0$ and in a finite amount of time (with time directed in such a way that $|g|$ decreases) the determinant of $|g|$ becomes zero. The last statement shows that in the synchronous system of coordinates the metric will inevitably have a singularity.

A general method for the construction of the synchronous system of coordinates consists of the following: an arbitrary three-dimensional space-like hypersurface is chosen; at each point of the hypersurface we consider all geodesics (time-like) orthogonal to the hypersurface; a family of space-like sections is determined by the conditions $t = $ const, where the time t coincides with the length along the geodesic and $t = 0$ on the initial hypersurface. In a general choice of the initial hypersurface ($t = 0$) the set of orthogonal geodesics is focused at the focal points. Note that a singularity of the metric in the constructed synchronous system of coordinates can be fictitious, i.e. it may disappear for a different choice of coordinates.

In the case of homogeneous cosmological models there is a special family of space-like sections, namely the orbits of the action of the three-dimensional group of isometries G. In the corresponding synchronous system of coordinates (which will be used from now on) the singularity of the metric of homogeneous cosmological models in generally intrinsic. Let us prove this fact for homogeneous models in the filled space (and in the absence of motion of matter). From the existence of the first integral of

$$H = \varepsilon(q_1 q_2 q_3)^{(1+k)/2}$$

(see (3.25)) we get $\varepsilon \to \infty$ for $|g| = q_1 q_2 q_3 \to 0$. Einstein's equations imply that the scalar curvature R equals

$$R = -T_{ij}g^{ij} = (3k - 1)\varepsilon.$$

Therefore $|R| \to \infty$ if $\varepsilon \to \infty$, so the singularity of the solution is not removable.

The monotone function F (3.46) in the phase space of the dynamical systems (3.41) and (3.43) has the form

$$F = \frac{d(q_1 q_2 q_3)^{1/6}}{dt} = \frac{p_1 q_1 + p_2 q_2 + p_3 q_3}{3(q_1 q_2 q_3)^{1/3}}, \qquad \frac{dF}{d\tau} \leqslant 0 \qquad (3.50)$$

and is used extensively in subsequent sections. Further analysis of homogeneous cosmological models will be done separately for classes A and B.

4. Transformation of the Hamiltonian System

I. Einstein's equations for homogeneous cosmological models of class A are equivalent (as shown in Sect. 3) to a Hamiltonian system in the phase space p_i, q_i $(i = 1, 2, 3)$ with the Hamiltonian

$$H = \frac{1}{(q_1 q_2 q_3)^{(1-k)/2}} [T(p_i q_i) + \tfrac{1}{4} V_G(g_i)],$$

where

$$T(p_i q_i) = 2 \sum_{i<j}^3 p_i p_j q_i q_j - \sum_{i=1}^3 p_i^2 q_i^2,$$

$$V_G(q_i) = 2 \sum_{i<j}^3 n_i n_j q_i q_j - \sum_{i=1}^3 n_i^2 q_i^2.$$

In the new time τ, which is related to the synchronous time t by

$$d\tau/dt = (q_1 q_2 q_3)^{-k/2},$$

the Hamiltonian system has the following form:

$$\dot{p}_i = -\frac{\partial H}{\partial q_i} = -\frac{1}{(q_1 q_2 q_3)^{(1-k)/2}}\left[2p_i(p_j q_j + p_k q_k - p_i q_i) \right.$$

$$\left. + \frac{1}{2}n_i(n_j q_j + n_k q_k - n_i q_i) - \left(\frac{1-k}{2}\right)\bar{H}/q_i \right], \tag{4.1}$$

$$\dot{q}_i = \frac{\partial H}{\partial p_i} = \frac{1}{(q_1 q_2 q_3)^{(1-k)/2}}[2q_i(p_j q_j + p_k q_k - p_i q_i)].$$

Here $i, j, k = 1, 2, 3$, $\bar{H} = T(p_i q_i) + 1/4V_G(q_i)$.

An important property of the Hamiltonian system (4.1) is the existence of the monotone function

$$F = \frac{d}{dt}(\det(g_{ij}))^{1/6} = \frac{1}{3}\frac{p_1 q_1 + p_2 q_2 + p_3 q_3}{(q_1 q_2 q_3)^{1/3}}. \tag{4.2}$$

By (4.1) we have

$$\frac{dF}{dt} = \frac{2}{3(q_1 q_2 q_3)^{5/6}}\left[-\frac{1}{3}(p_1 q_1 + p_2 q_2 + p_3 q_3)^2 + T(p_i q_i) - \frac{1+3k}{4}\bar{H}\right]. \tag{4.3}$$

Since the quadratic form

$$-\tfrac{1}{3}(p_1 q_1 + p_2 q_2 + p_3 q_3)^2 + T(p_i q_i)$$

is nonnegative definite (it has two negative eigenvalues and one zero eigenvalue corresponding to the eigenvector $p_1 q_1 = p_2 q_2 = p_3 q_3$) and $\bar{H} \geqslant 0$, then from (4.3) it follows that $dF/dt \leqslant 0$. Thus $d^2(\det g_{ij})^{1/6}/dt^2 \leqslant 0$, and consequently $\det(g_{ij})(t)$ can have only one local maximum and cannot have any local minima at all. Negative values of F correspond to a contraction of the volume $\det(g_{ij})(t)$ and positive values correspond to expansion. Both of these processes are described by the system (4.1) in the same region $p_1 q_1 + p_2 q_2 + p_3 q_3 < 0$, but with opposite directions of time (in view of this we will speak of the direction of time towards contraction $(d(\det g_{ij})/dt < 0)$ and towards expansion $(d(\det g_{ij})/dt > 0)$).

The trajectories of the system (4.1) intersect the set of zeroes of the derivative $dF/dt = 0$ $(p_1 q_1 = p_2 q_2 = p_3 q_3, \bar{H} = 0)$ transversally everywhere, except at the critical points T_k $(p_\alpha q_\alpha = 0, n_i q_i = n_j q_j, n_k q_k = 0)$. Therefore, the function F decreases monotonically along the trajectories of the system (4.1). This system (4.1) has no critical points in the physical region $q_i > 0$ (except for the previously mentioned critical points T_k for the models of types I and VII_0 corresponding to the flat Minkowski solution). This means that stationary homogeneous cosmological models, except for the Minkowski solution, do not exist.

All the critical points of the dynamical system (4.1) lie on the boundary of the physical region. The existence of a three-dimensional manifold of degenerate critical points of the system (4.1) $(q_1 = q_2 = q_3 = 0$, momenta p_i are arbitrary) and the fact that the phase space is not compact make it more difficult to study the behavior of this system near a cosmological singularity, i.e. near $q_1 q_2 q_3 = 0$.

To apply the methods of qualitative theory of differential equations to analyze the system (4.1) we will construct a compact manifold S with a dynamical system on it, which is equivalent to (4.1) in the interior of S and continues to the boundary Γ of S in such a way as to preserve the direction of time. The boundary Γ is not a smooth manifold as it consists of several faces, which intersect at corners of various dimensions. At the same time all critical points will have at least two non-zero eigenvalues corresponding to each.

Some critical points correspond to trajectories, which approach them from the physical region (the interior of the manifold S). Such points provide power asymptotics of the solutions near the singularity $q_1 q_2 q_3 = 0$. Other critical points, which lie on the corners of the boundary Γ, have no separatrices approaching them from the physical region. Their separatrices lie entirely on the boundary of the manifold S (more accurately on the various components of the boundary intersecting at the particular corner) and go from one critical point to another. The critical points and their separatrices form the "skeleton" of the dynamical system. The asymptotic behavior of solutions near a singularity is determined by the motion of the trajectory along the separatrices from a neighborhood of one unstable critical point to a neighborhood of another and so on, until it approaches a neighborhood of an attracting critical point (if the cosmological model in question has such critical points).

II. Let us now construct the compact manifold S and the dynamical system on it. In the general case (all $n_i \neq 0$) the manifold S has dimension 5 and is covered by two local charts W_1 and W_2. The local chart W_1 has coordinates $\bar{s}_i$, y_i:

$$\bar{s}_i = p_i q_i / G, \quad y_i = q_i / G, \quad G = (q_1^2 + q_2^2 + q_3^2)^{1/2}. \tag{4.4}$$

The coordinates $\bar{s}_i$ are those of the Euclidean space $\mathbb{R}^3$, whereas the coordinates y_i are those of the two-dimensional unit sphere S^2: $y_1^2 + y_2^2 + y_3^2 = 1$.

To compactify the Euclidean space $\mathbb{R}^3$ at infinity we introduce a local chart W_2 with coordinates y_i, s_i, w:

$$s_i = \bar{s}_i / (\bar{s}_1^2 + \bar{s}_2^2 + \bar{s}_3^2)^{1/2} = p_i q_i / P,$$

$$w = (\bar{s}^2 + \bar{s}_2^2 + \bar{s}_3^2)^{-1}, \tag{4.5}$$

$$P = (p_1^2 q_1^2 + p_2^2 q_2^2 + p_3^2 q_3^2)^{1/2}.$$

The coordinates s_i are those of the unit sphere S^2: $s_1^2 + s_2^2 + s_3^2 = 1$, and w runs along a ray $0 \leqslant w < \infty$. The two-dimensional sphere $s_1^2 + s_2^2 + s_3^2 = 1$, $w = 0$ corresponds to a sphere at infinity in the coordinates $\bar{s}_i$. Together the two systems of coordinates $\bar{s}_i$ and s_i, w cover the closed three-dimensional ball D^3. The compact five-dimensional manifold S is a subset of the cross-product $D^3 \times S_y^2$ determined by the physical conditions

$$y_i \geqslant 0, \quad w \geqslant 0, \quad H_2 \geqslant 0, \quad \bar{s}_1 + \bar{s}_2 + \bar{s}_3 \leqslant 0, \tag{4.6}$$

where

$$H_2 = \frac{1}{P^2} H(q_1 q_2 q_3)^{(1-k)/2} = 2 \sum_{i<j}^{3} s_i s_j - 1$$

$$+ \frac{1}{4} w \left(2 \sum_{i<j}^{3} n_i n_j y_i y_j - \sum_{i=1}^{3} n_i^2 y_i^2 \right).$$

The boundary Γ of S is not a smooth manifold. Γ consists of six four-dimensional components (faces) determined by the conditions

$$\Gamma_0: H_2 = 0, \quad \Gamma_i: y_i = 0, i = 1, 2, 3,$$

$$\Gamma_w: w = 0, \quad \Gamma_m: \bar{s}_1 + \bar{s}_2 + \bar{s}_3 = 0.$$

Components of the boundary Γ_0 and Γ_m lie in the "physical" region of the manifold S, i.e. these components correspond to nonsingular states of the metric, whereas the components $\Gamma_1, \Gamma_2, \Gamma_3, \Gamma_w$ are a compactification of the physical region of S by degenerate states of the metric and correspond to the cosmological singularity $q_1 q_2 q_3 = 0$. By (4.6), the component Γ_w is a cross-product $\Gamma_w = \Delta \times D^2$, where the coordinates y_i run over the triangle Δ ($y_i \geq 0$) and the coordinates s_i run over the disc D^2:

$$2 \sum_{i<j}^{3} s_i s_j - 1 \geq 0, \tag{4.7}$$

given as a subset of the sphere $s_1^2 + s_2^2 + s_3^2 = 1$, $w = 0$ by the condition $H_2 \geq 0$.

After a transformation into the local coordinates W_1 and a time change

$$\frac{d\tau_1}{d\tau} = \frac{G}{2(q_1 q_2 q_3)^{(1-k)/2}}$$

the Hamiltonian system (4.1) takes the form

$$\dot{\bar{s}}_i = -n_i y_i (n_j y_j + n_k y_k - n_i y_i) + (1-k) H_1 - 4\bar{s}_i (\bar{s}_\alpha - 2\bar{s}_\beta y_\beta^2),$$

$$\dot{y}_i = 8 y_i (y_\alpha^2 \bar{s}_\alpha - \bar{s}_i), \tag{4.8}$$

$$\dot{G} = 4 G (\bar{s}_\alpha - 2\bar{s}_\beta y_\beta^2),$$

where

$$H_1 = (q_1 q_2 q_3)^{(1-k)/2} H/G^2 = T(\bar{s}_i) + \tfrac{1}{4} V_G(y_i)$$

(in (4.8) and (4.9) summation is understood over the indices $\alpha, \beta, \gamma = 1, 2, 3$).

After a time change

$$\frac{d\tau_2}{d\tau} = \frac{P}{2(q_1 q_2 q_3)^{(1-k)/2}}$$

the Hamiltonian system (4.1) takes on the following form in the local coordinates W_2:

$$\dot{s}_i = w[-n_i y_i(n_j y_j + n_k y_k - n_i y_i)$$
$$+ s_i(n_\alpha y_\alpha s_\alpha(n_\beta y_\beta + n_\gamma y_\gamma - n_\alpha y_\alpha))]$$
$$+ (1-k)(1 - s_i(s_1 + s_2 + s_3))H_2,$$
$$\dot{w} = 2w[w(n_\alpha y_\alpha s_\alpha(n_\beta y_\beta + n_\gamma y_\gamma - n_\alpha y_\alpha)) \qquad (4.9)$$
$$- (1-k)(s_\alpha)H_2 + 4(s_\alpha - 2s_\beta y_\beta^2)],$$
$$\dot{y}_i = 8y_i(s_\alpha' y_\alpha^2 - s_i),$$
$$\dot{G} = 4G(s_\alpha - 2s_\beta y_\beta^2).$$

The local coordinates W_1 can be conveniently used to study the behavior of homogeneous cosmological models in a neighborhood of the state of maximal expansion $\bar{s}_1 + \bar{s}_2 + \bar{s}_3 = 0$. The local coordinates W_1 are used in a neighborhood of the cosmological singularity $q_1 q_2 q_3 = 0$.

Note that in the systems (4.8) and (4.9) the variable G can be separated. This is a consequence of the invariance of the Hamiltonian system (4.1) relative the group of scale transformations

$$\varphi_\lambda: q_i \to \lambda q_i, \quad \varphi_\lambda: p_i \to p_i, \quad \varphi_\lambda: \tau \to \lambda^{(1-3k)/2}\tau. \qquad (4.10)$$

Due to such a reduction of order the dynamical system (4.1) is equivalent to the dynamical system (4.8)–(4.9) on the compact five-dimensional manifold S. Obviously the newly obtained dynamical system can be smoothly continued to the components of the boundary Γ. Note that these components $\Gamma_0, \Gamma_1, \Gamma_2, \Gamma_3, \Gamma_w$ are invariant submanifolds of the dynamical system.

Various components of the boundary Γ pairwise intersect and thus form the three-dimensional corners of the boundary $Y_1, Y_2, Y_3, Y_4, Y_5, Y_6$. All these corners except Y_6 are invariant submanifolds of the dynamical system. On the corner $Y_0 = \Gamma_0 \cap (\Gamma_1 \cup \Gamma_2 \cup \Gamma_3)$ we have $H_2 = 0$ and the coordinates y_1, y_2, y_3 run along the sides of the spherical triangle Δ. The corners Y_1, Y_2, Y_3 project to the vertices of the triangle Δ. On Y_i we have $y_i = 1, y_j = y_k = 0$, i.e. $Y_i = \Gamma_j \cap \Gamma_k (i, j, k = 1, 2, 3)$. If $n_i = 0$, then Y_i is a solid cone

$$2\sum_{i<j}^{3} \bar{s}_i \bar{s}_j - \sum_{i=1}^{3} \bar{s}_i^2 \geq 0, \quad \bar{s}_1 + \bar{s}_2 + \bar{s}_3 \leq 0$$

completed by the disc D^2 (4.7). If $n_i = \pm 1$, then Y_i is a cut paraboloid

$$2\sum_{i<j}^{3} s_i s_j - 1 - \frac{1}{4}w \geq 0, \quad w \geq 0.$$

On the corner $Y_4 = (\Gamma_1 \cup \Gamma_2 \cup \Gamma_3) \cap \Gamma_w = \partial\Delta \times D^2$ the coordinates y_1, y_2, y_3 run along the sides of the spherical triangle Δ, $w = 0$ and the coordinates s_i are in the disc D^2 (4.7). The corner Y_4 is a solid three-dimensional torus. It consists of three cylinders $\Gamma_i \cap \Gamma_w$ glued pairwise at their bases. On the corner $Y_5 = \Gamma_0 \cap \Gamma_w = \Delta \times S^1$ the coordinates y_1, y_2, y_3 are on the spherical triangle Δ, $w = 0$ and the coordinates s_i are on the circle $S^1 = \partial D^2$. The corner $Y = \Gamma_0 \cap \Gamma_m$ lies entirely in the "physical" region of the manifold S, except for three points $T_i^0 \colon \bar{s}_k = 0$, $y_i = 0$, $y_j = y_k = 1/\sqrt{2}$.

The corners Y_i also intersect at points, where three components of the boundary meet $Y_{kl} = Y_k \cap Y_l$ $(k, l = 0, 1, \ldots, 6; \alpha = 1, 2, 3)$:

$$Y_{0\alpha} = \partial Y_\alpha \times x_\alpha, \quad Y_{04} = Y_{05} = \partial \Delta \times S^1, \quad Y_{06} = T_\alpha^0,$$

$$Y_{\alpha 4} = x_\alpha \times D^2, \quad Y_{\alpha 5} = x_\alpha \times S^1, \quad Y_{45} = Y_{045} = \partial \Delta \times S^1.$$

Here x_α are the three vertices of the spherical triangle Δ. There are also intersections of four components of the boundary or four corners

$$Y_{0\alpha 4} = Y_{0\alpha 5} = Y_{\alpha 45} = Y_{0\alpha 45} = x_\alpha \times S^1 = (\psi, \alpha).$$

The circle (ψ, α) (ψ is the angle on the circle and α is the number of the vertex x_α) has coordinates

$$w = 0, \quad y_i = \delta_{i\alpha}, \quad s_1 + s_2 + s_3 = -\sqrt{2},$$

$$s_1^2 + s_2^2 + s_3^2 = 1 \; (H_2((\psi, \alpha)) = 0)$$

and consists entirely of critical points of the dynamical system (4.9). All intersections of the corners of the boundary are invariant sets of the dynamical system on the manifold S.

The manifolds S (corresponding to various models) can have some components of the boundary Γ (or some corners), which are completely identical as well as the dynamical systems on them. For example the system on the component of a corner Y_i corresponding to the i-th vertex is determined by whether $n_i = \pm 1$ or $n_i = 0$ (see (4.9)). For all models the system on the boundary Γ_w ($w = 0$) is the same. All critical points, their eigenvalues and their separatrices (lying in Γ_w) are the same for all models of class A. For models with solvable groups (types I, II, VI_0, VII_0) the system (4.7) contains a closed subsystem of a smaller number of equations, so it is not necessary to consider these systems on the full five-dimensional manifold S. In fact all information about these models is contained in models of types VIII and IX, because the dynamical systems determining the evolution of homogeneous models with solvable groups coincide with systems on various components of the boundary Γ and corners Y_i for models of types VIII and IX.

By invariance relative to the scale transformations (4.10) the function F is also defined on the five-dimensional manifold S and in the coordinates (4.5) has the form

$$F = (s_1 + s_2 + s_3)/3w(y_1 y_2 y_3)^{1/3}.$$

Note that on the components of the boundary Γ_1, Γ_2, Γ_3, Γ_w we have $w(y_1 y_2 y_3)^{1/3} = 0$. On the components $\Gamma_i(y_i = 0)$ from the condition

$$H_2 = (s_1 + s_2 + s_3)^2 - 2 - \frac{w}{4}(n_j y_j - n_k y_k)^2 \geqslant 0$$

we obtain $(s_1 + s_2 + s_3) \leqslant -\sqrt{2}$. Since the same happens on Γ_w, $F = -\infty$ on the components Γ_w, Γ_1, Γ_2, Γ_3 for finite values of w. The only points on these components of the boundary at which $F \neq -\infty$ (and is not defined) are the exceptional points T_i^0 ($\bar{s}_i = 0$, $y_i = 0$, $y_j = y_k = 1/\sqrt{2}$). Since F is monotone (see (4.3)), it follows that all trajectories of the dynamical system (4.9) for models of types VIII and IX leave any compact region U inside the manifold S as well as the components of the boundary Γ_0, Γ_m for $\tau \to \infty$ (otherwise the function F, which must be bounded inside the region U, would grow indefinitely by absolute value). Consequently all trajectories of the dynamical system (4.9) for models of types VIII and IX approach the components of the boundary Γ_1, Γ_2, Γ_3, Γ_w for $\tau \to \infty$ while $F \to -\infty$.

Note that the local coordinates (4.4) and (4.5) are useful primarily for the construction of the compact manifold S. However the analysis of critical points (of the dynamical system (4.9) for example) for some models leads to unwieldy calculations. In fact we have a closed system of seven equations in the variables s_i, w, y_i subject to

$$s_2^1 + s_2^2 + s_3^2 = 1, \quad y_1^2 + y_2^2 + y_3^2 = 1.$$

In many cases in order to study the dynamical system (4.9) it is convenient to reduce it to a closed system of five equations by applying the above conditions. Indeed for $y_1 > 0$ let us introduce new coordinates V_1:

$$\bar{y}_2 = \frac{y_2}{y_1}, \quad \bar{y}_3 = \frac{y_3}{y_1},$$

$$\bar{u} = \frac{s_1}{s_2 + s_3}, \quad \bar{v}_2 = \frac{s_2 - s_3}{s_2 + s_3}, \quad \bar{w} = \frac{y_1}{2(s_2 + s_3)} w^{1/2} \tag{4.11}$$

and perform a time change

$$\frac{d\tau_3}{dt} = \frac{q_1}{(q_1 q_2 q_3)^{1/2} |\bar{w}|}.$$

For $y_2 \neq 0$ and $y_3 \neq 0$ let us introduce two analogous systems of coordinates V_2 and V_3 by cyclically permuting the indices 1, 2, 3. In the coordinates (4.11) and time τ_3 the dynamical system (4.1)–(4.9) has the following form:

$$\dot{\bar{u}} = n_1(n_2 \bar{y}_2 + n_3 \bar{y}_3 - n_1)\bar{w}^2 - \bar{u}\bar{w}^2(n_2 \bar{y}_2(n_1 + n_3 \bar{y}_3 - n_2 \bar{y}_2)$$
$$+ n_3 \bar{y}_3(n_1 + n_2 \bar{y}_2 - n_3 \bar{y}_3)) + (2\bar{u} - 1)\bar{H}_1,$$

$$\dot{\bar{v}} = [n_2 \bar{y}_2(n_1 - n_2 \bar{y}_2) - n_3 \bar{y}_3(n_1 - n_3 \bar{y}_3)]\bar{w}^2$$
$$- \bar{v}_2 \bar{w}^2(n_2 \bar{y}_2(n_1 + n_3 \bar{y}_3 - n_2 \bar{y}_2)$$
$$+ n_3 \bar{y}_3(n_1 + n_2 \bar{y}_2 - n_3 \bar{y}_3)) + 2\bar{v}_2 \bar{H}_1, \tag{4.12}$$

$$\dot{\bar{w}} = \bar{w}[\bar{u} - 1 - \bar{w}^2(n_2 \bar{y}_2(n_1 + n_3 \bar{y}_3 - n_2 \bar{y}_2)$$
$$+ n_3 \bar{y}_3(n_1 + n_2 \bar{y}_2 - n_3 \bar{y}_3)) + 2\bar{H}_1],$$

$$\dot{\bar{y}}_2 = \bar{y}_2(1 + \bar{v}_2 - 2\bar{u}), \quad \dot{\bar{y}}_3 = \bar{y}_3(1 - \bar{v}_2 - 2\bar{u}), \quad \dot{q}_1 = q_1(\bar{u} - 1),$$

where

$$\bar{H}_1 = H(q_1 q_2 q_3)^{(1-k)/2} \frac{\bar{w}^2}{q_1^2} = \frac{1-k}{4}[1 - (\bar{u} - 1)^2 - \bar{v}_2^2$$

$$+ \bar{w}^2(2n_1 n_2 \bar{y}_2 + 2n_1 n_3 \bar{y}_3 + 2n_2 n_3 \bar{y}_2 \bar{y}_3 - n_1^2 - n_2^2 - n_3^2 \bar{y}_3^2)].$$

The system of coordinates V_1 (4.11) is defined for $y_1 > 0$ and $s_2 + s_3 < 0$, so $\bar{w} < 0$. From $\bar{H}_1 > 0$ it follows that on the components of the boundary $\Gamma_1, \Gamma_2, \Gamma_3$, Γ_w of the manifold S, except for the three points T_i^0: $\bar{s}_k = 0$, $y_i = 0$, $y_j = y_k = 2^{-1/2}$ (if they belong to the manifold S for the model in question) we have that $\bar{w}$ is bounded by absolute value. Therefore the systems of coordinates V_1, V_2, V_3 cover a neighborhood of these components of the boundary Γ (except for a small neighborhood of the exceptional points T_i^0), so in order to analyze the behavior of the dynamical system in this neighborhood it is sufficient to study the dynamical system (4.12). Obviously the components of the boundary $\Gamma_2, \Gamma_3, \Gamma_w$ in the coordinates (4.11) are given by the equations $\bar{y}_2 = 0$, $\bar{y}_3 = 0$, $\bar{w} = 0$ respectively.

5. Cosmological Models of Types I and II

I. As mentioned in section 2, the solutions of the homogeneous model of type I for $k = 0$ $(p = 0)$ were integrated explicitly in the works [59, 88]. The dynamics of these solutions are completely determined by a closed subsystem of the system (4.9) for $n_i = 0$ (and arbitrary $0 \leqslant k < 1$):

$$\dot{s}_i = (1 - k)(1 - s_i(s_1 + s_2 + s_3))H_3,$$

$$H_3 = 2\sum_{i<j}^{3} s_i s_j - 1. \tag{5.1}$$

The system (5.1) is defined in the disc D^2 (4.7) $(H_3 \geqslant 0)$ on a two-dimensional sphere S^2: $s_1^2 + s_2^2 + s_3^2 = 1$ and is equivalent to the dynamical system (4.9) on the component of the boundary Γ_w (this system is the same for all homogeneous models of class A).

Thus, for a model of type I the manifold S is the disc D^2. The boundary of the disc is a circle

$$S^1: s_1^2 + s_2^2 + s_3^2 = 1, s_1 + s_2 + s_3 = -\sqrt{2}(H_3 = 0),$$

which consists of attracting critical points. These points correspond to Kasner solutions for vacuum

$$q_i = C_i t^{2p_i}, \quad p_i = 1 - \frac{2s_i}{s_1 + s_2 + s_3} = 1 + \sqrt{2}s_i. \tag{5.2}$$

The system (5.1) also has one repelling critical point $\Phi(s_i = -1/\sqrt{3})$, which

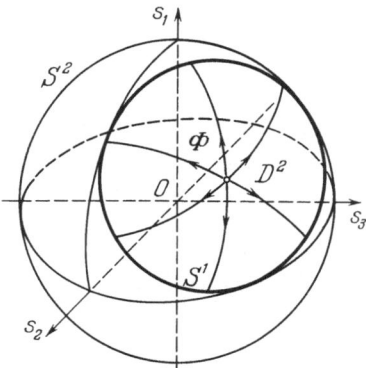

Fig. 7. Phase portrait of the dynamical system (5.1) defined in the disc D^2 on the unit sphere S^2 (*homogeneous cosmological model of type* I)

corresponds to the flat Friedmann solution

$$q_i = C_i t^{4/3(1+k)}. \tag{5.3}$$

By (5.1) we have $(s_1 - s_2)/(s_2 - s_3) = \text{const}$, so the trajectories of the system (5.1) are arcs of great circles on the sphere S^2 leaving (for $\tau \to -\infty$) the repelling critical point Φ and approaching (for $\tau \to +\infty$) attracting critical points on the circle S^1 (Fig. 7). Meanwhile the metric has asymptotic forms (5.3) and (5.2) respectively.

From this follows a well known statement that for the filled space for a model of type I, every solution has asymptotic forms of Kasner solutions towards contraction and asymptotic forms of the Friedmann solution towards expansion (i.e. it isotropizes).

II. Let us study the asymptotic behavior of solutions of the model of type II with $n_1 = 1$ and $n_2 = n_3 = 0$. For this model the system (4.12) has the following form:

$$\dot{\bar{u}} = -\bar{w}^2 + (2\bar{u} - 1)\bar{H}_2, \quad \dot{\bar{v}}_2 = 2\bar{v}_2\bar{H}_2, \quad \dot{\bar{w}} = \bar{w}(\bar{u} - 1 + 2\bar{H}_2),$$

$$\dot{\bar{y}}_2 = \bar{y}_2(1 + \bar{v}_2 - 2\bar{u}), \quad \dot{\bar{y}}_3 = \bar{y}_3(1 - \bar{v}_2 - 2\bar{u}), \quad \dot{q}_1 = q_1(\bar{u} - 1). \tag{5.4}$$

Here

$$\bar{H}_2 = \frac{(1-k)}{4}(1 - (\bar{u} - 1)^2 - \bar{v}_2^2 - \bar{w}^2).$$

The first three equations (5.4) form a closed subsystem, which coincides with the dynamical system (4.9) on the corner Y_1 ($\bar{y}_2 = \bar{y}_3 = 0$) of the boundary Γ. Therefore the manifold S for the model of type II is the corner Y_1 or one half of a ball $\bar{H}_2 \geqslant 0$ (Fig. 8):

$$(\bar{u} - 1)^2 + \bar{v}_2^2 + \bar{w}^2 \leqslant 1, \quad \bar{w} \leqslant 0. \tag{5.5}$$

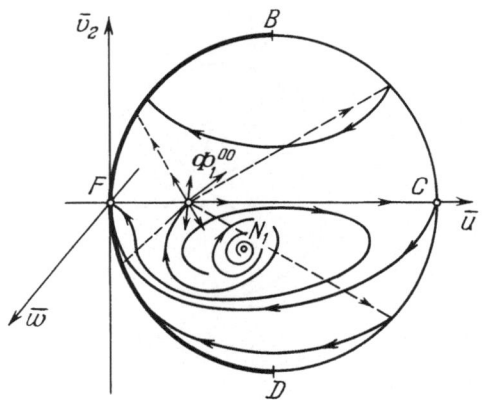

Fig. 8. Qualitative behavior of the trajectories of the dynamical system (5.4) on the closed three-dimensional manifold S for the homogeneous cosmological model of type II

On S the system (5.4) has the following invariant submanifolds:

1) $\bar{w} = 0$;   2) $\bar{v}_2 = 0$;   3) $\bar{H}_2 = 0$.

Let us enumerate the sets of critical points of the system (5.4):

1) circle: $\bar{w} = 0, (\bar{u} - 1)^2 + \bar{v}_2^2 = 1$;

2) point: Φ_1^{00}: $\bar{w} = 0, \bar{u} = \frac{1}{2}, \bar{v}_2 = 0$;

3) point: N_1: $\bar{u} = \dfrac{3 + k}{5 - k}, \bar{w} = -\dfrac{1}{5 - k}((1 - k)(1 + 3k))^{1/2}, \bar{v}_2 = 0$.

The eigenvalues of these critical points are as follows (next to the eigenvalues in parentheses we have included the variables corresponding to the appropriate eigenvectors):

1) $\lambda_1 = -\dfrac{(1 + k)}{2}(1 + \bar{u}), \lambda_2 = 0, (\bar{u}, \bar{v}_2); \quad \lambda_3 = \bar{u} - 1, (\bar{w})$.

(the zero eigenvalue corresponds to the fact that the dimension of the circle of critical points (1) is one)

2) $\lambda_1 = \dfrac{3}{8}(1 - k), (\bar{u}); \quad \lambda_2 = \dfrac{3}{8}(1 - k), (\bar{v}_2);$

$\lambda_3 = -\dfrac{1 + 3k}{8}, (w)$.

3) $\lambda_{1,2} = \dfrac{1 - k}{5 - k} \pm \dfrac{i}{5 - k}\left(\left(\dfrac{1 - k}{2}\right)(3 + 16k - 3k^2)\right)^{1/2}, (\bar{u}, w);$

$\lambda_3 = \dfrac{2(1 - k)}{5 - k}, (\bar{v}_2)$.

From this it follows that the arc DFB $(\bar{u} < 1$, see Fig. 8) consists of attracting critical points, the critical point N_1 is repelling and the critical points on the arc BCD $(\bar{u} > 1)$ and Φ_1^{00} are unstable.

Trajectories approaching the attracting critical points on the arc DFB correspond to the Kasner asymptotic form K_1, which generalizes the exact solutions (5.2):

$$q_1 \approx C_1 t^{2(1-\bar{u})/(1+\bar{u})}, \quad q_2 \approx C_2 t^{2(\bar{u}-\bar{v})/(1+\bar{u})}, \quad q_3 \approx C_3 t^{2(\bar{u}+\bar{v})/(1+\bar{u})}, \quad (5.6)$$

The arc FB corresponds to the asymptotic form K_2 $(p_2 < 0$ (see (5.2)), i.e. $q_2 \to \infty$, q_1 and $q_3 \to 0$ for $t \to 0$). The arc DF corresponds to the Kasner asymptotic form K_3 $(p_3 < 0)$. The repelling critical point N_1 corresponds to the following exact solution (also denoted by N_1), which is stable during the expansion of space:

$$q_1 = C_1 t^{(1-k)/(1+k)}, \quad q_2 = C_2 t^{(3+k)/2(1+k)}, \quad q_3 = C_3 t^{(3+k)/2(1+k)}. \quad (5.7)$$

The one-dimensional separatrix approaching the unstable critical point Φ_1^{00} corresponds to the asymptotic form Φ (5.3) (for $t \to 0$).

The asymptotic behavior of solutions of a homogeneous cosmological model of type II is completely described by the following theorem.

Theorem. *All metrics of a homogeneous model of type II in the filled space during the contraction of space have one of the following asymptotic forms:*

$$N_1, \Phi, K_2, K_3.$$

During the expansion of space, all metrics approach the exact solution N_1 (5.7). The solutions in vacuum have asymptotic forms K_2 and K_3 during contraction and K_1 during expansion.

To prove the first part of the theorem it is sufficient to show that in the physical region $w < 0$ every trajectory of the system (5.4) (which is not the stationary point N_1) approaches one of the critical points on the arc DFB or Φ_1^{00} for the direction of time towards contraction. Assume that at the initial point of the trajectory, $\bar{v}_2 \neq 0$. Since the function $\bar{v}_2$ is monotone along the trajectories of the system (5.4), this trajectory stays close to the boundary $\bar{H}_2 = 0$ of the manifold S (see Fig. 8). Critical points of the dynamical system (5.4), which lie on the boundary $\bar{H}_2 = 0$, form two acrs: BCD and DFB. Trajectories from the physical region $\bar{w} < 0$ cannot approach the unstable points on the arc BCD. Trajectories approaching these critical points lie on the boundary $\bar{w} = 0$ of the manifold S and therefore do not correspond to any physical solutions.

On the boundary $\bar{H}_2 = 0$ all trajectories approach the attracting critical points of the arc DFB. Therefore each trajectory approaching the boundary $\bar{H}_2 = 0$ (i.e. all trajectories in the interior of the manifold S) also approach one of the attracting critical points on the arc DFB. Suppose that at the initial point of the trajectory $\bar{v}_2 = 0$. Then in the plane $\bar{v}_2 = 0$ the system (5.4) has the form

$$\dot{\bar{u}} = -\bar{w}^2 + (2\bar{u} - 1)\bar{H}_2 = f_{\bar{u}},$$

$$\dot{\bar{w}} = w(\bar{u} - 1 + 2\bar{H}_2) = f_{\bar{w}}, \quad \bar{H}_2 = \frac{1-k}{4}(1 - (\bar{u} - 1)^2 - w^2) \tag{5.8}$$

and is defined on the half-disc $H_2 \geqslant 0$, $\bar{w} \leqslant 0$. It is easy to check that the following equation holds:

$$\frac{\partial}{\partial \bar{u}}(F_0 f_{\bar{u}}) + \frac{\partial}{\partial \bar{w}}(F_0 f_{\bar{w}}) = \frac{2}{\bar{w}} < 0, \quad \text{where} \quad F_0 = \frac{1}{\bar{w}\bar{H}_2}.$$

Therefore, according to the Dulac-Bendixon criterion the system (5.7) has no limit cycles. Thus there exists a unique separatrix, which leaves the point N_1 (in the plane $\bar{v}_2 = 0$ it is a repelling focus) as an unwinding spiral and approaches the point Φ_1^{00}. All other trajectories approach the point F.

By changing the direction of time we see that towards expansion all trajectories from the interior of the manifold S (i.e. in the completed space) approach the critical point N_1.

The trajectories on the boundary $\bar{H}_2 = 0$ (i.e. in vacuum) move in a circle in the plane $\bar{u}, \bar{w}$ with a constant value of $\bar{v}_2$. Thus after leaving the point $(\bar{u}, \bar{v}_2)$ on the arc BCD a trajectory approaches the point $(2 - \bar{u}, \bar{v}_2)$ on the arc DFB. Corresponding solutions have the asymptotic form (5.6) corresponding to the arc BCD (i.e. $K_1 (p_1 < 0)$) during the expansion of space and asymptotic forms (5.6) $K_2 (p_2 < 0)$ or $K_3 (p_3 < 0)$.

6. Cosmological Models of Type IX[2]

The homogeneous cosmological model of type IX with the group of isometries $SO(3)$ $(n_1 = n_2 = n_3 = 1)$ by virtue of its importance takes an exceptional place among other cosmological models. This is because the homogeneous model of type IX is the only possible homogeneous perturbation of the homogeneous and isotropic closed Friedmann solution. In the analysis of homogeneous models of type IX, the most important things are the study of various modes of behavior of the metric near singularities, the study of the most complex oscillatory mode of behavior of the metric, the statement and solution of the question about the typical states of the metric near singularities during contraction and expansion of space and the study of isotropization for the expansion of space.

I. Critical Points of the Dynamical System and Power Asymptotics of the Solutions. The dynamical system (4.9) has the following form on the closed manifold S for the model of type IX:

---

[2] The main results of this section were obtained in a joint work by the author and S. P. Novikov [12].

$$\dot{s}_i = w[-y_i(y_j + y_k - y_i) + s_i(s_\alpha y_\alpha(y_\beta + y_\gamma - y_\alpha))]$$
$$+ (1 - k)(1 - s_i(s_1 + s_2 + s_3))H_2,$$
$$\dot{w} = 2w[w(s_\alpha y_\alpha(y_\beta + y_\gamma - y_\alpha)) - (1 - k)(s_1 + s_2 + s_3)H_2$$
$$+ 4(s_\alpha - 2s_\beta y_\beta^2)],$$
$$\dot{y}_i = 8y_i(s_\alpha y_\alpha^2 - s_i),$$
$$\dot{G} = 4G(s_\alpha - 2s_\beta y_\beta^2),$$

(6.1)

where

$$H_2 = 2\sum_{i<j}^{3} s_i s_j - 1 + w\frac{1}{4}\left(2\sum_{i<j}^{3} y_i y_j - 1\right)$$

and there is implicit summation over the indices $\alpha, \beta, \gamma = 1, 2, 3$. From now on the time variable τ_2 in the system (6.1) will be denoted by τ.

As mentioned in section 4, from the existence of the monotone function

$$F = \frac{1}{3}\frac{(s_1 + s_2 + s_3)}{w(y_1 y_2 y_3)^{1/3}}$$

it follows that the dynamical system (6.1) has no critical points inside the physical region of the manifold S. Indeed all these critical points lie on the various components of the boundary Γ. The entire collection of critical points consists of sets of six types:

$$\Phi, N_i, T_i, A_i, B_i, (\psi, i)$$

(Fig. 9). We will list the coordinates of these critical sets as well as their eigenvalues.

1) The triangle of critical points Φ lies on the component of the boundary Γ_w and is determined by the conditions $w = 0$, $s_i = -1/\sqrt{3}$. The coordinates y_1, y_2, y_3 run along the triangle Δ, $H_2(\Phi) = 1$. The eigenvalues of the system (6.1) on the manifold S at these critical points are

$$\lambda_{1,2} = (1 - k)\sqrt{3} \qquad \text{(variables } s_i\text{)},$$
$$\lambda_3 = -2(1 + 3k)/\sqrt{3} \quad \text{(variable } w\text{)}, \qquad (6.2)$$
$$\lambda_{4,5} = 0 \qquad \text{(variables } y_i\text{)}.$$

Let Φ_i^0 denote the sides of the triangle Φ (on Φ_i^0 $y_i = 0$) and Φ_i^{00} denote the vertices of the triangle Φ (at $\Phi_i^{00} y_k = \delta_{ik}$).

2) Three isolated critical points N_i lie on the corners of the boundary $Y_i = \Gamma_j \cap \Gamma_k$ and have coordinates

$$s_i = -(3 + k)\sqrt{2}u_0, \quad s_j = s_k = -(5 - k)u_0/\sqrt{2},$$
$$w = 8(1 + 3k)(1 - k)u_0^2, \quad y_i = 1, \quad y_j = y_k = 0, \qquad (6.3)$$
$$u_0 = (43 + 2k + 3k^2)^{-1/2}.$$

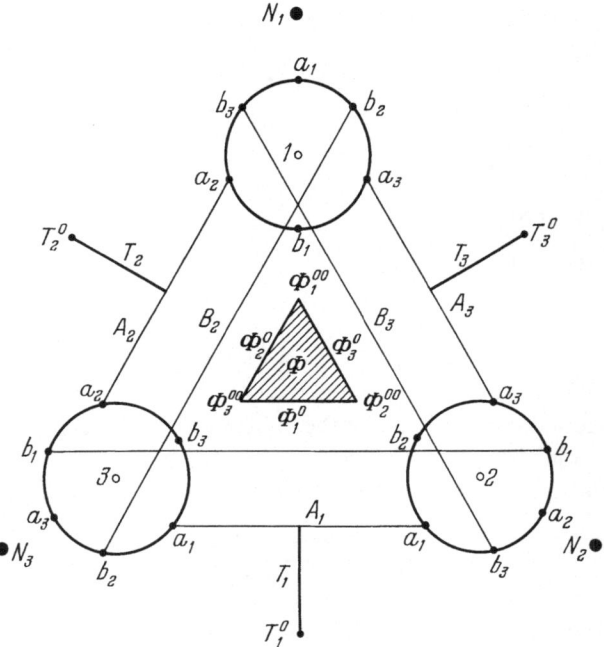

Fig. 9. The general distribution of critical points of the dynamical system (6.1) on the manifold S for the homogeneous cosmological model of type IX

The eigenvalues of the system (6.1) at the critical points N_i are

$$\lambda_1 = 8(1 - k)\sqrt{2u_0} \qquad \text{(variables } s_j),$$

$$\lambda_{2,3} = 4\left[(1 - k) \pm i\left(\left(\frac{1 - u_0}{2}\right)(3 + 16k - 3k^2)\right)^{1/2}\right]\sqrt{2u_0} \quad \text{(variables } s_j, w),$$

$$\lambda_4 = \lambda_5 = -4(1 + 3k)\sqrt{2u_0} \qquad \text{(variables } y_i).$$

$$(6.4)$$

3) Three segments of critical points T_i lie on the components of the boundary Γ_i and have coordinates

$$y_i = 0, \quad y_j = y_k = 1/\sqrt{2}, \quad s_i = 0, \quad s_j = s_k = -1/\sqrt{2}, \quad 0 \leqslant w \leqslant \infty, \quad H_2(T_i) = 0.$$

One endpoint of a segment T_i has coordinate $w = 0$ and lies in the intersection $\Gamma \cap \Gamma_w$. The other endpoint T_i^0 $(w = \infty)$ belongs to the local chart W_1 (4.2) with coordinates $\bar{s}_1 = \bar{s}_2 = \bar{s}_3 = 0$, i.e. lies in the intersection $\Gamma_i \cap \Gamma_0 \cap \Gamma_m$. At the critical points T_i the system (6.1) has the following eigenvalues:

$$\lambda_1 = -4\sqrt{2} \qquad \text{(variable } y_i),$$

$$\lambda_{2,3} = \pm 2i\sqrt{w}, \quad \lambda_4 = -2\sqrt{2}(1 - k), \quad \lambda_5 = 0 \quad \text{(variables } y_j, y_k, s_i, w).$$

$$(6.5)$$

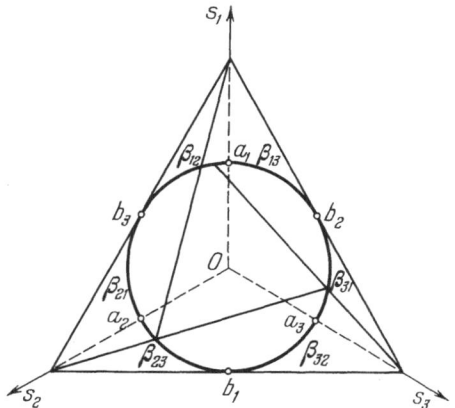

Fig. 10. The circle S^1 is divided into six arcs β_{ik}. Straight lines passing through the vertices of the triangle Δ define the image of the projection T on the circle S^1

4) Three circles of critical points are intersections of four components of the boundary $\Gamma_j \cap \Gamma_k \cap \Gamma_w \cap \Gamma_0$ and have coordinates $y_j = \delta_{ij}$, $w = 0$. The coordinates s_1, s_2, s_3 run over the circle S^1:

$$s_1^2 + s_2^2 + s_3^2 = 1, \quad s_1 + s_2 + s_3 = -\sqrt{2}.$$

At the critical points (ψ, i) the system (6.1) has the following eigenvalues:

$$\lambda_1 = 2(1 - k)(s_1 + s_2 + s_3) \quad \text{(variables } s_j\text{)},$$
$$\lambda_2 = 8(s_j + s_k - s_i) \quad \text{(variable } w\text{)}, \tag{6.6}$$
$$\lambda_3 = 8(s_i - s_j), \quad \lambda_1 = 8(s_i - s_k) \quad \text{(variables } y_j, y_k\text{)}.$$

The zero eigenvalue λ_5 corresponds (just as $\lambda_5 = 0$ for critical points on the segments T_i (see (6.5))) to the dimension of the set of critical points (ψ, i) being one.

The circle S^1 is separated by points

$$a_i(s_i = -2\sqrt{2}/3, \quad s_j = s_k = -1/3\sqrt{2})$$
$$b_i(s_i = 0, \quad s_j = s_k = -1/\sqrt{2})$$

into six arcs β_{ik} (see Fig. 10). On β_{ik} we have

$$s_i < s_k < s_j < 0.$$

It is also convenient to introduce arcs

$$\alpha_i = \beta_{ij} + \alpha_i + \beta_{ik}(s_i < s_j, s_k).$$

Arcs of critical circles (ψ, i) will be denoted by (α_i, k) and (β_{ij}, k), where $k = 1, 2, 3$ shows the number of circle.

Table 2. The signs of the eigenvalues of critical points on the arcs of circles (ψ, i)

	(β_{ik}, i)	(β_{ij}, i)	(β_{ki}, i)	(β_{kj}, i)	(β_{jk}, i)	(β_{ji}, i)
λ_1	−	−	−	−	−	−
$\lambda_2(w)$	+	+	−	−	−	−
$\lambda_3(y_j)$	−	−	−	+	+	+
$\lambda_4(y_k)$	−	−	+	+	+	−

In Table 2 we have given the signs of the eigenvalues (6.6). They depend on the position of a critical point on a circle (ψ, i) and do not change within each arc (β_{jk}, i). From the table it follows that all critical points on the circles (ψ, i) (except for the points a_i and b_i, where the eigenvalues (6.6) change sign) are non-degenerate and unstable.

5) Three segments of critical points A_i and three segments of critical points B_i lie in the intersections of the components of the boundary $\Gamma_i \cap \Gamma_w \cap \Gamma_0$. At these critical points we have

$$y_i = 0, \quad y_j^2 + y_k^2 = 1, \quad w = 0, \quad H_2(A_i) = H_2(B_i) = 0.$$

On the segments A_i we have

$$s_i = -2\sqrt{2}/3, \quad s_j = s_k = -1/3\sqrt{2}.$$

At the critical points of A_i the eigenvalues of the system (6.1) have the following form:

$$\lambda_1 = 4\sqrt{2} \qquad \text{(variable } y_i),$$
$$\lambda_2 = -2\sqrt{2}(1-k) \quad \text{(variables } s_j),$$
$$\lambda_3 = -16\sqrt{2}/3 \qquad \text{(variable } w),$$
$$\lambda_4 = \lambda_5 = 0.$$

$$(6.7)$$

On the segments B_i we have

$$s_i = 0, \quad s_j = s_k = -1/\sqrt{2}.$$

At these critical points the eigenvalues of the system (6.1) have the following form:

$$\lambda_1 = -4\sqrt{2} \qquad \text{(variable } y_i),$$
$$\lambda_2 = -2\sqrt{2}(1-k) \quad \text{(variables } s_j),$$
$$\lambda_3 = \lambda_4 = \lambda_5 = 0.$$

$$(6.8)$$

The segments of critical points A_i connect the critical points (a_i, k) and (a_i, j), whereas the segments B_i connect the critical points (b_i, k) and (b_i, j), which lie on the critical circles (ψ, k) and (ψ, j).

This calculation of the eigenvalues of critical points shows that the dynamical system on the manifold S does not have any attracting or repelling critical points whatsoever. Critical points of the sets Φ, N_i, T_i and (ψ, i) are nondegenerate (i.e. the number of zero eigenvalues is equal to the dimension of these sets) and unstable (because at each critical point there are eigenvalues with opposite signs of their real parts; note that the instability of critical points T_i appears only in the second order and follows from analyzing the behavior of the dynamical system on the boundary Γ_i (see below and Sect. IV)). The degenerate critical points of A_i and B_i are also unstable.

The separatrices approaching the critical points Φ, N_i and T_i pass through the physical region of the manifold S (at least almost all of them). The solutions, which correspond to these separatrices, have power asymptotics during the contraction of space $(t \to 0)$. Critical points in the set Φ are approached by a three-dimensional separatrix (corresponding to the eigenvalues λ_3, λ_4, λ_5 (6.2)), which corresponds to a quasi-isotropic asymptotic form [48] generalizing the asymptotic form of Friedmann solutions:

$$q_i \approx C_i t^{4/3(1+k)}. \tag{6.9}$$

The critical points N_i are approached by a two-dimensional separatrix (eigenvalues λ_4, λ_5 (6.4)), which corresponds to the asymptotic form due to Novikov [62]:

$$q_i \approx C_i t^{(1-k)/(1+k)}, \quad q_j \approx C_j t^{(3+k)/2(1+k)},$$
$$q_k \approx C_k t^{(3+k)/2(1+k)}. \tag{6.10}$$

The critical points T_i are approached by a three-dimensional separatrix (eigenvalues λ_1, λ_4, λ_5 (6.5)), which corresponds to the asymptotic form due to Taub [47]:

$$q_i \cong Ct^2, \quad q_j \cong q_k \cong C_1. \tag{6.11}$$

The power asymptotic forms of solutions during the contraction of space $(t \to 0)$ listed here are unstable, because the critical points Φ, N_i and T_i are unstable. A trajectory of the dynamical system, which moves in a neighborhood of the separatrices approaching the critical points Φ, N_i and T_i, diverges from the approaching separatrix as it nears the boundary Γ and starts moving along a separatrix leaving these (unstable) critical points. All separatrices leaving the critical points Φ, N_i and T_i lie on the various components of the boundary Γ. Therefore in order to study further movement of the trajectories along the boundary Γ it is necessary to study the behavior of the dynamical system (6.1) in the various components of the boundary Γ.

II. A Study of the Separatrices of the Critical Point Φ. On the component of the boundary Γ_w the dynamical system (6.1) has the following form:

$$\dot{s}_i = (1 - k)\big(1 - s_i(s_1 + s_2 + s_3)\big)H_3,$$

$$\dot{y}_i = 8y_i\left(\sum_{\alpha=1}^{3} s_\alpha y_\alpha^2 - s_i\right),$$

(6.12)

where

$$H_2 = 2\sum_{i<j}^{3} s_i s_j - 1.$$

By (6.12) we have $(s_1 - s_2)/(s_3 - s_2) = $ const in the region $H_3 > 0$. Therefore the trajectories of the system (6.12) in the coordinates s_i move on the unit sphere $s_1^2 + s_2^2 + s_3^2 = 1$ along arcs of great circles passing through the point $\Phi(s_i = -1/\sqrt{3})$. For $\tau \to -\infty$ these trajectories leave the critical point Φ and the coordinates y_i tend to some constants. Consequently all trajectories of the system (6.12) in the region $H_3 > 0$ are in fact separatrices leaving the critical points

$$\Phi(s_i = -1/\sqrt{3}, \quad y_1^2 + y_2^2 + y_3^2 = 1, \quad y_i \geqslant 0, \quad w = 0).$$

The critical points, which make up the triangle Φ, are repelling in the component of the boundary Γ_w and the separatrices leaving a critical point (y_1, y_2, y_3) form a two-dimensional surface.

Let us determine which critical points are approached by the separatrices leaving a critical point (y_1, y_2, y_3) in the triangle Φ as $\tau \to +\infty$. Each such separatrix in the coordinates s_i for $\tau \to +\infty$ tends to some (arbitrary) point s_i^0 on the circle

$$S^1: s_1^2 + s_2^2 + s_3^2 = 1, s_1 + s_2 + s_3 = -\sqrt{2}(H_3 = 0).$$

By (6.11) we have

$$(y_i/y_j)^{\cdot} = -(y_i/y_j)(s_i - s_j).$$

(6.13)

Suppose that a critical point (y_1, y_2, y_3) lies inside the triangle Φ. Then if $s_k^0 < s_i^0$, s_j^0, we have $y_k/y_j, y_k/y_i \to \infty$ for $\tau \to \infty$, i.e. the trajectory approaches the arc (α_k, k) on the circle of critical points (ψ, k). If however $s_k^0 = s_j^0 < s_i^0$ (in this case $s_i^0 = 0$), then $y_k/y_j = $ const and $y_k/y_i \to \infty$, so the trajectory tends to some point in the segment B_i for $\tau \to \infty$.

If the critical point (y_1, y_2, y_3) belongs to the side Φ_i^0 ($y_i = 0$) of the triangle Φ, then along the separatrices leaving this point we have $y_i \equiv 0$. At the same time if $s_j^0 < s_k^0$, then according to (6.13) $y_j/y_k \to \infty$ for $\tau \to \infty$, i.e. the trajectory approaches the arc $(\alpha_j, j) + (\beta_{ij}, j)$ ($s_j < s_k$) on the circle of critical points (ψ, j). If $s_j^0 = s_k^0$, then $y_j/y_k = $ const and for

$$s_i^0 = -2\sqrt{2}/3, \quad s_j^0 = s_k^0 = -1/3\sqrt{2}$$

the trajectory tends to a point x_i^0 on the segment A_i and for

$$s_i^0 = 0, \quad s_j^0 = s_k^0 = -1/\sqrt{2}$$

it tends to a point x_i^1 on the segment B_i.

If the critical point (y_1, y_2, y_3) is a vertex Φ_i^{00} $(y_k = \delta_{ki})$ of the triangle Φ, then the separatrices leaving this critical point lie on the corners of the boundary Γ $(y_k = \delta_{ki})$ and only the coordinates s_i change along them. The endpoints (for $\tau \to +\infty$) of these separatrices entirely fill the circle of critical points (ψ, i).

III. A Study of the Separatrices of Critical Points (ψ, i), A_i, B_i, N_i. According to (6.6) on the arc (β_{ji}, i) $(s_j < s_i < s_k)$ we have $\lambda_1 < 0$, $\lambda_2 < 0$, $\lambda_3 > 0$ and $\lambda_4 > 0$, so from each critical point (s_i^0) on the arc (β_{ji}, i) leaves a one-dimensional separatrix along which $y_k = 0$, $w = 0$, $s_i = s_i^0 = $ const and the fraction y_j/y_i grows without bound. The final point of this separatrix lies on the arc (β_{ji}, i) of the circle (ψ, j).

According to (6.6) on the arc (β_{ji}, i) $(s_j < s_k < s_i)$ we have $\lambda_1 < 0$, $\lambda_2 < 0$, $\lambda_3 > 0$ and $\lambda_4 > 0$, so from each critical point (s_i^0) on the arc (β_{jk}, i) leaves a two-dimensional separatrix along which $w = 0$, $s_i = s_i^0 = $ const $(H_3 = 0)$. The dynamics of the coordinates y_1, y_2, y_3 are determined by the system (6.12) and can be integrated explicitly:

$$y_i(\tau) = \frac{y_i^0 e^{-8s_i^0 \tau}}{\left(\sum\limits_{k=1}^{3} (y_k^0)^2 e^{-16s_k^0 \tau} \right)^{1/2}}. \tag{6.14}$$

All these separatrices (apart from one exceptional separatrix) approach the critical point (s_i^0) on the arc (β_{jk}, j) for $\tau \to \infty$. The exceptional separatrix, along which $y_j = 0$, approaches the point (s_i^0) on the arc (β_{jk}, k) (the next separatrix step from this point again leads to the point (s_i^0) on the arc (β_{jk}, j)).

Along the separatrices leaving the critical point (b_i, i) $(s_i = 0, s_j = s_k = -1/\sqrt{2})$ we have $y_k/y_j = $ const and y_k/y_i for $\tau \to \infty$. Therefore as $\tau \to \infty$ these separatrices approach some arbitrary points on the segment of critical points B_i.

Along the separatrices leaving the segment of critical points $A_i (s_i = -2\sqrt{2}/3$, $s_j = s_k = -1/3\sqrt{2})$ (see (6.7)) we have $y_j/y_k = $ const and $y_i/y_j \to \infty$ for $\tau \to \infty$. Therefore as $\tau \to \infty$ these separatrices approach the critical point (a_i, i).

According to (6.6), at the critical points of the arc $(\alpha_i, i) = (\beta_{ij}, i) + (a_i, i) + (\beta_{ik}, i)(s_i < s_j, s_k)$ we have λ_3, $\lambda_4 < 0$ and $\lambda_2 > 0$. Therefore from each critical point (s_i^0) on the arc (α_i, i) leaves a one-dimensional separatrix along which $y_k = \delta_{ik}$. The dynamics of the coordinates w, s_i are determined by the system (6.1) on the corner of the boundary $y_k = \delta_{ik}$:

$$\dot{s}_i = w(1 - s_i^2) + (1 - k)(1 - s_i(s_1 + s_2 + s_3))\bar{H}_2,$$

$$\dot{s}_j = w(-s_i s_j) + (1 - k)(1 - s_j(s_1 + {}^{\cdot}s_2 + s_3))\bar{H}_2,$$

$$\dot{s}_k = w(-s_i s_k) + (1 - k)(1 - s_k(s_1 + s_2 + s_3))\bar{H}_2,$$

$$\dot{w} = 2w\left(-s_i w - (1 - k)(s_1 + s_2 + s_3)\bar{H}_2 + 4\left(\sum_{\alpha=1}^{3} s_\alpha - 2s_i \right) \right),$$

$$\tag{6.15}$$

where

$$\bar{H}_2 = 2 \sum_{i<j}^{3} s_i s_j - 1 + w/4.$$

Because it is unique, the separatrix in question lies on the surface $\bar{H}_2 = 0$. According to (6.15), along this separatrix we have $s_j/s_k = \text{const}$ and $\dot{s}_i > 0$. Therefore, on the sphere $s_1^2 + s_2^2 + s_3^2 = 1$ the separatrix moves along an arc of the great circle passing through the point $s_i = 1$, $s_j = s_k = 0$ and through the initial point (s_i^0) on the arc α_i of the circle S^1. The final point of the separatrix (s_i^1) is the second point of intersection of this great circle and the circle S^1:

$$w = 0, \quad s_i^1 = -\frac{3s_i^0 + 2\sqrt{2}}{3 + 2\sqrt{2}s_i^0}, \quad s_j^1 = \frac{s_j^0}{3 + 2\sqrt{2}s_i^0}, \quad s_k^1 = \frac{s_k^0}{3 + 2\sqrt{2}s_i^0}. \quad (6.16)$$

We will give a simple geometric interpretation of the newly obtained image T of the circle S^1 (the initial point of the separatrix gets mapped to the final point). The circle S^1 lies in the plane

$$P: \ s_1 + s_2 + s_3 = -\sqrt{2}$$

and is inscribed in the equilateral triangle Δ, which is the intersection of the plane P with the faces of the quadrant $s_i \leqslant 0$. The points where S^1 touches Δ divide the circle into the three arcs α_i (see Fig. 10). Let l be the intersection of the plane $s_j/s_k = \text{const}$ (where the separatrix lies) and the plane P. Obviously l is a straight line which passes through the i-th vertex of the triangle Δ, the initial point of the separatrix (s_i^0) (lying on the arc α_i) and the final point of the separatrix (s_i^1). Consequently the image T is a projection of the arc (α_i, i) of the circle S^1 by rays leaving the i-th vertex of the triangle towards the other two arcs:

$$(\alpha_j, i) + (b_i, i) + (\alpha_k, i) = (\beta_{ji}, i) + (a_j, i) + (\beta_{jk}, i)$$
$$+ (b_i, i) + (\beta_{kj}, i) + (a_k, i) + (\beta_{ki}, i) \quad (6.17)$$

(see Fig. 10).

Let us study the separatrices of the critical points N_i. According to (6.4), the critical points N_i (6.3) have an approaching two-dimensional separatrix[3] in the physical region of the manifold S and a leaving three-dimensional separatrix[4] in the corner of the boundary $y_k = \delta_{ik}$. On this corner $y_k = \delta_{ik}$ the system (6.15) has one repelling critical point N_i, two saddle points on the arc (α_i, i) and at the vertices Φ_i^{00} of the triangle Φ and attracting critical points on the arc (6.17) of the circle of critical points (ψ, i). The system (6.15) is equivalent to the system (5.4) describing the dynamics of the homogeneous cosmological model of type II, which was analyzed (in other phase variables) in Sect. 5.

From the results of Sect. 5 it follows that all separatrices leaving a critical point N_i (except one) approach arbitrary critical points on the arc (6.17) for $\tau \to$

---

[3] *Translator's note:* a stable invariant two dimensional manifold
[4] *Translator's note:* an unstable invariant three-dimensional manifold

$+\infty$. The exceptional separatrix lying in the invariant manifold $s_j = s_k$ $(\bar{v}_2 = 0)$ leaves a critical point N_i in the form of an unwinding spiral and as $\tau \to \infty$ approaches the critical point Φ_i^{00}.

IV. A Study of the Separatrices of the Critical Points T_i. According to (6.5), the segment of critical points T_i is approached by a three-dimensional separatrix, which fills the invariant manifold V_i: $y_j = y_k$, $s_j = s_k$. The intersection of this manifold with the component of the boundary Γ_i $(y_i = 0)$ is a two-dimensional plane

$$\mathcal{L}: \; y_i = 0, \, y_j = y_k = 1/\sqrt{2}, \, s_j = s_k$$

filled by separatrices approaching the critical points T_i and corresponding to the eigenvalues λ_4, λ_5 (6.5). In the neighborhood of T_i all trajectories of the system (6.1) not lying on the manifold V_i rotate around V_i, because of the presence of purely imaginary eigenvalues. Let us show that on the component of the boundary Γ_i all trajectories move away from the invariant plane $\mathcal{L}$ (this shows the instability of the critical points of T_i). Indeed on Γ_i $(y_i = 0)$ the function H_2 (see (6.1)) has the following form:

$$H_2 = 2 \sum_{i<j}^{3} s_i s_j - 1 - \frac{w}{4}(y_j - y_k)^2 \geqslant 0,$$

so

$$2 \sum_{i<j}^{3} s_i s_j - 1 \geqslant 0,$$

and therefore $s_1 + s_2 + s_3 \leqslant 0$ implies that all $s_k \leqslant 0$. Consequently on Γ_i the following function F_i monotonically increases along the trajectories of the system (6.1):

$$F_i = \frac{4(s_k - s_j)^2 + w(y_k - y_j)^2}{w y_k y_j}, \quad \frac{dF_i}{d\tau} = -32 \frac{s_i(s_k - s_j)^2}{w y_k y_j} \geqslant 0. \tag{6.18}$$

Note that the trajectories of the system (6.1) intersect the surface $s_k = s_j$ (the set of zeroes of the derivative $dF/d\tau_2$) transversally everywhere outside the plane $\mathcal{L}$. Therefore outside the plane $\mathcal{L}$ every trajectory X leaves any bounded region U in the interior of Γ_i and consequently approaches the corners of the boundary

$$y_i = 0, \, w = 0 \quad \text{or} \quad y_i = y_j = 0, \, y_k = 1; \quad y_i = y_k = 0, \, y_j = 1$$

(otherwise the function F_i would grow indefinitely along the trajectory X in the region U). As it approaches the corners of the boundary the trajectory X starts moving along the trajectories in these corners of the boundary, which (as shown above) approach the critical points N_j, N_k, Φ_i^0, A_i, B_i, (ψ, j) and (ψ, k). After a finite number of steps along the separatrices of unstable critical points the trajectory X either approaches the attracting (on the component of the boundary Γ_i) critical points on the arcs (β_{ik}, k) and (β_{ij}, j) or the trajectory X is itself a separatrix of one of the unstable critical points (such trajectories are not typical and form submanifolds of smaller dimension).

Let us consider the behavior of the trajectories of the system (6.1) in the invariant plane $\mathscr{L}$. For

$$y_i = 0, \quad y_j = y_k = 1/\sqrt{2}, \quad s_j = s_k$$

the system (6.1) takes the following form:

$$\dot{s}_i = 2(1 - k)s_i s_j(s_j - s_i)(4s_j - s_i),$$
$$\dot{w} = 2ws_i[4 - (1 - k)(s_i + 2s_j)(4s_j - s_i)], \tag{6.19}$$

where $s_i^2 + 2s_j^2 = 1$. The system (6.19) is defined in the region $H_2 = s_i(4s_j - s_i) > 0$ or $4s_j < s_i < 0$. The system (6.19) obviously has a segment of attracting critical points T_i: $s_i = 0$, $0 \leqslant w \leqslant \infty$, an attracting critical point $s_i = -1/3\sqrt{2}$, $w = 0$ lying on the segment A_i and a saddle point $s_i = -1/\sqrt{2}$ lying on the segment Φ_i^0. As $\tau \to +\infty$ the trajectory $s_i = s_j$ approaches a critical point on the segment Φ_i^0, the trajectories in the region $s_j < s_i$ approach the segment T_i and the trajectories in the region $4s_j < s_i < s_j$ approach a critical point on the segment A_i. For $\tau \to -\infty$ all trajectories of the system (6.19) leave the degenerate critical point T_i^0 $(s_\alpha = 0, y_i = 0, y_j = y_k = 1/2)$.

V. The Separatrix Diagram During the Contraction of Space. The results of the above analysis of the separatrices of the critical points Φ, N_i, T_i, A_i, B_i and (ψ, i) can be collected in a separatrix diagram (Table 3). In the diagram we use the following notation: a square with an entry denotes a separatrix going (time is directed towards contraction) from one set of critical points in the top row to another set in the leftmost column, the number in the square is the dimension of the separatrix, an empty square denotes the absence of a separatrix and the letter T denotes the image of the projection (6.16).

For the direction of time towards contraction each trajectory of the system (6.1) (as shown in Sect. 4) by the existence of the monotone function $F = (s_1 + s_2 + s_3)/w(y_1 y_2 y_3)^{1/3}$ approaches the components of the boundary $\Gamma_1, \Gamma_2, \Gamma_3, \Gamma_w$ for $\tau \to \infty$. Trajectories (separatrices) approaching the critical points Φ, N_i, T_i correspond to solutions with power asymptotics of the metric near a singularity (6.9)–(6.11). All other trajectories for $\tau \to \infty$ start moving along the trajectories of the system (6.1) on the components of the boundary $\Gamma_1, \Gamma_2, \Gamma_3, \Gamma_w$, which (as shown in parts II–IV) go from one unstable critical point to another. Thus almost all trajectories of the system (6.1) for $\tau \to \infty$ move along sequences of separatrices of unstable critical points and therefore admit separatrix approximation.

According to the separatrix diagram (Table 3) a trajectory after a finite number of steps (not more than three) along the separatrices of the critical points Φ, N_i, T_i starts moving along the separatrices of the three critical circles (ψ, i), which form a closed system. Thus for $\tau \to \infty$ a trajectory can be approximated by an infinite sequence of separatrices

$$\cdots \rightarrow (\psi_0, \alpha_0) \rightarrow (\psi_1, \alpha_1) \rightarrow (\psi_2, \alpha_2) \rightarrow \cdots \qquad (6.20)$$

At the same time $(\psi_{s+1}, \alpha_{s+1})$ is a single-valued function of (ψ_s, α_s) if ψ_s belongs to the arcs (β_{ij}, i) and (β_{ji}, i). If however ψ_s belongs to the arc (β_{jk}, i), then the step to $(\psi_{s+1}, \alpha_{s+1})$ is two-valued:

$$(\beta_{jk}, k) \overset{\text{II}}{\leftarrow} (\beta_{jk}, i) \overset{\text{I}}{\rightarrow} (\beta_{jk}, j). \qquad (6.21)$$

Step I is along a two-dimensional departing separatrix and step II is along its one-dimensional boundary separatrix. At the same time if we leave from the end of step II, then the next step will lead us to the same place as step I would have in one leap.

The three circles (ψ, k) and the separatrices leaving them (with time directed towards the contraction of space) form a set invariant relative to the dynamical system (6.1), which is also a simplicial complex. The one-dimensional simplices of this set are the circles (ψ, k). The two-dimensional simplices are formed by the separatrices of the circles (ψ, k), which lie in the corners of the boundary Γ. The three-dimensional simplices are formed by the separatrices, which belong to the component of the boundary Γ_w and leave the arcs (β_{jk}, i) as part of step I in (6.21).

The set P is an attracting set for the dynamical system (6.1). Indeed, because of the existence of the monotone function F (4.2) all trajectories of the system (6.1) (for $\tau \rightarrow \infty$) eventually come arbitrarily close to the components of the boundary $\Gamma_1, \Gamma_2, \Gamma_3$ and Γ_w of the manifold S. Because of the existence of the monotone functions F_i (6.18) almost all trajectories of the dynamical system on the components of the boundary Γ_i $(i = 1, 2, 3)$ (apart from a few exceptional separatrices) tend to the set P as $\tau \rightarrow \infty$. More precisely they tend to the circles (ψ, k). On Γ_w the same property follows from explicit integration of the dynamical system on Γ_w (see part II). As a consequence of this, all trajectories of the dynamical system on the manifold S (except for separatrices approaching the critical points Φ, N_i, T_i) approach the set P, which therefore is indeed an attracting set.

Following modern terminology, an attracting set, which is not a smooth manifold, can be called a strange attractor. Let us list some properties of our strange attractor P. The set P is located on the boundary Γ, which is attached to the physical region of the manifold S. Therefore each trajectory on P by itself does not correspond to any exact solution of the initial dynamical system (4.1). The dynamics of the trajectories in the set P are quite simple: all trajectories are separatrices going from some critical point (ψ, α) to another critical point (ψ_1, α_1). However the trajectories of the dynamical system (6.1) inside the manifold S exhibit very complex behavior as they approach the attractor P: for $\tau \rightarrow \infty$ each trajectory can be approximated by a sequence of separatrices (6.20), which in general is everywhere dense in the set P. Indeed the sequential change of the points (s_i) on the critical circles (ψ, k) along the trajectory (6.20) is determined by the action of the mapping T on the circle S^1 (projection from the vertices of the circumscribed equilateral triangle Δ, see part III, (6.16) and Fig. 10). Obviously

Table 3. The separatrix diagram of the dynamical system (6.1) for the homogeneous cosmological model of type IX (for the contraction of space)

	S	T_i^0	Φ	T_i	N_i	Φ_i^0
Φ	3					
T_j	3 $j = 1, 2, 3$	2 $j = i$				
N_j	2 $j = 1, 2, 3$	1 $j = i \pm 1$				
Φ_j^0		2 $j = i$				
Φ_j^{00}					1 $j = i$	
(β_{xy}, x)			4 $x = 1, 2, 3$ $y = x \pm 1$			3 $x = i \pm 1$ $y = x \pm 1$
(β_{yx}, x)			4 $x = i \pm 1$ $y = i$	3 $x = i \pm 1$ $y = i$	3 $x = i$ $y = i \pm 1$	3 $x = i \pm 1$ $y = i$
(β_{yz}, x)				3 $x = i$ $y = i \pm 1$		
(a_x, x)			3 $x = 1, 2, 3$			2 $x = i \pm 1$
(b_x, x)					2 $x = i$	
A_j			3 $j = i$	2 $j = i$		2 $j = i$
B_j			3 $j = 1, 2, 3$			2 $j = i$

$\deg T = -2$ (T changes the orientation on S^1 and each point has two preimages). The map T has three fixed points, namely the three points where the circle S^1 and the triangle Δ touch. Obviously T increases lengths of arcs everywhere (only at the above three points $|dT(\varphi)/d\varphi| = 1$). As a consequence of this, there exists a countable everywhere dense in S^1 set of periodic (for T) points, each of which is unstable. For all remaining points x on S^1 the set $T^k(x)$ $(k = 1, 2, \ldots)$ is everywhere

Φ_i^{00}	(β_{ij}, i)	(β_{ji}, i)	(β_{jk}, i)	(a_i, i)	(b_i, i)	A_i	B_i
2 $x = i$ $y = i \pm 1$		2 $x = j$ $y = i$	3 $x = j$ $y = k$				
2 $x = i$ $y = i \pm 1$	2^T $x = i$ $y = j$		2 $x = k$ $y = j$			2 $x = i \pm 1$ $y = i$	2 $x = i \pm 1$ $y = i$
2 $x = i$ $y = i \pm 1$	2^T $x = i$ $y = j$						
1 $x = i$						2 $x = i$	
1 $x = i$				1^T $x = i$			
				2 $j = i$			

dense in S^1 and this is why a general sequence of separatrices (6.20) is everywhere dense in the attractor P. Note that there is a smooth invariant (with respect to T) measure on the circle S^1, which becomes singular at the above three points. Also note that the complete measure on S^1 turns out to be infinite.

The most important property of the attractor P for the dynamical system on the manifold S describing the homogeneous cosmological model of type IX is that in a neighborhood of P all trajectories of the dynamical system (despite their considerably complex behavior) can be analyzed in great detail, because in the separatrix approximation of the trajectories (6.20) all separatrix steps can be integrated explicitly. An analogous situation occurs also in some other dynamical systems, where the attractor is located on the boundary attached to the initial physical region (see Chapt. VIII).

Recall that the circles of critical points (ψ, k) are in the corners of the boundary Y_{0123} where $y_i = \delta_{ik}$. Therefore on the trajectory moving along the sequence of separatrices (6.20) we have $q_i \gg q_j$, q_k periodically, i.e. the metric g_{ij} changes in an oscillatory mode. Also for the direction of time towards contraction the trajectory approaches the attractor P and periodically goes nearer and nearer the critical circles (ψ, k). Consequently the amplitude of oscillations of the quantities q_i/q_k for $\tau \to \infty$ grows to infinity.

The general oscillatory mode of behavior of the metric near a cosmological singularity was discovered by Belinsky, Lifshitz and Khalatnikov [49, 50] and was subsequently studied by Misner [58].

In the works [49–52], the evolution of the metric (for example of the homogeneous cosmological model of type IX) near a singularity is broken into a series of so called "Kasner epochs." In each epoch the metric is approximated by Kasner solutions $q_i = C_i t^{2p_i}$, where the "Kasner exponents" p_i satisfy the following conditions:

$$p_1 + p_2 + p_3 = 1, \quad p_1^2 + p_2^2 + p_3^2 = 1. \tag{6.22}$$

When two neighboring "Kasner epochs" are glued together the exponents change and if in one epoch $p_1 < 0 < p_2 < p_3$, then in the next one

$$p_1' = \frac{-p_1}{1 + 2p_1}, \quad p_2' = \frac{2p_1 + p_2}{1 + 2p_1}, \quad p_3' = \frac{2p_1 + p_3}{1 + 2p_1}. \tag{6.23}$$

Often the numbers $p_1 < p_2 < p_3$ are represented in parametric form $(u > 1)$

$$p_1(u) = \frac{-u}{1 + u + u^2}, \quad p_2(u) = \frac{1 + u}{1 + u + u^2}, \quad p_3(u) = \frac{u(1 + u)}{1 + u + u^2}.$$

The law of change of Kasner exponents (6.23) looks particularly simple if the parameter u is used [50]: if in one Kasner epoch we had

$$p_l = p_1(u), \quad p_m = p_2(u), \quad p_n = p_3(u),$$

then in the next one

$$p_l' = p_2(u - 1), \quad p_m' = p_1(u - 1), \quad p_n' = p_3(u - 1).$$

Let us show that if in (6.21) we choose the single-valued step I, then the separatrix approximation of the trajectories (6.20) is equivalent to the description of the oscillatory mode obtained in [50]. Indeed as the trajectory moves along the separatrices of the critical points $(\beta_{jk}, i)(s_j < s_i)$ in $\Gamma_w (w = 0)$ the coordinates s_i are constant in the first approximation. The metric is approximated by the Kasner solution

$$q_i = C_i t^{2p_i}, \quad p_i = 1 + \sqrt{2s_i}. \tag{6.24}$$

Conditions (6.22) are satisfied, because on the circles of critical points (ψ, i) we have

$$s_1^2 + s_2^2 + s_3^2 = 1, \quad H_2(s) = 2\sum_{i<j}^{3} s_i s_j - \sum_{i=1}^{3} s_i^2 = 0.$$

According to the separatrix diagram (Table 3), after each separatrix step in the component of the boundary Γ_w $(w = 0)$ approaching the critical points (β_{jk}, i) there is a separatrix step going in the corner of the boundary $Y_i (y_k = \delta_k^i)$ and vice versa. In each such step the coordinates s_i change. The final coordinates (s_i^1) are related to the initial coordinates (s_i^0) by $(s_i^1) = T(s_i^0)$ (6.16). The step is equivalent to a "change of Kasner exponents." It is easy to check that under the conditions $p_i = 1 + \sqrt{2s_i}$ (6.24) the equations (6.23) follow from the equations (6.16).

Thus, the oscillatory mode corresponds to the movement of the trajectories of the dynamical system (6.1) in the neighborhood of the strange attractor P. The "long epoch" in which there is oscillation of two eigenvalues of the metric q_i, $q_j \gg q_k$ occurs when the projection of the trajectory into the coordinates s_i is near the points where the circle S^1 touches the triangle Δ (see Fig. 10). When time is directed towards the contraction of space all metrics of the homogeneous cosmological model of type IX (except for the metrics with power asymptotic forms (6.9)–(6.11)) enter the oscillatory mode, which therefore is the typical state of the metric during contraction.

VI. Typical States of the Metric in the Early Stages of the Expansion of Space. Modern astronomical observations show that the observed expanding universe is isotropic (with great precision). From the observation of background radiation, the isotropy of the universe began very early when the magnitude of the relative red shift became $z \gg 10^3$. Note that from the information about the chemical content of matter it follows that isotropization occurs for $z > 10^9$ (see [74]). Therefore in relativistic cosmology there is a problem of finding a theoretic basis for the isotropization of the metric in the early stages of the expansion of space.

As mentioned in [74] the area where the equations of the general theory of relativity apply is bounded by the Planck scale of time and length:

$$t_g = \sqrt{G\hbar/c^5} = 5.3 \times 10^{-44} \text{ sec}$$

$$l_g = \sqrt{G\hbar/c^3} = 1.6 \times 10^{-33} \text{ cm}$$

where G is the gravitational constant, $\hbar$ is Planck's constant and c is the speed of light. For $t < t_g$, $l < l_g$ quantum effects play a major role. One possible point of view is that the process of formation of an isotropic solution near a cosmological singularity can be described by a presently unknown gravitational quantum theory and at the moment when the classical general theory of relativity becomes applicable the solution is already isotropic with great precision, so it remains isotropic throughout the processes of further evolution.

However, it seems closer to the truth that at the moment when the general theory of relativity takes over, some unknown quantum mechanism of formation of the initial data causes some occurrence from the entire set of possible states of the metric, which are quite arbitrary and may be far from isotropic. Under this natural assumption, in order to study the possibility of early isotropization of the metric it is necessary to analyze the evolution of solutions of Einstein's equations with initial data from a selected set of states. In principle, after some time in the process of expansion of space, these solutions can end up being concentrated in neighborhoods of some special modes, which are quite naturally called the typical states of the metric during the expansion of space. If the isotropic Friedmann solution is among the typical states of the metric in the early stages of expansion, then in the framework of the classical general theory of relativity there is a mechanism, which ensures an arbitrarily early isotropization in a whole region of the space of solutions.

We now proceed with the program sketched out for the homogeneous cosmological model of type IX based on the full qualitative analysis of the dynamical system (6.1) defined on the compact manifold S. As mentioned in part V, it was shown in [49, 50] that the typical state of the metric near a cosmological singularity for the contraction of space is the oscillatory mode. However it does not yet follow from this that the oscillatory mode is a typical state of the metric in the early stages of expansion, because the typical properties of the solutions of the dynamical system can irreversibly depend on the direction of time.

Irreversibility of the typical properties of behavior of the solutions of the dynamical system (6.1) near the cosmological singularity $q_1 q_2 q_3 = 0$ (or near the components of the boundary $\Gamma_1, \Gamma_2, \Gamma_3, \Gamma_w$ of the manifold S) is easily seen from the comparison of the separatrix diagram for the contraction of space (Table 3) to the separatrix diagram for the expansion of space (obtained by reversing time). When time is directed towards the contraction of space, in general a trajectory of the dynamical system (6.1) after a finite number of steps along the separatrices of critical points Φ, N_i, T_i, A_i, B_i tends arbitrarily close to the attractor P and moves along the separatrices of critical points (ψ, α) as $\tau \to \infty$.

When time is directed towards expansion, the trajectories go through critical points in reverse order. Before leaving the singularity or the boundary Γ the only steps in the separatrix diagram (reversed in time) allowed are the ones in Fig. 11. The multiple dots (in Fig. 11) denote the critical points of types (ψ, α), A_i and B_i. Straight arrows denote the separatrices. The numbers above the straight arrows are the dimensions of the corresponding separatrices. Wavy arrows denote the steps by continuities, which occur because one critical set can lie on the boundary of another. Exceptional points T_i must be counted as part of the physical region even though they belong to the closure of the boundary, because the system can remain near these points for a long time (even up to the moment of maximal expansion). Figure 11 shows that a trajectory can leave the boundary Γ only by passing near the critical points Φ, N_i and T_i. Because of the absence of separatrices going from the corners of the boundary Γ into the physical region the oscillatory

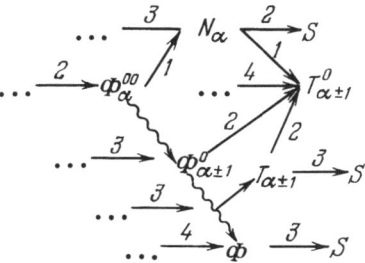

Fig. 11. The steps in the separatrix diagram towards expansion before leaving the singularity

mode, which is based on the separatrices in these corners, can (in principle) disappear arbitrarily early. Note that some of the steps in Fig. 11 were also found (by other (mainly numerical) methods) in [70].

To pose the question about the typical states of the metric in the early stages of the expansion of space exactly, we need the following property of the monotone function $F = d(\det g_{ij})^{1/6}/dt$ (see (4.2)): when time is directed towards the expansion of space the function $|F|$ decreases from ∞ to 0 along each solution and $F = 0$ at the moment of maximal expansion. It is natural to call the states of the metric, for which $F \gg 1$, the early stage of expansion. Note that the function F is invariant relative to the scale transformations (4.10) and has a simple physical meaning. The function F is the rate of change of the mean radius of the universe defined by $|g|^{1/6}$ ($|g|^{1/6}$ has the dimensions of length). For the Friedmann solution $F = \dot{a}$, where a is the radius of the three-dimensional sphere (the function $F = \partial(|g|^{1/6})/\partial t$ is monotone along any (e.g. inhomogeneous) solution of Einstein's equations in the synchronous system of coordinates (see the end of Sect. 3).

Definition of the typical states of the metric in the early stage of the expansion of space consists of the following. Suppose that for $|F| = F_1 \gg 1$ in the phase space with the coordinates g_{ij} and momenta p^{ij} we are somehow given initial conditions (for example we can take a uniform distribution on the surface $|F| = F_1$). By Einstein's equations these initial conditions are shifted in the phase space and for some value of $|F| = F_2 \gg 1$ $(F_2 < F_1)$ can concentrate in small neighborhoods of some critical points of the phase space. Meanwhile the metric will be approximated by some special modes, which we will call the typical states of the metric in the early stage of the expansion of space.

In the model we have considered, the condition $|F| = F_1 \gg 1$ means that the initial data on the manifold S are in a small neighborhood of the boundary Γ. Therefore the trajectories leaving these initial conditions move along the separatrices of critical points of the dynamical system (6.1) lying on the boundary Γ until they arrive in the neighborhood of critical points, which have separatrices going into the physical region of the manifold S. These critical points belong to the sets Φ, N_i and T_i (see Fig. 11). Thus the starting distribution of the initial data changes into a distribution concentrated near the critical points Φ, N_i and T_i. At the same time the trajectories remain close to the boundary Γ, so $|F| \gg 1$. As the trajectories move along the separatrices of the critical points Φ, N_i and T_i entering the physical region of the manifold S the metric is approximated by the power modes (6.9), (6.10) and (6.11). These modes are the typical states of the metric of the homogeneous cosmological model of type IX in the early stage of the expansion of space. Meanwhile the function $|F|$ decreases to $|F| \sim 1$.

Note that the time t_0 when the metric enters one of these power modes (t_0 is greater than the time of the oscillatory mode) can be arbitrarily small and largely depends on the solution itself (this is obvious from the existence of scale transformations (4.10)). For all solutions we have $t_0 \ll t_m$, where t_m is the time, which passes from the singularity ($t = 0$) to the moment of maximal expansion. Notice that the time t_0 when the metric takes on a quasi-isotropic asymptotic form (6.9) can also be arbitrarily small, i.e. the isotropization of solutions for the homogeneous model of type IX can begin (after an anisotropic oscillatory mode) arbitrarily early.

An analysis of the typical states of the metric for all other homogeneous cosmological models in the early stages of the expansion of space can be done in an analogous fashion based on the material in chapters II and III. Such analysis shows that the typical states of the metric in the early stages of the expansion of space depend very little on the type of the model and on the conditions of the motion of matter. More precisely the power asymptotic forms (6.10) and (6.11) may change form for various models, but in all models one of the typical states of the metric is the quasi-isotropic asymptotic form (6.9). This fact makes it likely that the quasi-isotropic asymptotic form (6.9) is a typical state of the metric in the early stage of the expansion of space in the general inhomogeneous case.

VII. Geometrical Model of the Oscillatory Mode. We will make another description of the oscillatory mode for the homogeneous cosmological model of type IX by means of separatrix approximation of the trajectories of the dynamical system. Let us transform the system (4.1) into the coordinates $P_i = 2p_i q_i$, q_i and make a time change

$$d\tau_0/dt = 1/q_1 q_2 q_3.$$

We obtain the following system:

$$\dot{P}_i = -q_i(q_j + q_k - q_i) + H_0, \quad \dot{q}_i = q_i(P_j + P_k - P_i), \tag{6.25}$$

where

$$H_0 = \frac{1-k}{4}(T(P_i) + V(q_i)), \quad (i,j,k) = (1,2,3).$$

Critical points of the system (6.25) form four two-dimensional sets:

1) Three planes of critical points

$$T_i: \quad q_i = 0, \, q_j = q_k, \, P_i = 0, \, P_j = P_k.$$

The planes T_i belong to $H_0 = 0$. The eigenvalues of these critical points are

$$\lambda_{1,2} = 0, \quad \lambda_3 = P_i(1-k), \quad \lambda_{4,5} = \pm 2iq_1, \quad \lambda_6 = 2P_i.$$

2) A cone of critical points

$$q_i = 0, \quad V(P_i) = 2\sum_{i<j}^{3} P_i P_j - \sum_{i=1}^{3} P_i^2 = 0,$$

lying in the negative quadrant of the coordinates P_1, P_2, P_3. The eigenvalues of the system (6.25) at the critical points of the cone K are

$$\lambda_{1,2} = 0, \quad \lambda_3 = P_1 + P_2 + P_3 < 0,$$

$$\lambda_{q_1} = P_2 + P_3 - P_1 = x_1, \quad \lambda_{q_2} = P_1 + P_3 - P_2 = x_2,$$

$$\lambda_{q_3} = P_1 + P_2 - P_3 = x_3.$$

It is easy to check that

$$V(P_i) = x_1 x_2 + x_2 x_3 + x_3 x_1 = 0$$

$$P_1 + P_2 + P_3 = x_1 + x_2 + x_3 < 0.$$

From this it follows that one of the x_i is positive and the other two are negative. If $x_1 = 0$, then either $x_2 = 0$ or $x_3 = 0$ (because $V(P_i) = 0$), so extra zero eigenvalues appear only when $P_i = 0$, $P_j = P_k$, i.e. when the cone intersects the planes T_i (as it should be).

From each point P_1^0, P_2^0, P_3^0 on the cone (except the intersections with the planes T_i) leaves a one-dimensional separatrix corresponding to the unique positive eigenvalue. Suppose that $x_1 > 0$. Then along this separatrix we have $q_2 = q_3 = 0$ and $H_0 = 0$, whereas the equations (6.25) take the following form:

$$\dot{P}_2 = 0, \quad \dot{P}_3 = 0, \quad \dot{P}_1 = q_1^2, \quad \dot{q}_1 = q_1(P_2 + P_3 - P_1).$$

Therefore along the separatrix we have

$$P_2 = P_2^0 = \text{const}, \quad P_3 = P_3^0 = \text{const}.$$

By making a time change

$$d\tau_1/d\tau_0 = q_1$$

we obtain

$$\dot{P}_1 = q_1, \quad \dot{q}_1 = P_2^0 + P_3^0 - P_1.$$

The solutions of the above equations are

$$q_1 = (P_2^0 + P_3^0 - P_1^0)\sin\tau_1,$$

$$P_1 = P_2^0 + P_3^0 - (P_2^0 + P_3^0 - P_1^0)\cos\tau_1,$$

where $0 \leqslant \tau_1 \leqslant \pi$.

Thus the maximum value of the coordinate q_1 along this separatrix is

$$q_1^{\max} = P_2^0 + P_3^0 - P_1^0 = x_1^0.$$

The separatrix approaches some other critical point on the cone K, whose coordinates can be obtained from the coordinates of the initial point by means of the projection T_0:

$$P_1^1 = 2(P_2^0 + P_3^0) - P_1^0, \quad P_2^1 = P_2^0, \quad P_3^1 = P_3^0.$$

Obviously this is the second point of intersection of the straight line $P_2 = P_2^0, P_3 = P_3^0$ with the cone $V(P_i) = 0$. At this point we have $x_1^1 = -x_1^0 < 0$, so along the separatrix, which leaves it, either q_2 or q_3 grows etc. For a trajectory moving along some sequence of separatrices we periodically have $q_i \gg q_j, q_1$ and the maximum value of q_i over one period of oscillation is approximately equal to $x_i^{(k)}$. Consequently on the cone $V(P_i) = 0$ the map T_0 defines another model of the oscillatory mode.

The map T_0 is a linear map, so for all points of the cone lying on a generatrix of the cone the law of succession of the maximum values of q_i is the same. The amplitude of oscillations depends on the position on the generatrix and decreases on the whole, because the sequence of points $(P_i^k) = T_0(P_i^0)$ converges to the vertex of the cone $P_i = 0$. By linearity the map T_0 determines the map T on the circle S^1, obtained by slicing the cone with the plane $P_1 + P_2 + P_3 = -\sqrt{2}$. Obviously the map T coincides with the mapping (6.16) obtained in part III in the analysis of separatrices of critical points on the circles (ψ, i) (see Fig. 10). Thus the approximation of trajectories of the dynamical system (6.25) by the separatrices of critical points lying on the cone $V(P_i) = 0$ leads to the earlier geometrical model of the oscillatory mode and adds to it a description of the subsequent change of the amplitude of oscillations of the maximum values of q_i. This description holds for any finite number of oscillations on the trajectories of the system (6.25), which start sufficiently close to the critical points of the cone K.

VIII. Asymptotic Form of the Taub Model with Matter Near a Singularity. The axisymmetric $(q_2 \equiv q_3)$ cosmological model of type IX (the Taub model) has the group of isometries $SO(2) \times SO(3)$. It was integrated by Taub in the case of vacuum $(\varepsilon = 0)$ [47]. We shall study the asymptotic behavior of this model in the general case $\varepsilon > 0$.

Theorem. For $0 \leqslant k < 1$ near the singularity $q_1 q_2^2 = 0$ the metric of the Taub model has (for $t \to 0$) one of the following asymptotic forms:

$$q_1 \approx C_1 t^{4/3(1+k)}, \quad q_2 \approx C_2 t^{4/3(1+k)}, \tag{6.26}$$

$$q_1 \approx C_1 t^{(1-k)/(1+k)}, \quad q_2 \approx C_2 t^{(3+k)/2(1-k)}, \tag{6.27}$$

$$q_1 \approx C_1 t^2, \quad q_2 \approx \text{const} \tag{6.28}$$

(obviously the asymptotic forms (6.26)–(6.28) are special cases of the forms (6.9)–(6.11)).

The axisymmetric cosmological model of type IX is described by the system (4.1) on the invariant manifold $V: q_2 \equiv q_3, p_2 \equiv p_3$. On the manifold V in the coordinates

$$\bar{u} = \frac{p_1 q_1}{2p_2 q_2}, \quad w = \frac{q_1}{4p_2 q_2}, \quad v = \left(\frac{q_2}{q_1}\right)^{1/2} w, \quad q_1$$

and time

$$\tau_2: \frac{d\tau_2}{dt} = -\frac{w}{q_1^{1/2} v^2}$$

the system (4.1) has the following form:

$$\dot{\bar{u}} = -w^2 + 2v^2 - 2\bar{u}v^2 + (2\bar{u} - 1)H_2,$$

$$\dot{w} = w(\bar{u} - 1 - 2v^2 + 2H_2),$$

$$\dot{v} = \frac{1}{2}v\left(-k - (1 - k)(\bar{u} - 1)^2 - (1 - k)w^2 - 4kv^2\right), \qquad (6.29)$$

$$\dot{q}_1 = q_1(\bar{u} - 1),$$

$$H_2 = \frac{1-k}{4}\left(1 - (\bar{u} - 1)^2 - w^2 + 4v^2\right).$$

The first three equations (6.29) form a closed system defined on the single-cavity hyperboloid $H_2 \geqslant 0$. The trajectories of the system (6.29) in the region $p_2 q_2 > 0$ $(w > 0, v > 0)$ and time $\tau_3 = -\tau_2 \, (d\tau_3/dt > 0)$ go to infinity in finite time in the coordinates w, v and $p_2 q_2(w, v)$ changes sign. Therefore when time is directed towards contraction, it is enough to consider the system (6.29) in the invariant region $w < 0, v < 0$.

By (6.29) $\dot{v}/v < 0$, so the trajectory moves inside $H_2 \geqslant 0$ and approaches the manifold $v = 0$. In particular the sign of v is conserved. The asymptotic behavior of solutions is thus determined by the system on the manifold

$$v = 0, \quad (\bar{u} - 1)^2 + w^2 \leqslant 1, \quad w \leqslant 0. \qquad (6.30)$$

Here is a list of the critical points of the system (6.29) and their eigenvalues (at all points we have $v = 0, q_1 = 0$):

$$T: \bar{u} = 0, \ w = 0, \ \lambda_{\bar{u}} = -\frac{1-k}{2}, \ \lambda_v = -\frac{1}{2}, \ \lambda_w = -1,$$

$$\lambda_{q_1} = -\frac{1}{2},$$

$$\Phi: \bar{u} = \frac{1}{2}, \ w = 0, \ \lambda_{\bar{u}} = \frac{3}{8}(1 - k),$$

$$\lambda_v = -\frac{1 + 3k}{8}, \lambda_w = -\frac{1 + 3k}{8}, \lambda_{q_1} = -\frac{1}{4},$$

$$N_1: \bar{u} = \frac{3 + k}{5 - k}, w = -\frac{1}{5 - k}\sqrt{(1 + 3k)(1 - k)},$$

$$\lambda_v = -\frac{1 + 3k}{2(5 - k)}, \lambda_{q_1} = -\frac{1 - k}{5 - k},$$

$$\lambda_{\bar{u}, w} = \frac{1 - k \pm i\sqrt{\dfrac{(1 - k)}{2}(3 + 16k - 3k^2)}}{5 - k},$$

$$C: \bar{u} = 2, w = 0, \lambda_{\bar{u}} = -\frac{3}{2}(1 - k),$$

$$\lambda_v = -\frac{1}{2}, \lambda_w = 1, \lambda_{q_1} = \frac{1}{2}.$$

The only stable critical point is T. It corresponds to the asymptotic form (6.28), which therefore holds for an entire region in the space of solutions. The critical point Φ is approached by a three-dimensional separatrix, which corresponds to the asymptotic form (6.26). The asymptotic form (6.27) corresponds to a two-dimensional separatrix approaching the critical point N_1. Separatrices approaching the critical point C belong to the non-physical boundary $w \equiv 0$. There are no physical asymptotic forms corresponding to the point C.

Let us prove that all trajectories of the system (6.29) (except for the separatrices of critical points Φ, N_1, C) approach the critical point T. Since every trajectory of the system (6.29) approaches the manifold (6.30), our statement follows from the fact that all trajectories on the manifold (6.30) (except for the separatrices of the above critical points) approach the critical point T. To prove that last statement it is sufficient to show that the system (6.29) on the manifold (6.30) (it coincides with the system (5.8)) has no limit cycles. This was already proved in Sect. 5 (for the system (5.8)).

7. Analysis of Cosmological Models of Types VIII, VII$_0$ and VI$_0$

Type VIII. The compact five-dimensional manifold S with the dynamical system (4.9) on it for the model of type VIII $(n_1 = -1, n_2 = n_3 = 1)$ was constructed in Sect. 4. The component of the boundary $\Gamma_m (\bar{s}_1 + \bar{s}_2 + \bar{s}_3 = 0)$ of the manifold S degenerates into one point

$$T_1^0: \bar{s}_i = 0, y_1 = 0, y_2 = y_3 = 1/\sqrt{2})$$

because (see (4.8))

$$H_1 = (\bar{s}_1 + \bar{s}_2 + \bar{s}_3)^2 - 2(\bar{s}_1^2 + \bar{s}_2^2 + \bar{s}_3^2)$$
$$+ \tfrac{1}{4}(-2y_1 y_2 - 2y_1 y_3 - y_1^2 - (y_2 - y_3)^2) \geqslant 0$$

Thus the point T_1^0 is the only point of "maximal expansion" in the model of type VIII. When time is directed towards expansion, since the function $F = (\bar{s}_1 + \bar{s}_2 + \bar{s}_3)/(y_1 y_2 y_3)^{1/3}$ (see Sect. 4) is monotone, each trajectory of the dynamical system (4.9) leaves any compact region inside the manifold S (and also in the neighborhood of Γ_0 lying in the physical region). Consequently it approaches the components of the boundary Γ_m, Γ_1, Γ_2, Γ_3, Γ_w. Since the function F grows monotonically (during the expansion of space) and at all points of the boundary $\Gamma_1, \Gamma_2, \Gamma_3, \Gamma_w$ (except T_1^0) $F = -\infty$, the trajectory can approach only the critical point T_1^0. Thus during the endless expansion of space in the model of type VIII the isotropization of the components of the metric does not occur.

When time is directed towards contraction, the function F monotonically decreases and the trajectories of the dynamical system (4.9) approach the components of the boundary Γ_1, Γ_2, Γ_3, Γ_w. On the components of the boundary Γ_1 ($y_1 = 0$) and Γ_w ($w = 0$) the dynamical system (4.9) coincides with the dynamical system (6.1) on the same components of the boundary. This last system was studied in detail in Sect. 6. The dynamical systems (4.9) on the corners of the boundary Y_i ($y_k = \delta_{ik}$) for models of types IX and VIII are also identical. On the components of the boundary Γ_2 ($y_2 = 0$) and Γ_3 ($y_3 = 0$) the systems (4.9) for the model of type VIII are equivalent, i.e. one becomes the other when the coordinates q_2 and q_3 are transposed. Therefore in order to analyze the behavior of the metric for the model of type VIII near a cosmological singularity (i.e. the dynamical system (4.9) in the neighborhood of the components of the boundary $\Gamma_1, \Gamma_2, \Gamma_3, \Gamma_w$) it remains to consider the dynamical system on the component of the boundary Γ_2.

It is convenient to analyze the dynamical system on Γ_2 in the local coordinates V_1 (4.11). For the cosmological model of type VIII the system (4.12) has the following form:

$$\dot{\bar{u}} = \bar{w}^2(-\bar{y}_2 - \bar{y}_3 - 1) - \bar{u}\bar{w}^2(\bar{y}_2(\bar{y}_3 - y_2 - 1)$$
$$+ \bar{y}_3(\bar{y}_2 - \bar{y}_3 - 1)) + (2\bar{u} - 1)H_2,$$
$$\dot{\bar{v}}_2 = \bar{w}^2(-\bar{y}_2(1 + \bar{y}_2) + \bar{y}_3(1 + \bar{y}_3))$$
$$- \bar{v}_2\bar{w}^2(\bar{y}_2(\bar{y}_3 - \bar{y}_2 - 1) + \bar{y}_3(\bar{y}_2 - \bar{y}_3 - 1)) + 2\bar{v}_2 H_2, \quad\quad (7.1)$$
$$\dot{\bar{w}} = \bar{w}[\bar{u} - 1 - \bar{w}^2(\bar{y}_2(\bar{y}_3 - \bar{y}_2 - 1) + \bar{y}_3(\bar{y}_2 - \bar{y}_3 - 1)) + 2H_2],$$
$$\dot{\bar{y}}_2 = \bar{y}_2(1 + \bar{u}_2 - 2\bar{u}), \quad \dot{\bar{y}}_3 = \bar{y}_3(1 - \bar{v}_2 - 2\bar{u}), \quad \dot{q}_1 = q_1(\bar{u} - 1).$$

where

$$H_2 = \frac{1-k}{4}[1 - (\bar{u} - 1)^2 - \bar{v}_2^2 + \bar{w}^2(-(\bar{y}_2 - \bar{y}_3)^2 - 2(\bar{y}_2 - \bar{y}_3) - 1)].$$

The condition that the energy is positive $H_2 \geqslant 0$ slices out a compact ellipsoid over each point of the spherical triangle Δ in the corresponding coordinates $\bar{u}, \bar{v}_2$, $\bar{w}$. Only over the point $y_1 = 0, y_2 = y_3 = 1/\sqrt{2}$ the value of $\bar{w}$ is unbounded (at the point $T_1^0\bar{w} = -\infty$). The component of the boundary Γ_2 is diffeomorphic to $D^3 \times I$, where D^3 is a half $(\bar{w} \leqslant 0)$ of an ellipsoid in the coordinates $\bar{u}, \bar{v}_2$ and I is the segment of definition of $\bar{y}_3$. If $0 \leqslant \bar{y}_3 < \varepsilon$, then we use the coordinate $\bar{y}_3$ and the local chart V_1. If $\bar{y}_3 \geqslant \varepsilon$, then the coordinate $q_1/q_3 = 1/\bar{y}_3$ and the local chart V_3 are used. The bases of $\Gamma_2 (D^3 \times 0$ and $D^3 \times 1)$ coincide with the corners Y_1 and Y_3. On Y_1 and Y_3 the dynamical system was analyzed earlier (see Sects. 5 and 6). In particular we found that on Y_i there are repelling (in Y_i) critical points N_1, N_3 and there is a one-dimensional separatrix approaching them from Γ_2. It turns out that inside Γ_2 there is a unique critical point, which is repelling in this manifold.

Here is the system (7.1) with the coordinate $x = 2\bar{u} + \bar{v}_2 - 1$ used instead of $\bar{v}_2$:

$$\dot{x} = x\bar{w}^2[\bar{y}_3(\bar{y}_3 + 1) + \bar{y}_2(\bar{y}_2 + 1 - 2\bar{y}_3)] + 2xH_3$$
$$+ 2\bar{w}^2[\bar{y}_3^2 - 1 - \bar{y}_2(\bar{y}_2 + 1)],$$

$$\dot{\bar{y}}_3 = -\bar{y}_3 x,$$

$$\dot{\bar{u}} = \bar{w}^2(-\bar{y}_2 - \bar{y}_3 - 1) - \bar{u}\bar{w}^2(\bar{y}_2(\bar{y}_3 - \bar{y}_2 - 1)$$
$$+ \bar{y}_3(\bar{y}_2 - \bar{y}_3 - 1)) + (2\bar{u} - 1)H_2,$$

$$\dot{\bar{w}} = \bar{w}[\bar{u} - 1 - \bar{w}^2(\bar{y}_2(\bar{y}_3 - \bar{y}_2 - 1) + \bar{y}_3(\bar{y}_2 - \bar{y}_3 - 1)) + 2H_2], \qquad (7.2)$$

$$\dot{\bar{y}}_2 = \bar{y}_2(2 + x - 4\bar{u}), \quad \dot{q}_1 = q_1(\bar{u} - 1),$$

$$H_2 = \frac{(1 - k)}{4}[1 - (\bar{u} - 1)^2 - (x - 2\bar{u} + 1)^2$$
$$+ \bar{w}^2(-(\bar{y}_2 - \bar{y}_3)^2 - 2(\bar{y}_2 + \bar{y}_3) - 1)].$$

For $\bar{y}_3 \neq 0, \bar{w} \neq 0$ the critical points of the system (7.2) must belong to the invariant manifold $x = 0, \bar{y}_3 = 1, \bar{y}_2 = 0$. Consider the system (7.2) on this manifold:

$$\dot{\bar{u}} = 2\bar{w}^2(\bar{u} - 1) + (2\bar{u} - 1)\bar{H}_2 = f_{\bar{u}},$$
$$\dot{\bar{w}} = \bar{w}(\bar{u} - 1 + 2\bar{w}^2 + 2\bar{H}_2) = f_{\bar{w}}. \qquad (7.3)$$

where

$$\bar{H}_2 = \frac{(1 - k)}{4}\left(\frac{4}{5} - 5\left(\bar{u} - \frac{3}{5}\right)^2 - 4\bar{w}^2\right) \geqslant 0.$$

For $\bar{w} \neq 0$ the system (7.3) has a unique critical point M_2 with coordinates:

$$\bar{u} = \frac{3 + k}{5 - k}, \quad \bar{w} = -\frac{(1 + 2k - 3k^2)^{1/2}}{5 - k}.$$

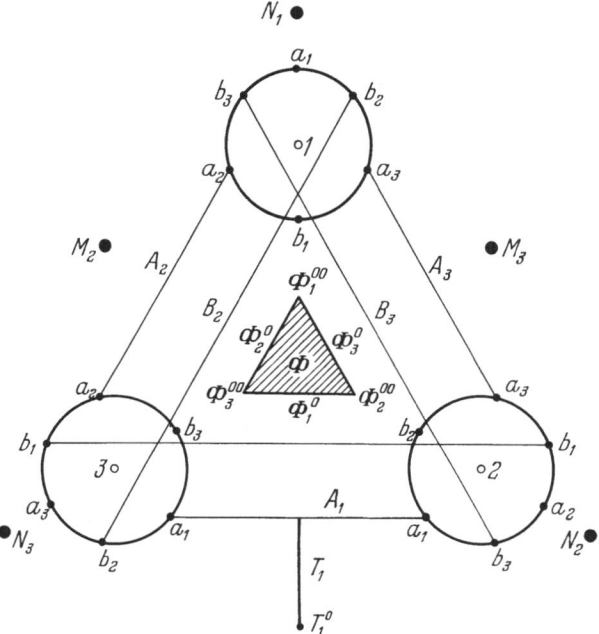

Fig. 12. The critical points of the dynamical system (4.9) on the manifold S for the homogeneous cosmological model of type VIII

Here are the eigenvalues of the system (7.2) at this critical point (in parentheses are the corresponding eigenvectors):

$$\lambda_{1,2} = \frac{1-k}{5-k}(1 \pm i(1 + 6k)^{1/2}), (\bar{u}, \bar{w}); \quad \lambda_3 = -2\frac{1+3k}{5-k}, (\bar{y}_2);$$

$$\lambda_{4,5} = \frac{1-k}{5-k} \pm i\frac{((1-k)(3+13k))^{1/2}}{5-k}, (x, \bar{y}_3); \tag{7.4}$$

$$\lambda_6 = -2\frac{1-k}{5-k}, (q_1).$$

Thus in the component of the boundary Γ_2 the critical point M_2 is repelling. Negative eigenvalues λ_3, λ_6 correspond to a two-dimensional unstable separatrix approaching the critical point M_2. For $t \to 0$ the corresponding solutions have the following asymptotic form:

$$q_2 \approx C_2 t^2, \quad q_1 \approx q_3 \approx C_1 t^{(1-k)/(1+k)}. \tag{7.5}$$

In this asymptotic form q_2 varies as in the Taub asymptotic form (6.11), whereas q_1 and q_3 vary as in the asymptotic form (6.10).

Figure 12 shows the critical points of the dynamical system on the compact

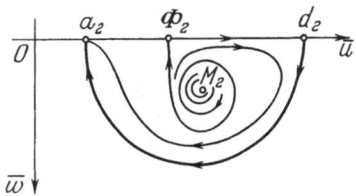

Fig. 13. The phase portrait of the dynamical system (7.3) in the region determined by physical conditions $\bar{H}_2 > 0, \bar{w} < 0$

manifold S for the model of type VIII. Instead of the segments of critical points T_2 and T_3 as in the model of type IX, here we have the isolated points M_2 and M_3. The remaining critical points along with all their eigenvalues are the same for both models. Therefore many separatrices coincide as well. The only exceptions are the separatrices inside Γ_2 and Γ_3. In Γ_2 besides the repelling critical point M_2, unstable critical points N_1, N_2 and some arcs of circles there are also completely attracting (in Γ_2) critical points of the arcs $(\beta_{21}, 1)$ and $(\beta_{23}, 3)$. Of course this does not imply that the separatrices leaving M_2 approach the critical points on these arcs, because generally speaking there can exist invariant manifolds inside Γ_2 with separatrices winding onto them. For the model of type IX the absence of such invariant manifolds in Γ_i was guaranteed by the existence of monotone functions F_i (6.18). For the model of type VIII we will prove a weaker assertion: in Γ_2 there does not exist a smooth three-dimensional invariant (relative to the system (7.1)) manifold W containing the critical point M_2. If such manifold W existed, then it would intersect the two-dimensional invariant manifold $x = 0$, $\bar{y}_3 = 1$, $\bar{y}_2 = 0$ passing through the point M_2 along some cycle (a closed trajectory of the system (7.3)). However it easy to prove that the system (7.3) considered in the region $\bar{H}_2 \geqslant 0$, $\bar{w} \leqslant 0$ has no cycles. Indeed the following equality holds:

$$\frac{\partial (F_0 f_{\bar{u}})}{\partial \bar{u}} + \frac{\partial (F_0 f_{\bar{w}})}{\partial \bar{w}} = \frac{2\bar{w}}{\bar{H}^2} + \frac{2}{\bar{w}} < 0, \quad \text{where } F_0 = \frac{1}{\bar{w}\bar{H}_2}.$$

Therefore according to the Dulac-Bendixon criterion, cycles do not exist.

Note that the phase portrait of the dynamical system (7.3) defined on the half-ellipse $\bar{H}_2 \geqslant 0$, $\bar{w} \leqslant 0$ is qualitatively the same as the phase portrait of the system (5.8) (Fig. 13). In the region we have considered, the system (7.3) has a repelling critical point M_2, an attracting critical point a_2 (contained in the segment A_2) with coordinates $\bar{u} = 1/5$, $\bar{w} = 0$ and two unstable critical points: Φ_2 $(\bar{u} = 1/2, \bar{w} = 0)$ contained in the segment Φ_2^0 and d_2 $(\bar{u} = 1, \bar{w} = 0)$ contained in the segment B_2.

All separatrix steps between the critical points in the components of the boundary Γ_1 and Γ_w for the model of type VIII are identical to those of the model of type IX. In particular all separatrix steps between the critical circles (ψ, k) are the same as in the separatrix diagram (Table 3). Therefore all conclusions about the properties of the oscillatory mode of behavior of the metric near a singularity for the model of type IX made in Sect. 6 equally apply to the model of type VIII.

As in part VIII of Sect. 6 we will prove the following theorem:

Theorem. *Near the singularity $q_1 q_2 q_3 = 0$ the axisymmetric metrics $(q_2 \equiv q_3)$ of the model of type VIII possess one of the asymptotic forms $(6.26)-(6.28)$ for $t \to 0$. During the expansion of space $(t \to \infty)$ all axisymmetric metrics have the following asymptotic form:*

$$q_1 \approx C_1, \quad q_2 = q_3 \approx C_2 t^2. \tag{7.6}$$

The considered class of metrics is described by the system (7.1) on the invariant manifold $\bar{y}_2 = \bar{y}_3, \bar{v}_2 = 0$. In new coordinates

$$\bar{u}, \bar{w}, v = \bar{y}_2^{1/2} \bar{w}, q_1$$

and time τ_2:

$$\frac{d\tau_2}{dt} = -\frac{1}{\bar{w}(q_1 y_2 y_3)^{1/2}} = -\frac{\bar{w}}{q_1^{1/2} v^2}$$

this system has the following form:

$$\begin{aligned}
\dot{\bar{u}} &= -\bar{w}^2 - 2v^2 + 2\bar{u}v^2 + (2\bar{u} - 1)\bar{H}_2, \\
\dot{\bar{w}} &= \bar{w}(\bar{u} - 1 + 2v^2 + 2\bar{H}_2), \\
\dot{v} &= \tfrac{1}{2}v(-k(1 - 4v^2) - (1 - k)(\bar{u} - 1)^2 - (1 - k)w^2), \\
\dot{q}_1 &= q_1(\bar{u} - 1).
\end{aligned} \tag{7.7}$$

where

$$\bar{H}_2 = \frac{1 - k}{4}(1 - (\bar{u} - 1)^2 - \bar{w}^2 - 4v^2).$$

Condition $\bar{H}_2 \geqslant 0$ determines an ellipsoid $(\bar{u} - 1)^2 + \bar{w}^2 + 4v^2 \leqslant 1$, while $\bar{w} \leqslant 0$, $v \leqslant 0$.

By (7.7) the function $|v|$ decreases monotonically, so towards contraction each trajectory approaches the plane $v = 0$. In the plane $v = 0$ the system (7.7) identically coincides with the system (6.29) for $v = 0$. All eigenvalues of the critical points also coincide, so the assertion about the asymptotic forms towards contraction is proved in part VIII of Sect. 6.

In the other direction of time, $|v|$ monotonically grows. Therefore all physical trajectories $(v \neq 0, w \neq 0)$ approach the attracting (towards expansion) critical point $v = -1/2, \bar{u} = 1, \bar{w} = 0$ (there are no other critical points for $v \neq 0$). At this critical point we have

$$\lambda_{\bar{u}} = -1/2, \quad \lambda_v = -k, \quad \lambda_{\bar{w}} = -1/2, \quad \lambda_{q_1} = 0.$$

After changing to the synchronous time t we obtain the asymptotic form (7.6). If we include the axisymmetric metrics into the general metrics of the model of type VIII, then the attracting (towards expansion) critical point $v = -1/2, \bar{u} = 1$, $\bar{w} = 0$ corresponds to the attracting critical point T_1^0.

Table 4. The separatrix diagram of the dynamical system (4.9) for the homogeneous cosmological models of types VII_0 and VI_0 (for the contraction of space)

	M_3	Φ_3^0	Φ_i^{00}	N_i	(β_{i3}, i)	(β_{ij}, i)
Φ_3^0	2					
Φ_j^{00}				1 $j=i$		
N_j	2 $j=1,2$					
(β_{x3}, x)		3 $x=1,2$	2 $x=i$			
(β_{xy}, x)		3 $x=1,2$	2 $x=i$			
(β_{yx}, x)			2 $x=i$	3 $x=i$		2^T $x=i$ $y=j$
(β_{3x}, x)	4 $x=1,2$	3 $x=1,2$	2 $x=i$	3 $x=i$	2^T $x=i$	
(β_{y3}, x)			2 $x=i$	3 $x=i$		2^T $x=i$ $y=j$
(β_{3y}, x)			2 $x=i$	3 $x=i$	2^T $x=i$	
A_3	3	2				
B_3		2				
T_3						

Type VIII_0. For the cosmological model of type VII_0 $(n_1 = 1, n_2 = 1, n_3 = 0)$ the dynamical system (4.9) contains a closed subsystem, which coincides with the system (6.1) for the model of type IX on the component of the boundary Γ_3, which therefore is the compact manifold S for the model of type VII_0. The critical points of the dynamical system (6.1) on Γ_3 are shown in the appropriate part of Fig. 9. Two arcs $(\beta_{31}, 1)$ and $(\beta_{32}, 2)$ on the circles $(\psi, 1)$ and $(\psi, 2)$ consist of attracting (in Γ_3) critical points (see Table 2 of eigenvalues). For $t \to 0$ the solutions, which correspond to the trajectories of the dynamical system on Γ_3 approaching these attracting critical points, have the stable Kasner asymptotic form:

(β_{ji}, i)	(β_{3i}, i)	(β_{j3}, i)	(β_{3j}, i)	A_3	B_3	T_3	T_3^0
							2
							1 $j = 1, 2$
		2 $x = j$					
2 $x = j$ $y = i$							
			2 $x = j$	2 $x = 1, 2$	2 $x = 1, 2$	3 $x = 1, 2$	4 $x = 1, 2$
						2	3
							2

$$q_i \cong C_i t^{2p_i}, \quad p_i = 1 + \sqrt{2}s_i. \tag{7.8}$$

On the arcs $(\beta_{31}, 1)$ and $(\beta_{32}, 2)$ we have $s_3 < -1/\sqrt{2}$, so $p_3 < 0$ and in the above asymptotic forms $q_1, q_2 \to 0$ and $q_3 \to \infty$.

When time is directed towards contraction the trajectories of the dynamical system on Γ_3 (as shown in part IV of Sect. 6) approach the boundary of the manifold Γ_3 (which consists of three components determined by $y_i = \delta_{i1}, y_i = \delta_{i2}$; $w = 0$) and start moving along the separatrices of critical points. The separatrix diagram for the model of type VII_0 (and also type VI_0) is shown in Table 4 (the notation is the same as in the separatrix diagram in Table 3, column M_3 pertains to the model of type VI_0). Notice that every sequence of separatrices of unstable critical points in Γ_3 after a finite number of steps ends in one of the attracting critical points, which belong to the arcs $(\beta_{31}, 1)$ and $(\beta_{32}, 2)$ (see Table 4). Therefore for the contraction of space, almost all solutions for the model of type VII_0 have the stable Kasner asymptotic form (7.8). The role of the transitional asymptotic form for the solutions of the model of type VII_0 is played by some finite number of oscillations in the "long epoch," where the eigenvalues of the metric q_1 and $q_2 \gg q_3$ oscillate in a precisely identical fashion to the models of types IX and VIII. The geometrical model of the oscillatory mode (see Fig. 10) makes it evident that these oscillations terminate after a finite number of steps, because any point of the circle S^1 ends up in the arcs β_{31} or β_{32} after a repeated application of the mapping (projection) T.

Trajectories of the dynamical system on the invariant manifold $s_1 = s_2$, $y_1 = y_2 = 1/\sqrt{2}$ describe the axisymmetric metrics for the model of type VII_0 (see (6.19)). Among them there are exact solutions: the flat Friedmann solution

$$q_1 = q_2 = C_1 t^{4/3(1+k)}, \quad q_3 = C_3 t^{4/3(1+k)},$$

approaching a critical point in the segment Φ_3^0, the Kasner solution

$$q_1 = q_2 = C_1 t^{4/3}, \quad q_3 = C_3 t^{-2/3},$$

approaching a critical point in the segment A_3, the Taub solution corresponding to the critical points of T_3 and isometric to the Minkowski solution

$$q_1 = q_2 = C_1, \quad q_3 = C_3 t^2.$$

In all axisymmetric solutions of the model of type VII_0 the three-dimensional spatial sections $(t = \text{const})$ are flat according to the formulas for the Ricci curvature (3.28).

When time is directed towards expansion, by the existence of the monotone function F_3 (6.18) all trajectories of the dynamical system on Γ_3 (except for the critical points of T_3) approach the (degenerate) critical point T_3^0 as $t \to \infty$ (see part IV of Sect. 6). However as the above exact solutions show, the asymptotic forms of the metric here can vary.

Type VI_0. For the cosmological model of type VI_0 $(n_1 = 1, n_2 = -1, n_3 = 0)$ the dynamical system (4.9) is equivalent to the dynamical system on the component of the boundary Γ_3 (or Γ_2) for the model of type VIII. Therefore the compact manifold S for the model of type VI_0 is the four-dimensional manifold Γ_3 (as shown above Γ_3 is the product of a half-ball (corner Y_1) and the segment I). The critical points of the dynamical system on Γ_3 are shown in the appropriate part of Fig. 12. All these critical points and their eigenvalues coincide with the analogous critical points for the model of type VII_0, except that instead of the segment T_3 and the point T_3^0 here we have a repelling critical point M_3. In accordance with

Table 5. Power asymptotics of the metric for the
homogeneous cosmological models of class A for
the contraction of space

Model type	Power asymptotics for $t \to 0$
II	K_2, K_3, Φ, N_1
VI_0	K_3, Φ, M_3, N_1, N_2
VII_0	K_3, Φ, T_3, N_1, N_2
VIII	$\Phi, T_1, M_2, M_3, N_1, N_2, N_3$
IX	$\Phi, T_1, T_2, T_3, N_1, N_2, N_3$

this we must discard the rows and columns T_3 and T_3^0 in the separatrix diagram
(Table 4).

The behavior of solutions of the model of type VI_0 for the contraction of space
is analogous to the behavior of solutions of type VII_0. After a finite number of
oscillations in the "long epoch" almost all solutions take on the stable Kasner
asymptotic form (7.8).

When time is directed towards expansion, the critical point M_3 is attracting.
In the model of type VI_0 it corresponds to the following exact solution:

$$q_3 = C_3 t^2, \quad q_1 = q_2 = C_1 t^{(1-k)/(1+k)}, \tag{7.9}$$

which is stable towards expansion. For $t \to \infty$ this asymptotic form (7.9) holds for
an entire region in the space of solutions. Therefore in the model of type VI_0 the
isotropization of the metric for the endless expansion of space does not occur.

In conclusion, we list the power asymptotics towards contraction, which
occur in homogeneous cosmological models of class A (see Table 5). In this table
the power asymptotic forms are designated by the corresponding critical points.
K_i are the Kasner asymptotic forms (7.8) in which the negative Kasner exponent
p_i corresponds to the variable q_i (i.e. for $t = 0$ we have $q_i \to \infty$ and $q_j, q_k \to 0$).

8. Transformation of the Dynamical System for Homogeneous Cosmological Models of Class B

I. General Properties of the Dynamical System. For the homogeneous cosmological models of class B Einstein's system of equations reduces to the following
dynamical system in the phase space $p_i, q_i, p_\varphi, \varphi$ (as was shown in Sect. 3):

$$\frac{dp_i}{d\tau} = -\frac{\partial H}{\partial q_i} - h_i, \quad \frac{dp_\varphi}{d\tau} = -\frac{\partial H}{\partial \varphi} - h_\varphi,$$

$$\frac{dq_i}{d\tau} = \frac{\partial H}{\partial p_i}, \quad \frac{d\varphi}{d\tau} = \frac{\partial H}{\partial p_\varphi}. \tag{8.1}$$

The system (8.1) is defined on the subspace defined by Einstein's equation $R_3^0 = 0$:

$$R = 2p_3q_3 - p_1q_1 - p_2q_2 - \frac{1}{2a}P_\varphi = 0, \qquad (8.2)$$

and is considered in the region $H = \varepsilon(q_1q_2q_3)^{(1+k)/2} \geqslant 0$. Time τ is related to the synchronous time t by

$$d\tau/dt = (q_1q_2q_3)^{-k/2}.$$

The Hamiltonian H and the summands h_i, h_φ have the following form:

$$H = \frac{1}{(q_1q_2q_3)^{(1-k)/2}} \left(T(p_iq_i) + V_G(q_i) \right),$$

$$T = 2 \sum_{i<j}^{3} p_iq_ip_jq_j - \sum_{i=1}^{3} p_i^2q_i^2 - \frac{p_\varphi^2 q_1q_2}{(n_1q_1 - n_2q_2)^2},$$

$$V_G(q_i) = -\frac{1}{4}\left(12a^2q_1q_2 + (n_1q_1 - n_2q_2)^2\right),$$

$$h_1 = \frac{1}{(q_1q_2q_3)^{(1-k)/2}} a^2q_2, \quad h_2 = \frac{1}{(q_1q_2q_3)^{(1-k)/2}} a^2q_1, \qquad (8.3)$$

$$h_3 = \frac{1}{(q_1q_2q_3)^{(1-k)/2}}\left(-2a^2 \frac{q_1q_2}{q_3}\right),$$

$$h_\varphi = \frac{1}{(q_1q_2q_3)^{(1-k)/2}} a(n_1q_1 - n_2q_2)^2.$$

The constants a, n_1, n_2 determine the type of the model in question according to Table 1. As shown in Sect. 3, all metrics of the model of type V $(a = 1, n_1 = n_2 = 0)$ (whose Lie algebra has a six-dimensional group of inner automorphisms) can be diagonalized. Therefore for the model of type V we have $p_\varphi \equiv 0$ and the condition (8.2) becomes $2p_3q_3 = p_1q_1 + p_2q_2$. As a result, the system (8.1) is significantly simplified. It can be reduced to a two-dimensional system (see Sect. 10 below).

Let us verify that (8.1) implies the following:

$$\frac{dR}{d\tau} = 0, \quad \frac{dH}{d\tau} = -\frac{4a^2q_1q_2}{(q_1q_2q_3)^{1-k}} R, \qquad (8.4)$$

(in particular the constraint $R = 0$ (8.2) is preserved and on the level $R = 0$ the Hamiltonian $H \geqslant 0$ is preserved). Let us find $dH/d\tau$:

$$\frac{dH}{d\tau} = \frac{\partial H}{\partial p_i}\left(-\frac{\partial H}{\partial q_i} - h_i\right) + \frac{\partial H}{\partial q_i}\frac{\partial H}{\partial p_i}$$

$$+ \frac{\partial H}{\partial p_\varphi}\left(-\frac{\partial H}{\partial \varphi} - h_\varphi\right) + \frac{\partial H}{\partial \varphi}\frac{\partial H}{\partial p_\varphi}$$

$$= -h_i \frac{\partial H}{\partial p_i} - h_\varphi \frac{\partial H}{\partial p_\varphi}$$

$$= -\frac{2a^2}{(q_1 q_2 q_3)^{1-k}} \Bigg[q_1(p_3 q_3 + p_2 q_2 - p_1 q_1)q_2$$

$$+ q_2(p_1 q_1 + p_3 q_3 - p_2 q_2)q_1 + q_3(p_1 q_1 + p_2 q_2 - p_3 q_3)\left(-\frac{2q_1 q_2}{q_3} \right)$$

$$- \frac{p_\varphi q_1 q_2}{(n_1 q_1 - n_2 q_2)^2} \frac{(n_1 q_1 - n_2 q_2)^2}{a} \Bigg]$$

$$= -\frac{4a^2 q_1 q_2}{(q_1 q_2 q_3)^{1-k}} R.$$

Here is the system (8.1) in an explicit form:

$$\dot{p}_1 = -\frac{1}{(q_1 q_2 q_3)^{(1-k)/2}} \Bigg[2p_1(p_2 q_2 + p_3 q_3 - p_1 q_1) + \frac{p_\varphi^2(n_1 q_1 + n_2 q_2)}{(n_1 q_1 - n_2 q_2)^3} q_2$$

$$- 2a^2 q_2 - \frac{1}{2} n_1(n_1 q_1 - n_2 q_2) - \left(\frac{1-k}{2} \right) \frac{H_0}{q_1} \Bigg],$$

$$\dot{p}_2 = -\frac{1}{(q_1 q_2 q_3)^{(1-k)/2}} \Bigg[2p_2(p_1 q_1 + p_3 q_3 - p_2 q_2) - \frac{p_\varphi^2(n_1 q_1 + n_2 q_2)}{(n_1 q_1 - n_2 q_2)^3} q_1$$

$$- 2a^2 q_1 + \frac{1}{2} n_2(n_1 q_1 - n_2 q_2) - \frac{1-k}{2} \frac{H_0}{q_2} \Bigg],$$

$$\dot{p}_3 = -\frac{1}{(q_1 q_2 q_3)^{(1-k)/2}} \Bigg[2p_3(p_1 q_1 + p_2 q_2 - p_3 q_3) \hspace{3cm} (8.5)$$

$$- 2a^2 \frac{q_1 q_2}{q_3} - \frac{1-k}{2} \frac{H_0}{q_3} \Bigg],$$

$$\dot{q}_i = \frac{1}{(q_1 q_2 q_3)^{(1-k)/2}} [2q_i(p_j q_j + p_k q_k - p_i q_i)], \quad i, j, k = 1, 2, 3,$$

$$\dot{p}_\varphi = -\frac{1}{(q_1 q_2 q_3)^{(1-k)/2}} a(n_1 q_1 - n_2 q_2)^2,$$

$$\dot{\varphi} = -\frac{1}{(q_1 q_2 q_3)^{(1-k)/2}} \frac{2p_\varphi q_1 q_2}{(n_1 q_1 - n_2 q_2)^2}.$$

where $H_0 = T + V_G$. In new coordinates $P_i = p_i q_i$ and time τ_0 determined by

$$d\tau_0/d\tau = (q_1 q_2 q_3)^{-(1-k)/2} \ (\text{or } d\tau_0/dt = (q_1 q_2 q_3)^{-1/2})$$

the system (8.5) has the following form:

$$\dot{P}_1 = -\frac{p_\varphi^2(n_1 q_1 + n_2 q_2)}{(n_1 q_1 - n_2 q_2)^3} q_1 q_2 + 2a^2 q_1 q_2$$

$$+ \frac{1}{2} n_1 q_1(n_1 q_1 - n_2 q_2) + \frac{1-k}{2} H_0,$$

$$\dot{P}_2 = +\frac{p_\varphi^2(n_1 q_1 + n_2 q_2)}{(n_1 q_1 - n_2 q_2)^3} q_1 q_2 + 2a^2 q_1 q_2$$

$$- \frac{1}{2} n_2 q_2(n_1 q_1 - n_2 q_2) + \frac{1-k}{2} H_0, \qquad (8.6)$$

$$\dot{P}_3 = 2a^2 q_1 q_2 + \frac{1-k}{2} H_0,$$

$$\dot{q}_i = 2q_i(P_j + P_k - P_i),$$

$$\dot{p}_\varphi = -a(n_1 q_1 - n_2 q_2)^2,$$

$$\dot{\varphi} = -\frac{2p_\varphi q_1 q_2}{(n_1 q_1 - n_2 q_2)^2}.$$

where

$$H_0 = T + V_G$$

$$= (P_1 + P_2 + P_3)^2 - 2\sum_{j=1}^{3} P_j^2 - \frac{p_\varphi^2 q_1 q_2}{(n_1 q_1 - n_2 q_2)^2} \qquad (8.7)$$

$$- \frac{1}{4}(12a^2 q_1 q_2 + (n_1 q_1 - n_2 q_2)^2).$$

Now it is not difficult to verify that the function $R = 2P_3 - P_1 - P_2 - (1/2a)p_\varphi$ is the first integral of the system (8.1)–(8.6):

$$\frac{dR}{d\tau_0} = 2\dot{P}_3 - \dot{P}_1 - \dot{P}_2 - \frac{1}{2a}\dot{p}_\varphi$$

$$= -\frac{1}{2}(n_1 q_1 - n_2 q_2)^2 - \frac{1}{2a}(-a)(n_1 q_1 - n_2 q_2)^2 = 0.$$

From the condition $H_0 \geqslant 0$ (see (8.7)) it follows that $T(P_i) > 0$ in the physical region $(q_i > 0)$. Therefore all P_i have the same sign and $P_i \neq 0$. Consequently for the homogeneous cosmological models of class B the volume $(q_1 q_2 q_3)^{1/2}$ varies monotonically (since $(\ln(q_1 q_2 q_3))' = P_1 + P_2 + P_3 \neq 0$). When time is directed towards contraction we have $P_1 + P_2 + P_3 < 0$, so all $P_i < 0$. The system (8.6) has two monotonically increasing functions $P_1 + P_2 < 0$ and $P_3 < 0$ and two monotonically decreasing functions $q_1 q_2$ and $p_\varphi = 2a(2P_3 - P_1 - P_2)$. Also as in the system (4.1) for the homogeneous models of class A there is a monotonically decreasing function

$$F = \frac{d}{dt}(\det g_{ij})^{1/6} = \frac{1}{3}\frac{P_1 + P_2 + P_3}{(q_1 q_2 q_3)^{1/3}},$$

$$\frac{dF}{dt} = \frac{2}{3(q_1 q_2 q_3)^{5/6}}\left[-\frac{1}{3}(P_1 + P_2 + P_3)^2 + T(P_i)\right.$$

$$\left. - \frac{p_\varphi^2 q_1 q_2}{(n_1 q_1 - n_2 q_2)^2} - \frac{1 + 3k}{4}H_0\right].$$

As a consequence of the presence of these monotone functions all critical points of the dynamical system (8.6) belong to the boundary of the physical region.

II. Reduction of Order of the Dynamical System. There are a few important (for later) properties of the dynamical system (8.6). Firstly the coordinate φ is a cyclic coordinate of this system. Secondly with the preserved identity $R = 0$ in mind, the momentum $p_\varphi = 2a(2P_3 - P_1 - P_2)$ can be eliminated from the equations (8.6). And thirdly after these eliminations the system (8.6) is invariant relative to the scale transformations

$$P_i \to \lambda P_i, \quad q_i \to \lambda q_i, \quad \tau_0 \to \frac{1}{\lambda}\tau_0$$

and contains a closed subsystem of five equations in the variables P_1, P_2, P_3, q_1, q_2. Thus the order of the system (8.6) is reduced to four.

Here we introduce the scale-invariant coordinates (analogous to the coordinates used in Sect. 4):

$$y_1 = \frac{q_1}{(q_1^2 + q_2^2)^{1/2}}, \quad y_2 = \frac{q_2}{(q_1^2 + q_2^2)^{1/2}},$$

$$s_i = \frac{P_i}{(P_1^2 + P_2^2 + P_3^2)^{1/2}}, \quad i = 1, 2, 3, \quad w = \frac{q_1^2 + q_2^2}{P_1^2 + P_2^2 + P_3^2}. \tag{8.8}$$

The coordinates y_i and s_i satisfy the following two relations:

$$y_1^2 + y_2^2 = 1, \quad s_1^2 + s_2^2 + s_3^2 = 1. \tag{8.9}$$

In the coordinates (8.8) and time τ_1 defined by the relation

$$\frac{d\tau_1}{dt} = \frac{(P_1^2 + P_2^2 + P_3^2)^{1/2}}{2(q_1 q_2 q_3)^{1/2}},$$

the system (8.6) has the following form:

$$\dot{s}_1 = 2\frac{\bar{p}_\varphi^2(n_1 y_1 + n_2 y_2)}{(n_1 y_1 - n_2 y_2)^3}y_1 y_2(-1 + s_1(s_1 - s_2))$$

$$+ w[4a^2 y_1 y_2(1 - s_1(s_1 + s_2 + s_3)) + (n_1 y_1 - n_2 y_2)$$

$$\times (n_1 y_1 - s_1(s_1 n_1 y_1 - s_2 n_2 y_2))] + (1 - k)H_2(1 - s_1(s_1 + s_2 + s_3)),$$

$$\dot{s}_2 = 2\frac{\overline{p}_\varphi^2(n_1 y_1 + n_2 y_2)}{(n_1 y_1 - n_2 y_2)^3} y_1 y_2 (1 + s_2(s_1 + s_2))$$

$$+ w[4a^2 y_1 y_2 (1 - s_2(s_1 + s_2 + s_3)) + (n_1 y_1 - n_2 y_2)$$

$$\times (-n_2 y_2 - s_2(s_1 n_1 y_1 - s_2 n_2 y_2))] + (1 - k)H_2(1 - s_2(s_1 + s_2 + s_3)),$$

$$\dot{s}_3 = 2\frac{\overline{p}_\varphi^2(n_1 y_1 + n_2 y_2)}{(n_1 y_1 - n_2 y_2)^3} y_1 y_2 (s_3(s_1 - s_2))$$

$$+ w[4a^2 y_1 y_2 (1 - s_3(s_1 + s_2 + s_3)) + (n_1 y_1 - n_2 y_2)$$

$$\times (-s_3)(s_1 n_1 y_1 - s_2 n_2 y_2)] + (1 - k)H_2(1 - s_3(s_1 + s_2 + s_3)),$$

$$\dot{w} = 2w\left[4(s_1 + s_2 + s_3 - 2y_i^2 s_i) + 2(s_1 - s_2)\frac{\overline{p}_\varphi^2(n_1 y_1 + n_2 y_2)}{(n_1 y_1 - n_2 y_2)^3} y_1 y_2 \right.$$

$$- w\{4a^2 y_1 y_2(s_1 + s_2 + s_3) + (s_1 n_1 y_1 - s_2 n_2 y_2)(n_1 y_1 - n_2 y_2)\}$$

$$\left. - (1 - k)H_2(s_1 + s_2 + s_3)\right], \tag{8.10}$$

$$\dot{y}_i = 8y_i(y_1^2 s_1 + y_2^2 s_2 - s_i), \quad i = 1, 2,$$

$$\dot{\varphi} = -\frac{4\overline{p}_\varphi y_1 y_2}{(n_1 y_1 - n_2 y_2)^2}, \quad \overline{p}_\varphi = 2a(2s_3 - s_1 - s_2),$$

$$\dot{q}_3 = 4q_3(s_1 + s_2 - s_3).$$

where

$$H_2 = \frac{H_0}{(P_1^2 + P_2^2 + P_3^2)} = 2\sum_{i<j}^{3} s_i s_j - \sum_{i=1}^{3} s_i^2 - \frac{\overline{p}_\varphi^2 y_1 y_2}{(n_1 y_1 - n_2 y_2)^2}$$

$$- \frac{w}{4}(12a^2 y_1 y_2 + (n_1 y_1 - n_2 y_2)^2). \tag{8.11}$$

The dynamical system (8.10) contains a closed subsystem in the coordinates $s_1, s_2, s_3, w, y_1, y_2$ satisfying the two relations (8.9) and thus defined on some four-dimensional manifold. All critical points of the system (8.10) for the models of types III and IV ($n_1 = 1, n_2 = -1, n_3 = 0, a > 0$) are non-degenerate. Therefore the dynamics of the models of types III and IV can be conveniently studied on the constructed four-dimensional manifold S. In the system of coordinates W_1 (8.8) the manifold S is determined by the natural conditions

$$y_1 \geqslant 0, \quad y_2 \geqslant 0, \quad H_2 \geqslant 0, \quad w \geqslant 0, \quad s_1 + s_2 + s_3 \leqslant 0,$$

$$y_1^2 + y_2^2 = 1, \quad s_1^2 + s_2^2 + s_3^2 = 1. \tag{8.12}$$

For each value of y_1, y_2 the conditions $H_2 \geqslant 0$ and $w \geqslant 0$ determine (in the coordinates s_i, w) a bounded three-dimensional set P, which is diffeomorphic to a half of a three-dimensional ball. Therefore the manifold S is diffeomorphic to $P \times I$, where I is the segment determined by $y_i \geqslant 0$ in the circle $y_1^2 + y_2^2 = 1$.

III. Construction of the Compact Four-Dimensional Manifold for the Homogeneous Model of Type VII. For the model of type VII $(n_1 = 1, n_2 = 1, n_3 = 0)$ the three-dimensional set P for $y_1 = y_2$ contracts to a two-dimensional disc $(p_\varphi = 0)$, which consists entirely of degenerate singularities of the system (8.10). To resolve these singularities we introduce a new system of coordinates W_2:

$$s_1, s_2, y_1, y_2, w, z = \frac{2s_3 - s_1 - s_2}{y_1 - y_2}. \tag{8.13}$$

For the model of type VII the compact manifold S is covered by local charts W_1 and W_2 and is determined in them by the conditions (8.12) (by $H_2 \geqslant 0$ the coordinates z and w are bounded in the manifold S). In the coordinates W_2 and time τ_3:

$$d\tau_3/d\tau_2 = 1/(y_1 - y_2) \text{ (assume that } y_1 > y_2)$$

the system (8.10) has the following form:

$$\dot{s}_1 = 8a^2z^2(y_1 + y_2)y_1 y_2(-1 + s_1(s_1 - s_2))$$
$$+ w(y_1 - y_2)[4a^2y_1 y_2(1 - s_1Z)$$
$$+ (y_1 - y_2)(y_1 - s_1(s_1 y_1 - s_2 y_2))] + (1 - k)(y_1 - y_2)H_2 \times (1 - s_1Z),$$
$$\dot{s}_2 = 8a^2z^2(y_1 + y_3)y_1 y_2(1 + s_2(s_1 - s_2))$$
$$+ w(y_1 - y_2)[4a^2y_1 y_2(1 - s_2Z)$$
$$+ (y_1 - y_2)(-y_2 - s_2(s_1 y_1 - s_2 y_2))] + (1 - k)(y_1 - y_2)H_2 \times (1 - s_2Z),$$
$$\dot{z} = 8a^2z^3(y_1 + y_2)y_1 y_2(s_1 - s_2)$$
$$+ wz[-4a^2y_1 y_2(y_1 - y_2)Z - (y_1 - y_2)^2(s_1 y_1 - s_2 y_2)]$$
$$- (1 - k)H_2z(y_1 - y_2)Z - w(y_1 - y_2)^2$$
$$- 8z[(y_1 - y_2)(y_1^2s_1 + y_2^2s_2) - (y_1s_1 - y_2s_2)],$$
$$\dot{w} = 2w(y_1 - y_2)[4(Z - 2y_k^2s_k)$$
$$- w\{4a^2y_1 y_2Z + (s_1 y_1 - s_2 y_2)(y_1 - y_2)\}$$
$$- (1 - k)H_2Z] + 16wa^2z^2(y_1 + y_2)y_1 y_2(s_1 - s_2),$$
$$\dot{y}_1 = 8(y_1 - y_2)y_1(y_1^2s_1 + y_2^2s_2 - s_1),$$
$$\dot{y}_2 = 8(y_1 - y_2)y_2(y_1^2s_1 + y_2^2s_2 - s_2),$$
$$\dot{\varphi} = -8azy_1 y_2,$$
$$\dot{q}_3 = 2q_3(y_1 - y_2)(s_1 + s_2 - z(y_1 - y_2)). \tag{8.14}$$

where we use the following notation:

$$Z = (s_1 + s_2 + s_3) = \frac{1}{2}(3(s_1 + s_2) + z(y_1 - y_2)),$$

$$H_2 = -(s_1 - s_2)^2 + \frac{1}{4}(s_1 + s_1 + z(y_1 - y_2))(3(s_1 + s_2)) \qquad (8.15)$$

$$- z(y_1 - y_2)) - 4a^2 z^2 y_1 y_2 - \frac{w}{4}(12a^2 y_1 y_2 + (y_1 - y_2)^2).$$

The coordinates W_2 (8.13) on the manifold S are used in a neighborhood of points $y_1 = y_2 = 1/\sqrt{2}$. For all y_1, y_2 the condition $H_2 \geqslant 0$ determines some compact three-dimensional set P in the coordinates s_i, z, w subject to

$$s_1^2 + s_2^2 + s_3^2 = s_1^2 + s_2^2 + \tfrac{1}{4}(z(y_1 - y_2) + s_1 + s_2)^2 = 1).$$

IV. Construction of the Compact Manifold S for the Homogeneous Model of Type IV. For the model of type IV $(n_1 = 1, n_2 = n_3 = 0, a > 0)$ the three-dimensional set P, determined in the coordinates s_i, w by the condition $H_2 \geqslant 0$ (see (8.11)) for $(y_1, y_2) \to (0, 1)$, is stretched to a two-dimensional strip. This strip $2s_1 - s_2 - s_3 = 0$, $s_1^2 + s_2^2 + s_3^2 = 1$, $0 \leqslant w < \infty$ consists entirely of degenerate singularities of the system (8.10). To resolve these critical points we introduce new coordinates W_3:

$$s_1, s_2, \bar{y}_1 = \bar{y}_1^{1/2}, \quad y_2, \quad w_1 = w y_1, \quad z_1 = \frac{2s_3 - s_1 - s_2}{y_1^{1/2}}. \qquad (8.16)$$

In the coordinates W_3 the function H_2 (8.11) has the following form:

$$H_2 = -(s_1 - s_2)^2 + \frac{1}{4}(s_1 + s_2 + z_1 \bar{y}_1)(3(s_1 + s_2) - z_1 \bar{y}_1)$$

$$- 4a^2 z_1^2 y_2 - \frac{w_1}{4}(12a^2 y_2 + \bar{y}_1^2). \qquad (8.17)$$

Obviously in a neighborhood of the points $\bar{y}_1 = 0$, $y_2 = 1$ the condition $H_2 \geqslant 0$ determines (in the coordinates s_1, s_2, z_1, w_1) a compact three-dimensional set subject to

$$s_1^2 + s_2^2 + s_3^2 = s_1^2 + s_2^2 + \tfrac{1}{4}(z_1 \bar{y}_1 + s_1 + s_2)^2 = 1).$$

For the model of type IV the compact four-dimensional manifold S is covered by the local charts W_1 and W_2. In them it is determined by the conditions (8.12). In the coordinates W_3 the dynamical system (8.10) has the following form:

$$\dot{s}_1 = 8a^2 z_1^2 y_2(-1 + s_1(s_1 - s_2))$$

$$+ w_1[4a^2 y_2(1 - s_1 Z_1) + (1 - s_1^2)\bar{y}_1^2] + (1 - k)H_2(1 - s_1 Z_1),$$

$$\dot{s}_2 = 8a^2 z_1^2 y_2(1 + s_2(s_1 - s_2)) + w_1[4a^2 y_2(1 - s_2 Z_1) - s_1 s_2 \bar{y}_1^2]$$

$$+ (1 - k)H_2(1 - s_2 Z_1),$$

$$\dot{z}_1 = 8a^2 z_1^3 y_2 (s_1 - s_2)$$
$$\quad + w_1 z_1 [4a^2 y_2 (-Z_1) - s_1 \bar{y}_1^2] - (1 - k) H_2 z_1 Z_1$$
$$\quad - w_1 \bar{y}_1 - 4 z_1 (\bar{y}_1^4 s_1 + y_2^2 s_2 - s_1),$$
$$\dot{w}_1 = 2w_1 \{ 4(Z_1 - 2\bar{y}_1^4 s_1 - 2y_2^2 s_2) + 8(s_1 - s_2) a^2 z_1^2 y_2$$
$$\quad - w_1 [4a^2 y_2 Z_1 + s_1 \bar{y}_1^2] - (1 - k) H_2 Z_1 \qquad (8.18)$$
$$\quad + 4(\bar{y}_1^4 s_1 + y_2^2 s_2 - s_1) \},$$
$$\dot{\bar{y}}_1 = 4\bar{y}_1 (\bar{y}_1^4 s_1 + y_2^2 s_2 - s_1), \quad \dot{y}_2 = 8y_2 (\bar{y}_1^4 s_1 + y_2^2 s_2 - s_2),$$
$$\dot{\varphi} = -8a z_1 \frac{y_2}{\bar{y}_1}, \quad \dot{q}_3 = 2q_3 (s_1 + s_2 - z_1 \bar{y}_1).$$

where

$$Z_1 = (s_1 + s_2 + s_3) = \tfrac{1}{2}(3(s_1 + s_2) + z_1 \bar{y}_1).$$

9. Several General Properties of the Dynamics of Homogeneous Cosmological Models of Types III, IV, VI and VII

I. Some Exact Solutions and Asymptotic Forms of the Metric for Models of Class B. Compact four-dimensional manifolds S constructed in Sect. 8 for the models of types III, IV, VI and VII are products $P \times I$, where P is a bounded three-dimensional set in coordinates s_i, w or s_i, z, w; s_i, z_1, w diffeomorphic to a three-dimensional half-ball and I is a segment on a circle: $y_1^2 + y_2^2 = 1$, $y_i \geq 0$. The boundary Γ of the manifold S consists of four components: Γ_0 $(H_2 = 0)$, Γ_w $(w = 0)$, Y_1 $(y_1 = 1, y_2 = 0)$ and Y_2 $(y_1 = 0, y_2 = 1)$. At the intersection of these components we have the corners of the boundary. The components of the boundary Γ and all their intersections are invariant manifolds of the dynamical system (8.10) (respectively (8.14) or (8.18)) in the manifold S.

It is useful to compare the system (8.10) with the analogous system (4.9) for homogeneous models of class A. As observed in Sect. 4, the systems (4.9) for all models of class A with $w = 0$ are completely identical. The opposite happens in class B. There the system (8.10) is entirely different (this is a consequence of the fact that for the models of class B the metric is not diagonal) and depends on the type of the model (because the metric can be diagonalized by various transformations (see Sect. 3)). However some critical points of the systems (8.10) and (4.9) for $w = 0$ do coincide. For example the critical points Φ $(s_i = -1/\sqrt{3}, w = 0)$ are the same and so are the eigenvalues at these critical points. Therefore in all the models of class B there are also quasi-isotropic asymptotic forms for $t \to 0$ of codimension 2:

$$q_i \approx C_i t^{4/3(1+k)}, \tag{9.1}$$

where $\varphi \to$ const.

On the components of the boundary Y_i for $n_i \neq 0$ the system (8.10) is identical to the system (4.9) (or (6.1)) on the appropriate corners Y_i of the manifold S for the models of class A. Therefore for the models of class B the manifold S contains circles of critical points

$$(\psi, i)(s_1 + s_2 + s_3 = -\sqrt{2}, s_1^2 + s_2^2 + s_3^2 = 1, y_k = \delta_{ik}, w = 0),$$

the isolated critical points N_i (6.3) and the endpoints Φ_i^{00} of the segment Φ $(y_1^2 + y_2^2 = 1)$ (where $i = 1, 2$ for the models of types III, VI and VII and $i = 1$ for the model of type IV). The eigenvalues at these critical points coincide as well. Therefore the models of class B also have (stable) asymptotic forms N_i of codimension 3:

$$q_i \approx C_1 t^{(1-k)/(1+k)}, \quad q_j \approx C_2 t^{(3+k)/2(1+k)},$$
$$q_3 \approx C_3 t^{(3+k)/2(1+k)}, \quad i, j = 1, 2, \tag{9.2}$$

where $\varphi \to$ const for $t \to 0$.

Just like in the models of types VI_0 and VII_0 (see Sect. 7) the two arcs $(\beta_{31}, 1)$ $(s_3 < s_1 < s_2)$ and $(\beta_{32}, 2)(s_3 < s_2 < s_1)$ on the circles of critical points $(\psi, 1)$ and $(\psi, 2)$ (for the model of type IV only the arc $(\beta_{31}, 1)$) consist of attracting (towards contraction) critical points. Solutions corresponding to the trajectories approaching these critical points have the stable Kasner asymptotic form of the metric for $t \to 0$:

$$q_i \approx C_i t^{2p_i}, \quad p_i = 1 + \sqrt{2}s_i, \quad \varphi \to \text{const.} \tag{9.3}$$

The listed critical points belong to the boundary of the physical region of the manifold S, so they correspond not to exact solutions, but to asymptotic forms of the metric. A number of exact solutions of homogeneous models of class B can be found by studying the critical points in the physical region $w \neq 0$, $y_i \neq 0$ of the manifold S. In further calculations it is convenient to resolve the relations (8.9) and (in the physical region) change the coordinates to

$$u = 2\frac{P_1 - P_2 - P_3}{q_1} = 2\frac{s_1 - s_2 - s_3}{y_1 w^{1/2}},$$

$$v = 4\frac{P_1 - P_2}{q_1} = 4\frac{s_1 - s_2}{y_1 w^{1/2}}, \tag{9.4}$$

$$x = 4\frac{2P_3 - P_1 - P_2}{q_1} = 4\frac{2s_3 - s_1 - s_2}{y_1 w^{1/2}}, \quad y = \frac{q_2}{q_1} = \frac{y_2}{y_1}.$$

In the coordinates (9.4) and time $\tau_1' = 2\tau_1$ the dynamical system (8.6)–(8.10) has the following form:

$$\dot{y} = yv,$$

$$\dot{v} = uv + 2\left(n_1^2 - n_2^2 y^2\right) - 2a^2 \frac{x^2(n_1 + n_2 y) y}{(n_1 - n_2 y_2)^3},$$

$$\dot{x} = xu - 2(n_1 - n_2 y)^2,$$

$$\dot{u} = u^2 + n_1^2 - n_1^2 y^2 - 4a^2 y - a^2 \frac{x^2(n_1 + n_2 y) y}{(n_1 - n_2 y)^3} + H_3.$$

(9.5)

where

$$H_3 = \frac{1-k}{4}\left[3u^2 + \frac{1}{2}v^2 - \frac{1}{2}vx - 3uv + ux \right.$$

$$\left. - \frac{a^2 x^2 y}{(n_1 - n_2 y)^2} - 12a^2 y - (n_1 - n_2 y)^2 \right].$$

For $y \neq 0$ the system (9.5) has a one-dimensional set of critical points L_y with coordinates (with parameter y):

$$v = 0, \quad x = \frac{(n_1 - n_2 y)^2}{a\sqrt{y}}, \quad u = 2a\sqrt{y}.$$

(9.6)

At these critical points we have $H_3 = 0$.

For the homogeneous model of type IV, there are no other critical points in the region $y \neq 0$, $w \neq 0$ for the system (9.5)–(8.10). For the models of types III, VI and VII, besides the line L_y, there are also critical points for $y = 1$ (they will be considered separately in Sect. 10). The critical points of L_y determine the following exact vacuum solutions for the models of types III, IV, VI and VII:

$$q_1 = q_1^0 t^{2/(x^2 + 1)}, \quad q_2 = y q_1, \quad \varkappa = \frac{n_1 - n_2 y}{2a\sqrt{y}},$$

(9.7)

$$q_3 = q_3^0 t^2, \quad \varphi = -\frac{\ln t}{2\sqrt{q_3^0}} + \varphi_0.$$

For models of types III, VI and VII these exact solutions were first discovered in [70], where approximate estimates of the region of their stability were also listed.

It is possible to analyze in full the stability of the solutions (9.7) (and find the exact boundaries (in the parameter y) of the region of stability) based on the calculation of eigenvalues of the critical points (9.6). Here is the characteristic matrix of the system (9.5)

$$A_i^j(\lambda) = \left| \frac{\partial f_i}{\partial x_j} - \lambda \delta_j^i \right|$$

at these critical points:

	y	v	x	u
y	$-\lambda$	y	0	0
v	$-\dfrac{2}{y}m_1$	$2a\sqrt{y}-\lambda$	$-\dfrac{4am_3\sqrt{y}}{m_2}$	0
x	$4n_2\cdot m_2$	0	$2a\sqrt{y}-\lambda$	$\dfrac{m_2^2}{a\sqrt{y}}$
u	$\begin{aligned}&-4a^2+\dfrac{1-k}{4}\\&\times\left(-12a^2-\dfrac{(m_2)^2}{y}\right)\\&-\dfrac{1}{y}m_1\end{aligned}$	$\begin{aligned}&\dfrac{1-k}{8}\left(\dfrac{(m_2)^2}{a\sqrt{y}}\right.\\&\left.+12a\sqrt{y}\right)\end{aligned}$	$-\dfrac{2am_3\sqrt{y}}{m_2}$	$\begin{aligned}&4a\sqrt{y}-\lambda\\&-\dfrac{1-k}{4}\left(12a\sqrt{y}\right.\\&\left.+\dfrac{(m_2)^2}{a\sqrt{y}}\right)\end{aligned}$

where

$$m_1 = n_1^2 + 4n_1 n_2 y + 3n_2^2 y^2, \quad m_2 = n_1 - n_2 y, \quad m_3 = n_1 + n_2 y.$$

It is convenient to calculate the determinant of this matrix by first making a few simplifications by means of the following elementary operations (each subsequent operation is applied to the result of the preceding one):

1) multiply row 2 by $(-1/2)$ and add to row 4,
2) multiply column 4 by $1/2$ and add to column 2,
3) multiply row 1 by $(-a/\sqrt{y})$ and add to row 4,
4) multiply column 4 by $(a/\sqrt{y})$ and add to column 1.

As a result of these operations (not changing the determinant) the initial matrix becomes

	y	v	x	u
y	$-\lambda$	y	0	0
v	$-\dfrac{2}{y}m_1$	$2a\sqrt{y}-\lambda$	$-\dfrac{4am_3\sqrt{y}}{m_2}$	0
x	$4n_2 m_2 + \dfrac{(m_2)^2}{2a\sqrt{y}}$	$\dfrac{(m_2)^2}{2a\sqrt{y}}$	$2a\sqrt{y}-\lambda$	$\dfrac{(m_2)^2}{a\sqrt{y}}$
u	0	0	0	$\begin{aligned}&4a\sqrt{y}-\lambda-\dfrac{1-k}{4}\\&\times\left(12a\sqrt{y}+\dfrac{(m_2)^2}{a\sqrt{y}}\right)\end{aligned}$

The characteristic roots (eigenvalues) λ_k of the matrix so obtained $(\det(A_i^j - \lambda_k) = 0)$ are the eigenvalues of the critical points L_y:

$$\lambda_1 = 0, \quad \lambda_{2,3} = 2a\sqrt{y} \pm 2i(n_1 + n_2 y),$$

$$\lambda_4 = 4a\sqrt{y} - \frac{1-k}{4}\left(12a\sqrt{y} + \frac{(n_1 - n_2 y)^2}{a\sqrt{y}}\right). \tag{9.8}$$

The zero eigenvalue $\lambda_1 = 0$ corresponds to the fact that the dimension of the set of critical points L_y is one. The eigenvalues $\lambda_{2,3}$ correspond to solutions in vacuum $(H_3 = 0)$. Since the real parts of $\lambda_{2,3}$ are positive, the critical points of L_y (9.6) and the corresponding solutions (9.7) are stable (for all $y > 0$) in vacuum when time is directed towards expansion.

The type of stability of the critical points (9.6) and the corresponding solutions (9.7) in the completed space is determined by the sign of λ_4. For a sufficiently small y, it is obvious that $\lambda_4 < 0$ for all considered models, i.e. the critical points (9.6) and solutions (9.7) are unstable. The boundary of the region of stability is given by the equation $\lambda_4 = 0$ or

$$\left(\frac{n_1}{\sqrt{y_0}} - n_2\sqrt{y_0}\right)^2 = \frac{4a^2(1 + 3k)}{(1 - k)}. \tag{9.9}$$

For the model of type VII $(n_1 = 1, n_2 = 1)$ and all values of the parameters a, k the equation (9.9) has two roots: $y_0 < 1$ and $1/y_0$. For the model of type IV $(n_1 = 1, n_2 = 0)$ this equation has one root: $y_1 = (1 - k)/4a^2(1 + 3k)$. For the models of types III and VI $(n_1 = 1, n_2 = -1)$ the equation (9.9) has no real roots if

$$\frac{a^2(1 + 3k)}{1 - k} < 1; \tag{9.10}$$

If the converse holds, then (9.9) has two roots: $y_0 < 1$ and $1/y_0$.

The solutions (9.7) (critical points of L_y (9.6)) which are stable towards expansion are determined by $\lambda_4 > 0$. Specifically, for the model of type VII they are determined by $y_0 < y < 1/y_0$, for type IV by $y_1 < y$ and for types III and VI by $y_0 < y < 1/y_0$, if $a^2(1 + 3k)/(1 - k) > 1$. If the converse is true, then all solutions (9.7) are unstable.

Note that for $\varkappa \neq 1$ and $t \to \infty$ the exact solutions (9.7) are obviously anisotropic. Therefore the existence of stable (towards expansion) asymptotic forms (9.7) for the homogeneous models of class B (except the model of type V) means that in these models isotropization of the metric is possible only in the intermediate stages of the evolution of the solution. According to (9.7) for $t \to \infty$ the metric becomes anisotropic.

Solutions (9.7) also produce asymptotics towards contraction. According to the formulas for the eigenvalues (9.8) the asymptotics (9.7) for which y belongs to the region of instability $\lambda_4 < 0$ have codimension 2. Asymptotics for which y belongs to the region of stability $\lambda_4 > 0$ (towards contraction) have codimension 3.

II. The Oscillatory Mode in the Long Epoch for the Models of Class B. First let us observe some properties of the dynamical system (8.10) related to the fact that the metric

$$g_{ij} = \begin{pmatrix} g_{11} & g_{12} & 0 \\ g_{21} & g_{22} & 0 \\ 0 & 0 & g_{33} \end{pmatrix} \tag{9.11}$$

for the models of class B is not diagonal. We will show that in the case of a model of type VII almost all trajectories of the system (8.10) never intersect the surface $y_1 = y_2$. Let us write Einstein's equations as a dynamical system in an eight-dimensional phase space $g_{ij}, \dot{g}_{ij}$. A trajectory of the system (8.10) intersecting the surface $y_1 = y_2$ becomes a trajectory intersecting the manifold W_0 of matrices with two equal eigenvalues. For the model of type VII the coordinates q_1, q_2, q_3 are the eigenvalues of the metric g_{ij}, because the metric can be diagonalized by means of an orthogonal rotation through an angle φ (see Sect. 3). The manifold W_0 of matrices with equal eigenvalues $q_1 = q_2$ has dimension two. Therefore in the $g_{ij}, \dot{g}_{ij}$ space it corresponds to a six-dimensional manifold W_1 (the velocities $\dot{g}_{ij}$ are arbitrary). The trajectories of the dynamical system passing through the manifold W_1 form some submanifold of dimension seven of the eight-dimensional phase space $g_{ij}, \dot{g}_{ij}$. This means that almost all trajectories of the dynamical system do not lie in this manifold, i.e. do not intersect the surface $y_1 = y_2$. Consequently in order to study general solutions for the model of type VII it is sufficient to analyze the system (8.10) in the region $y_1 > y_2$.

In the case of homogeneous models of types III, IV and VI the metric g_{ij} can be diagonalized by means of commutator preserving transformations which are not orthogonal (for example for the models of types III and VI the transformation is a hyperbolic rotation). Here the coordinates q_1 and q_2 are not the eigenvalues of the matrix g_{ij} (9.11). It is easily verified that the submanifold of matrices g_{ij}, for which $q_1 = q_2$, has dimension three, so an entire region in the space of solutions intersects the manifold $q_1 = q_2$. This explains the significant difference between the dynamics of the specified models and the model of type VII. Here the amplitude of oscillations of the value of $y = q_2/q_1$ can be unbounded, whereas for the model of type VII we have $y < 1$.

In the homogeneous models of class B that we have considered, just like in the models of types VI_0 and VII_0 of class A, there is an intermediate mode of behavior of the metric before (for the contraction of space) the stable Kasner asymptotic form (9.3) is assumed. This is the oscillatory mode in the long epoch (q_1, q_2, q_3 oscillate) and it is determined by the drift of the trajectories of the system (8.10) along the separatrices of unstable critical points contained in the circles $(\psi, 1)$ and $(\psi, 2)$. However in the class B, because of the reasons outlined above, the oscillatory mode of change of the quantities q_1, q_2, q_3 principally depends on the type of the model in question.

Separatrices of the unstable critical points of (ψ, i) belong to the components of the boundary Y_i and $\Gamma_w (w = 0)$. On the components of the boundary Y_i (except

Y_2 for the model of type IV) the system (8.10) is identical to the system for the models of class A, which was analyzed in Sect. 6. On the component of the boundary Γ_w the system (8.10) has the following form:

$$\dot{s}_1 = Z_2(-1 + s_1(s_1 - s_2)) + (1 - k)H_2(1 - s_1(s_1 + s_2 + s_3)),$$
$$\dot{s}_2 = Z_2(1 + s_2(s_1 - s_2)) + (1 - k)H_2(1 - s_2(s_1 + s_2 + s_3)),$$
$$\dot{s}_3 = Z_2(s_3(s_1 - s_2)) + (1 - k)H_2(1 - s_3(s_1 + s_2 + s_3)),$$
$$\dot{y}_1 = 8y_1(y_1^2 s_1 + y_2^2 s_2 - s_1), \quad \dot{y}_2 = 8y_2(y_1^2 s_1 + y_2^2 s_2 - s_2).$$

$$(9.12)$$

where

$$H_2 = 2\sum_{i<j}^{3} s_i s_j - \sum_{i=1}^{3} s_i^2 - \frac{4a^2(2s_3 - s_1 - s_2)^2 y_1 y_2}{(n_1 y_1 - n_2 y_2)^2},$$

$$Z_2 = \frac{8a^2(2s_3 - s_1 - s_2)^2(n_1 y_1 + n_2 y_2)y_1 y_2}{(n_1 y_1 - n_2 y_2)^3}.$$

$$(9.13)$$

To be specific consider the separatrices of a critical point of the circle $(\psi, 1)$. For the models of types III, VI and VII the system (8.10) does not change if the indices 1 and 2 are transposed and for the model of type IV there is only one circle $(\psi, 1)$. According to Table 2 of the signs of eigenvalues, from each point (s_1, s_2, s_3) of the arc $s_2 < s_1$ $((\beta_{21}, 1) + (\beta_{23}, 1) + (\beta_{32}, 1))$ of the circle $(\psi, 1)$ leaves a one-dimensional separatrix going along the corner of the boundary $\Gamma_w \cap \Gamma_0$: $w = 0$, $H_2 = 0$. By (9.12) along this separatrix we have

$$\left(\frac{s_1 + s_2}{s_3}\right)^{\cdot} = 0, \quad \dot{y} = 8y(s_1 - s_2), \quad y = \frac{y_2}{y_1}.$$

$$(9.14)$$

Consequently in the coordinates s_1, s_2, s_3 the separatrix moves on the sphere $s_1^2 + s_2^2 + s_3^2 = 1$ along the arc γ_c of the great circle cut out by the plane $s_1 + s_2 = cs_3$ (Fig. 14). The direction of this movement is determined by the sign of $\dot{s}_1 = Z_2(-1 + s_1(s_1 - s_2))$. Since $H_2 > 0$ we have $s_i < 0$, so on the manifold S we have $-1 + s_1(s_1 - s_2) = -s_2^2 - s_3^2 - s_1 s_2 < 0$. Therefore the direction of movement is determined by the sign of Z_2 or $(n_1 y_1 + n_2 y_2)/(n_1 y_1 - n_2 y_2)^3$. According to (9.14) the value of $y = y_2/y_1$ increases up to the intersection of the arc γ_c with the arc $\varkappa$ $(s_1 = s_2)$ and decreases afterwards.

For a model of type VII $(n_1 = n_2 = 1)$ for $c \neq 2$ on all separatrices $s_1 + s_2 = cs_3$ we have $y < 1$. Otherwise for $y = 1$, the function H_2 would be negative $(H_2 = -\infty)$. Therefore on these separatrices we have $Z_2 > 0$, the motion along the arc γ_c occurs only in one direction $(\dot{s}_1 < 0)$ and the value of y goes through the maximum $y_m < 1$. The final point of this separatrix is the image of the initial point (s_1^0, s_2^0, s_3^0) under the map θ in the plane $s_1 = s_2$. We get (s_2^0, s_1^0, s_3^0). If at the initial point we have $2s_3^0 - s_1^0 - s_2^0 = 0$, then along the separatrix leaving it we have $s_i \equiv s_i^0$ and the separatrix approaches the same point (s_1^0, s_2^0, s_3^0) on the circle $(\psi, 2)$.

For a model of type IV $(n_1 = 1, n_2 = 0)$ we have $Z_2 > 0$ (except $2s_3^0 - s_1^0 -$

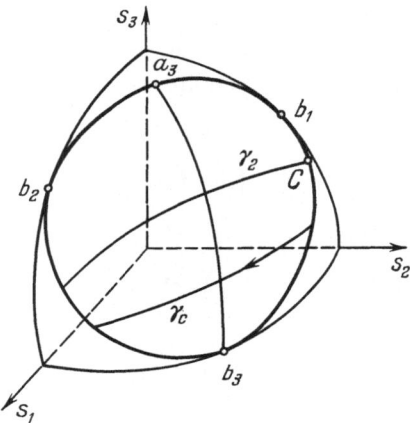

Fig. 14. The movement of separatrices of critical points of the circle $(\psi, 1)$ in the coordinates s_1, s_2, s_3

$s_2^0 = 0$). Therefore the behavior of the separatrices is the same as in the model of type VII. However the value y_m can be arbitrary.

For models of types III and VI $(n_1 = 1, n_2 = -1)$ there are three kinds of separatrices. The first kind are the ones along which $y < 1$. These separatrices are like those of model VII. The second kind are the ones where the value of y goes through 1 while the value of Z_2 changes sign and the motion along the arc γ_c reverses direction, y increases to infinity and the separatrix approaches the point (s_1^0, s_2^0, s_3^0) on the circle $(\psi, 2)$. The third kind are the two exceptional separatrices approaching the critical points j_1 and j_2, where $y = 1$, $w = 0$.

Let us find the maximum value y_m which is attained on the separatrices returning to a point on the initial circle $(\psi, 1)$. The maximum y_m is achieved at the point (s, s, s_3') of intersection of the arc γ_c with the arc $\varkappa$, where

$$\frac{s_1^0 + s_2^0}{s_3^0} = \frac{2s}{s_3^1}. \tag{9.15}$$

Along the separatrix we have $H_2 = 0$, so by substituting (9.15) into (9.13) we get

$$\frac{y}{(n_1 - n_2 y_m)^2} = \frac{T(s_i^1)}{4a^2 (2s_3^1 - s_1^1 - s_2^1)^2} = \frac{s_3^0}{4a^2} \frac{(2(s_1^0 + s_2^0) - s_3^0)}{(s_1^0 + s_2^0 + 2s_3^0)^2}. \tag{9.16}$$

This formula implies that as the initial point of the separatrix (s_1^0, s_2^0, s_3^0) on the circle $(\psi, 1)$ tends to the points

$$a_3\left(\frac{-1}{3\sqrt{2}}, \frac{-1}{3\sqrt{2}}, \frac{-4}{3\sqrt{2}}\right) \quad \text{or} \quad b_3\left(\frac{-1}{\sqrt{2}}, \frac{-1}{\sqrt{2}}, 0\right) \quad \text{(see Fig. 14)}$$

we have $y_m \to 0$. For a model of type VII the function $y_m/(1 - y_m)^2$ is monotone so as the initial point (s_1^0, s_2^0, s_3^0) tends to the point C on the arc γ_2 $(s_1^0 + s_2^0 = 2s_3^0)$ the value y_m monotonically increases to 1. For a model of type IV under the same conditions $y_m \to \infty$.

For models of types III and VI $y_m/(1 + y_m)^2 \leqslant 1/4$ for all y_m. Therefore if the initial point of the separatrix belongs to the arc I_1 $(s_2 < s_1)$, i.e.

$$\frac{s_3^0(2(s_1^0 + s_2^0 - s_3^0))}{4a^2(s_1^0 + s_2^0 - 2s_3^0)^2} > \frac{1}{4}, \qquad (9.17)$$

then along such a separatrix the function $y = y_2/y_1$ has no maximum and monotonically increases to infinity. It approaches the critical point (s_1^0, s_2^0, s_3^0) on the circle $(\psi, 2)$. Therefore it is the same separatrix step as in the models of types VI_0 and VII_0 of class A. If the initial point (s_1^0, s_2^0, s_3^0) does not belong to the arc I_1, then the separatrix leaving it is the same as in the model of type VII ($y_m < 1$). If the point (s_1^0, s_2^0, s_3^0) is one of the endpoints of the arc I_1, then the separatrix leaving it approaches the critical point j_1 or j_2 ($y = 1, w = 0$). Note that the arc I_1 contains the critical point C $(s_1^0 + s_2^0 = 2s_3^0)$ (see Fig. 14). For $a \to 0$ the arc I_1 is stretched to the entire arc $s_2 < s_1$.

Let us describe the properties of the oscillatory mode in the long epoch for the homogeneous models of class B based on the analysis of separatrices of the critical circles $(\psi, 1)$ and $(\psi, 2)$. The corresponding trajectories of the system (8.10) are approximated by the sequence of critical points and their separatrices along which the trajectories move:

$$\cdots \to (\psi_l, i_l) \to (\psi_{l+1}, i_{l+1}) \to (\psi_{l+2}, i_{l+2}) \to \cdots \qquad (9.18)$$

For the models of class B all separatrix steps between the critical circles are single-valued. Each separatrix in the component of the boundary Y_1 (or Y_2) is followed by a separatrix in the component of the boundary Γ_w ($w = 0$) and vice versa.

The sequence of critical points (ψ_i, i_1) in (9.8) is determined by the following mappings (in parentheses we have the components of the boundary where the appropriate separatrices belong; for models IV and VII $i_{l+1} = i_l = i$, so assume that $i = 1$):

$$K(\psi_l, 1) = (\psi_{l+1}, 1) = \begin{cases} (T_1\psi_l, 1), \ \psi \text{ on arc } (\alpha_1, 1), \ (Y_1), \\ (\theta\psi_l, 1), \ \psi \text{ on arc } s_2 < s_1, \ (\Gamma_w), \\ (\psi_l, 1), \ \psi \text{ on arc } (\beta_{31}, 1). \end{cases} \qquad (9.19)$$

where T_1 is the projection (6.16) of the arc α_1 from the first vertex of the triangle $\varDelta$ and θ is the reflection $\theta(s_1, s_2, s_3) = (s_2, s_1, s_3)$. When the point $(\psi_s, 1)$ ends up in the arc $(\beta_{31}, 1)$, which consists of attracting critical points, the approximation (9.18) is terminated and the stable Kasner asymptotic form (9.3) is assumed.

For models of types III and VI we have

$$K(\psi_l, 1) = (\psi_{l+1}, i_{l+1}) = \begin{cases} (\psi_l, 2), \ \psi \text{ on arc } I_1, \ (\Gamma_w), \\ (T_1\psi_l, 1), \ \psi \text{ on arc } (\alpha_1, 1), \ (Y_1), \\ (\theta\psi_l, 1), \ \psi \text{ on arc } s_2 < s_1 \text{ outside } I_1, \ (\Gamma_w), \\ (\psi_l, 1), \ \psi \text{ on arc } (\beta_{31}, 1). \end{cases} \qquad (9.20)$$

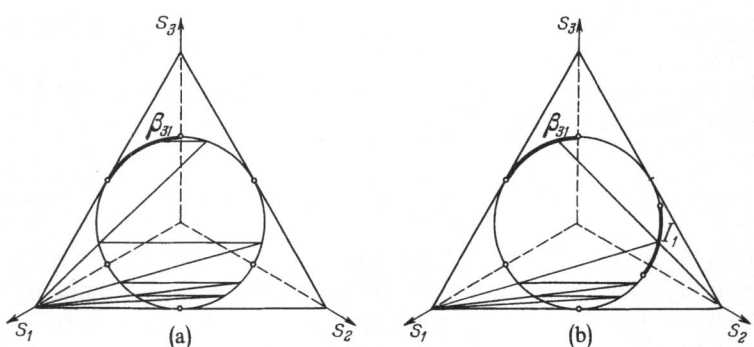

Fig. 15. The sequential transformation of a critical point on the circle S^1 for the homogeneous cosmological models of types IV and VII (a) and types III and VI (b)

The mapping $K(\psi_1, 2)$ is obtained from (9.20) by transposing the indices $1 \to 2$, $2 \to 1$. The exceptional separatrices leaving the boundary arcs I_1 and I_2 approach the unstable critical points j_1 and j_2 (see Sect. 10) and do not affect the oscillatory mode.

The action of the mappings (9.19) and (9.20) on a point (s_1, s_2, s_3) of the circle S^1 $\left(s_1 + s_2 + s_3 = -\sqrt{2}, s_1^2 + s_2^2 + s_3^2 = 1\right)$ is shown in Fig. 15 (parts (a) and (b) respectively). These pictures show that in the long epoch the oscillations terminate after a finite number of steps when the point ψ_s ends up in the arc $(\beta_{31}, 1)$ (or $(\beta_{32}, 2)$). It is also obvious that for models of types III and IV the oscillations inside the segment I_1 (see Fig. 15(b)) are the same as for models of types VI_0 and VII_0 of class A.

When the trajectories (9.18) are in the neighborhood of a critical point (s_1, s_2, s_3) on the circle (ψ_l, i_l) the metric is approximated by the Kasner solution (9.3). Here, as opposed to the models of class A, the Kasner exponents p_i are approximately constant only in the neighborhood of the circles (ψ_l, i_l) and vary along both types of separatrices (in Y_i and Γ_w). The Kasner exponents are parametrized by a parameter u:

$$p_1(u) = \frac{-u}{1 + u + u^2}, \quad p_2(u) = \frac{1 + u}{1 + u + u^2}, \quad p_3(u) = \frac{u(1 + u)}{1 + u + u^2} \quad (9.21)$$

For models of types IV and VII in one full oscillation of the value of $y = q_2/q_1$ the parameter u decreases to 1, whereas in the models of class A it decreases to 2. Indeed one oscillation in the models of class B corresponds to the transformations θT_1 of a point on the circle S^1, while in the models of class A it corresponds to the transformations $T_2 T_1$. Under the projections T_1 and T_2 the parameter u decreases to 1 (see part V of Sect. 6), whereas under the reflection θ it does not change.

The amplitude of oscillations of the value of $y = q_2/q_1$ in the long epoch can be calculated from (9.6). After a substitution of expressions (9.21) and $p_i = 1 +$

$\sqrt{2}s_i$ it can be expressed in terms of the parameter u of the Kasner exponents as follows:

$$\frac{y_m}{(n_1 - n_2 y_m)^2} = \frac{1}{4a^2}\left(\frac{1 + 2u}{1 - 2u - 2u^2}\right)^2.$$ (9.22)

where the value of the parameter u is taken at the beginning of a given oscillation and is distributed outside the segment I_1 (9.17) (for models III and IV). For $u \to (\sqrt{3} - 1)/2$ we have $y_m \to \infty$ for the model of type IV and $y_m \to 1$ for the model of type VII.

According to the system (8.10), the variation of the angle φ through one full oscillation (at this time the trajectory goes along one separatrix (9.18) in Y_1 and one separatrix in Γ_w) has the following form:

$$\Delta\varphi = \oint \dot\varphi \, d\tau_1 = -8a \int \frac{(2s_3 - s_1 - s_2)y}{(n_1 - n_2 y)^2} d\tau_1.$$ (9.23)

The step along a separatrix in the component of the boundary Y_1 ($y = 0$) does not contribute to this integral. For a model of type IV the integral (9.23) along a separatrix in the component of the boundary Γ_w can be calculated explicitly. Indeed according to the system (9.12) for $H_2 = 0$ (on the separatrix) we have:

$$\Delta\varphi = -8a \int (2s_3 - s_1 - s_2)y \, d\tau_1 = -a \int \frac{2s_3 - s_1 - s_2}{s_1 - s_2} dy$$

$$= \frac{1}{2a}\left(\int_0^{y_m} + \int_{y_m}^0\right)\left(\frac{d}{dy}\right)\left(\frac{s_1 - s_2}{2s_3 - s_1 - s_2}\right)dy$$ (9.24)

$$= \frac{1}{2a}\left(\frac{s_1 - s_2}{2s_3 - s_1 - s_2}\right)\bigg|_0^1 = -\frac{1}{a}\left(\frac{s_1^0 - s_2^0}{2s_3^0 - s_1^0 - s_2^0}\right).$$

This last equality takes into account that the final point of the considered separatrix is obtained from the initial point by the mapping θ: $(s_1^1, s_2^1, s_3^1) = (s_1^0, s_2^0, s_3^0)$. Substituting expressions $p_i = 1 + \sqrt{2}s_i$ and (9.21) into the obtained formula gives an expression of the angle of rotation of the axes q_1, q_2 in terms of the parameter u of the Kasner exponents:

$$\Delta\varphi = -\frac{1}{a}\frac{1 + 2a}{1 - 2u - 2u^2}.$$ (9.25)

For the homogeneous models of types III, VI and VII this formula (9.25) holds to the order of magnitude for large u. From (9.25) it follows that the complete rotation of the axes formally calculated over the entire long epoch

$$\left(\sum_{i=1}^{N} \Delta\varphi(u + i)\right)$$

diverges like $(1/a)\ln N$ for $N \to \infty$.

10. Analysis of Several Special Properties of Homogeneous Cosmological Models of Types V, VII, III, VI and IV

Type V. The homogeneous cosmological model of type V $(a = 1, n_1 = n_2 = n_3 = 0)$ contains the open Friedmann solution $q_1 \equiv q_2 \equiv q_3$ as a special case, so it is used to study the dynamics of perturbations of this solution. Analysis of the homogeneous model of type V is simpler than of the other models of class B, because all metrics of the model of type V without the motion of matter are diagonalizable (see Sect. 3) and therefore $p_\varphi = 0$. In this case the system (8.6) has the following form:

$$\dot{P}_i = 2q_1 q_2 + \frac{1 - k}{2} H_1, \quad \dot{q}_i = 2q_i(P_j + P_k - P_i). \tag{10.1}$$

where

$$H_1 = 2\sum_{i<j}^3 P_i P_j - \sum_{i=1}^3 P_i^2 - 3q_1 q_2.$$

The condition (8.2) $R = 0$ gives $P_3 = (P_1 + P_2)/2$.

In the scale-invariant coordinates

$$r = \frac{(q_1 q_2)^{1/2}}{P_1 + P_2}, \quad v = \frac{2(P_1 - P_2)}{P_1 + P_2}$$

and new time

$$\tau_1: \; d\tau_1/d\tau_0 = -(P_1 + P_2) = -2P_3 > 0$$

the system (10.1) takes the following form:

$$\dot{r} = r\left(-1 + \left(\frac{1-k}{4}\right) H_2 + 4r^2\right), \quad \dot{v} = v\left(\frac{1-k}{4} H_2 + 4r^2\right),$$

$$\dot{q}_1 = q_1(v - 1), \quad \dot{q}_2 = q_2(-v - 1), \quad \dot{q}_3 = -q_3, \tag{10.2}$$

$$\dot{P}_3 = -P_3\left(\frac{1-k}{4} H_2 + 4r^2\right).$$

where

$$H_2 = 4H_1/(P_1 + P_2)^2$$
$$= 4[-(P_1 - P_2)^2 - P_3(P_3 - 2(P_1 + P_2)) - 3q_1 q_2]/(P_1 + P_2)^2$$
$$= -v^2 + 3 - 12r^2.$$

The first two equations (10.2) form a closed system defined on a half-ellipse:

$$v^2 + 12r^2 \leqslant 3 \; (H_2 \geqslant 0, r \leqslant 0).$$

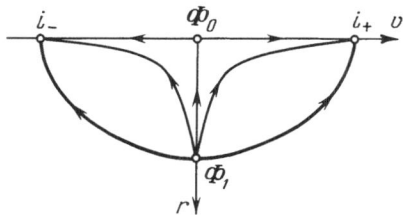

Fig. 16. Phase portrait of the dynamical system (10.2) on the two-dimensional manifold S for the homogeneous cosmological model of type V

The phase portrait of this two-dimensional system is shown in Fig. 16. There are two attracting critical points $i_\pm$ $(r = 0, v = \pm\sqrt{3})$. They correspond to stable (towards contraction) Kasner asymptotic forms

$$q_i \approx C_i t^{2p_i}, \quad p_1^\pm = \frac{1 \mp \sqrt{3}}{2}, \quad p_2^\pm = \frac{1 \pm \sqrt{3}}{2}, \quad p_3^\pm = \frac{1}{3}. \tag{10.3}$$

The unstable critical point Φ_0 $(r = v = 0)$ corresponds to a quasi-Friedmann asymptotic form

$$q_i \approx C_i t^{4/3(1+k)}.$$

The repelling critical point Φ_1 $(r = -1/2, v = 0)$ corresponds to an exact solution in vacuum:

$$q_i = C_i t^2. \tag{10.4}$$

The open Friedmann solution in the model of type V corresponds to the separatrix $v = 0$ going from the critical point Φ_1 to the critical point Φ_0. Since the coordinate v varies monotonically, the following well-known ([61]) assertion holds: when time is directed towards expansion all metrics for the model of type V isotropize and for $t \to \infty$ have the asymptotic form of the open Friedmann solution (10.4). All trajectories of the system (10.2) approach the attracting (for this direction of time) critical point Φ_1. For the direction of time towards contraction all metrics have for $t \to 0$ either the stable Kasner asymptotic (10.3) or the unstable quasi-Friedmann asymptotic.

Type VII. Just like the model of type V the model of type VII $(n_1 = n_2 = 1, n_3 = 0, a > 0)$ contains the open Friedmann solution as a special case. It corresponds to the exceptional trajectory of the system (8.5)–(8.10):

$$q_1 = q_2 = q_3, \quad p_\varphi^2/(q_1 - q_2)^3 \equiv 0, \quad s_1 = s_2 = s_3.$$

As shown in Sect. 9, for the model of type VII for all a there exists a segment of attracting (for the direction of time towards expansion) critical points L_y (9.6) on the manifold S. These critical points correspond to the stable (towards expansion) exact solutions (9.7). A special case of these solutions for $y = 1$ is the exact solution in vacuum:

$$q_1 = q_2 = C_1 t^2, \quad q_3 = a^2 t^2, \quad \varphi = \text{const.} \tag{10.5}$$

For $C_1 = a^2$ this solution (10.5) is isotropic and for $t \to \infty$ has the asymptotic form of the open Friedmann solution. All other stable solutions (9.7) for $y \neq 1$ become anisotropic for $t \to \infty$. Therefore the isotropization of solutions for the model of type VII is possible only in the intermediate stage of the evolution.

Let us consider the last remaining unanalyzed critical points of the dynamical system (8.10)–(8.14) on the manifold S. At these critical points we have $y_1 = y_2$. On the invariant manifold $y_1 = y_2 = 1/\sqrt{2}$ the system (8.14) has the following form:

$$\begin{aligned}
\dot{s}_1 &= 4\sqrt{2}a^2 z^2 \left(-1 + s_1(s_1 - s_2)\right), \\
\dot{s}_2 &= 4\sqrt{2}a^2 z^2 \left(1 + s_2(s_1 - s_2)\right), \\
\dot{z} &= 4\sqrt{2}a^2 z^3 (s_1 - s_2) + 4\sqrt{2}z(s_1 - s_2), \\
\dot{w} &= 8\sqrt{2}wa^2 z^2 (s_1 - s_2).
\end{aligned} \tag{10.6}$$

This system (10.6) is considered on the level of the relation

$$s_1^2 + s_2^2 + s_3^2 = s_1^2 + s_2^2 + \tfrac{1}{4}(s_1 + s_2)^2 = 1$$

in the region

$$H_2 \geq 0, \quad H_2 = -(s_1 - s_2)^2 + \tfrac{3}{4}(s_1 + s_2)^2 - 2a^2 z^2 - \tfrac{3}{2}a^2 w, \quad w \geq 0.$$

The critical points of the system (10.6) form a two-dimensional disc W: $y_1 = y_2$, $z = 0$. The eigenvalues of the system (8.14) at these critical points are

$$\lambda_w = 0, \quad \lambda_{s_i} = 0, \quad \lambda_z = 4\sqrt{2}(s_1 - s_2), \quad \lambda_{y_i} = -4\sqrt{2}(s_1 - s_2). \tag{10.7}$$

Thus the critical points in W with $s_1 \neq s_2$ are non-degenerate and unstable. The segment $s_1 = s_2$, $0 < w < 2/3a^2$ in W consists of degenerate critical points and joins the critical points Φ ($y_1 = y_2$) and L_y ($y = 1$).

The trajectories of the system (8.14) can intersect the invariant manifold $y_1 = y_2$ only by entering some critical point in it, i.e. a critical point in W. The set of trajectories (separatrices) approaching the saddle points of W is a three-dimensional subset of the four-dimensional manifold S. It is not difficult to verify that as these separatrices approach the critical points, the expressions q_i tend to non-zero constants, i.e. these are the exceptional separatrices which have initial data in the manifold W_0 of codimension two (see part II of Sect. 9). All other trajectories of the system (8.14) never intersect the manifold $y_1 = y_2$, so for almost all trajectories we either have $y_1 > y_2$ or $y_1 < y_2$.

Here is a list of critical points of the system (8.10)–(8.14) on the compact manifold S (Fig. 17):

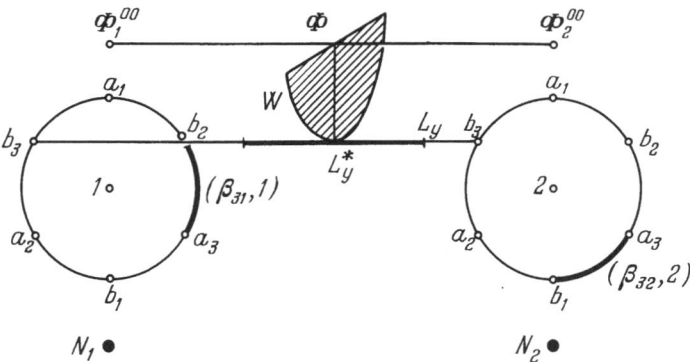

Fig. 17. The general distribution of critical points of the dynamical system (8.10)–(8.14) on the manifold S for the homogeneous cosmological model of type VII

1) On the components of the boundary Y_1 $(y_1 = 1, y_2 = 0)$ and Y_2 $(y_1 = 0,$ $y_2 = 1)$ the critical points are the same as in the models of class A. There are two critical circles $(\psi, 1)$ and $(\psi, 2)$, two isolated critical points N_1 and N_2 and the critical points Φ_1^{00} and Φ_2^{00}. The arcs $(\beta_{31}, 1)$ and $(\beta_{32}, 2)$ in the circles (ψ, i) are attracting for the contraction of space.

2) For $w = 0$, $y_1 \neq y_2 \neq 0$ we have a line of critical points Φ.

3) In the region $w \neq 0$, $y_1 \neq y_2 \neq 0$ we have a line of critical points L_y (9.6). The line L_y contains a segment L_y^* of attracting (towards expansion) critical points.

4) For $y_1 = y_2$ we have the disc W of unstable critical points.

The separatrix steps between the critical points are systematized in the separatrix diagram (Table 6). Here because of invariance of the region $y_1 > y_2$ noted above we consider only a half of the manifold $S(y_1 \geqslant y_2)$. The set of critical points W is divided into two parts:

$$W_+ \ (\lambda_z > 0, \ s_1 > s_2) \text{ and } W_- \ (s_1 < s_2).$$

Types III and VI. Homogeneous models of types II and VI are defined by the conditions $n_1 = 1$, $n_2 = -1$, $n_3 = 0$, $a > 0$. The model of type III is defined by $a = 1$ and in this case the Lie algebra of the corresponding Lie group G has a one-dimensional derived subalgebra (the subalgebra generated by the commutators). For $a \neq 1$ the derived subalgebra is two-dimensional. The model of type III is considered below as a special case of the model of type VI. It does however have a few differences in the properties of the asymptotics.

All critical points of the dynamical system (8.10) on the corresponding compact manifold S for $y_1 \neq y_2$ are listed in Sect. 9. For $y_1 = y_2$, $w \neq 0$ it is convenient to analyze the critical points in the coordinates (9.4). These critical points must belong to the invariant manifold $y = 1$, $v = 0$. On it the dynamical system (9.5) takes the following form:

Table 6. The separatrix diagram of the dynamical system (8.10)–(8.14) for the homogeneous cosmological model of type VII (for the contraction of space)

	L_y^*	L_y	Φ	Φ_1^{00}	N_1	W_+
L_y	2					
Φ	2					
Φ_1^{00}					1	
N_1	1					
W_+	3					
W_-						3
$(\beta_{13},1)$			3	2		
$(\beta_{12},1)$			3	2		
$(\beta_{21},1)$				2	3	
$(\beta_{31},1)$	4	3		2	3	
$(\beta_{23},1)$				2	3	
$(\beta_{32},1)$				2	3	

$$\dot{x} = xu - 8, \quad \dot{u} = u^2 - 4a^2 - H_3,$$

$$H_3 = \frac{1-k}{4}\left[3u^2 + ux - \frac{a^2 x^2}{4} - 12a^2 - 4\right]. \tag{10.8}$$

In the region $H_3 \geqslant 0$ the system (10.8) has two critical points. The first one is L_1: $u = 2a$, $x = 4/a$. It lies on the line of critical points L_y (9.6) for $y = 1$, $H_3(L_1) = 0$. The second one is M:

$$M: u = 2\sqrt{\frac{1-k}{1+3k}}, \quad x = 4\sqrt{\frac{1+3k}{1-k}},$$

$$H_2(M) = 4\left(\frac{1-k}{1+3k} - a^2\right).$$

Note that the coordinates of this point do not depend on a. The characteristic matrix of the system (9.5) at the critical point M breaks down into two blocs, so its eigenvalues are easily calculated:

W_-	$(\beta_{13},1)$	$(\beta_{12},1)$	$(\beta_{21},1)$	$(\beta_{31},1)$	$(\beta_{23},1)$	$(\beta_{32},1)$
					2^θ	
			2^θ			
		2^T				
3	2^T					2^θ
		2^T				
	2^T					

$$\lambda_{y,v} = \sqrt{\frac{1-k}{1+3k}} \pm \sqrt{\frac{1-k}{1+3k} - 4\left(1 - a^2\frac{1+3k}{1-k}\right)},$$

$$\lambda_{x,u} = \sqrt{\frac{1-k}{1+3k}} \pm \sqrt{\frac{1-k}{1+3k} - 2(1 - k - a^2(1+3k))}. \tag{10.9}$$

An analogous critical point already appeared earlier in models of types VIII and VI$_0$. The critical point M belongs to the physical region $H_3 \geqslant 0$ only when

$$a \leqslant \sqrt{\frac{1-k}{1+3k}}. \tag{10.10}$$

In particular for $a > 1$ the point M does not belong to the physical region for any value k. For

$$a = \sqrt{\frac{1-k}{1+3k}}$$

the point M coincides with the point L_1 on the line (9.6) of critical points which were already analyzed. Thus the value $a = 1$ (model of type III) is characterized by the fact that for $a \leqslant 1$ and some values of k (see (10.10)) the model of type VI has an asymptotic form corresponding to M, whereas for $a > 1$ such asymptotic forms do not exist.

When the point M belongs to the physical region $H_3 \geqslant 0$, all four eigenvalues (10.9) have positive real parts (because (10.10) holds), i.e. the critical point M is attracting towards expansion. This critical point corresponds to an exact solution, which is stable towards the expansion of space:

$$q_1 = q_2 = q_1^0 t^{(1-k)(1+k)}, \quad q_3 = q_3^0 t^2,$$

$$\varphi = -a \sqrt{\frac{1 + 3k}{1 - k}} (q_3^0)^{-1/2} \ln t. \tag{10.11}$$

Note that the condition (10.10) for the existence of an attracting critical point M towards expansion in the physical region coincides with the condition (9.10) of absence of a segment of attracting critical points on the line L_y (9.6). Therefore if $a < 1$, then as the parameter k varies from 0 to 1 the solutions in the completed space at first for $t \to \infty$ have the asymptotic form (10.11), whereas in vacuum the solutions have the asymptotic form (9.7). Later on for $a > \sqrt{(1 - k)/(1 + 3k)}$ all solutions have the asymptotic form (9.7). Thus a change of the parameter k of the equation of state of matter corresponds to a qualitative change of asymptotic properties of the solutions for the homogeneous cosmological model of type VI. However in any of the stable asymptotic forms of the metric (9.7) or (10.11) for the models of types III and VI there is no isotropization for $t \to \infty$.

Let us analyze the critical points of the dynamical system (8.10) $(n_1 = 1, n_2 = -1)$ for $y_1 = y_2$, $w = 0$. At these critical points we must have $s_1 = s_2$ and either $H_2 = 0$ or $s_i = -1/\sqrt{3}$. Therefore for $y_1 = y_2$, $w = 0$ the system (8.10) has three critical points. The first critical point Φ_1 $(s_i = -1/\sqrt{3})$ belongs to the segment Φ. The other two critical points are j_+ and j_-, where $s_1 = s_2$, $H_2(j_\pm) = 0$. The coordinates $s_1 = s_2 = s$ and s_3 are determined by the conditions

$$2s^2 + s_3^2 = 1, \quad \frac{s_3}{s} = 2\frac{1 + 2a^2 \pm \sqrt{1 + 3a^2}}{1 + 4a^2}.$$

The eigenvalues of the system (8.10) at these critical points have the following form:

$$\lambda_w = 8s_3 < 0, \quad \lambda_{1,2} = \pm 2\sqrt{2}a(2s_3 - s_1 - s_2),$$

$$\lambda_3 = -\frac{1 - k}{2a} \sqrt{1 + 3a^2} |\lambda_1|.$$

Thus the critical points j_+ and j_- are non-degenerate and unstable. From the critical points $j_\pm$ leave one-dimensional separatrices which belong to the inva-

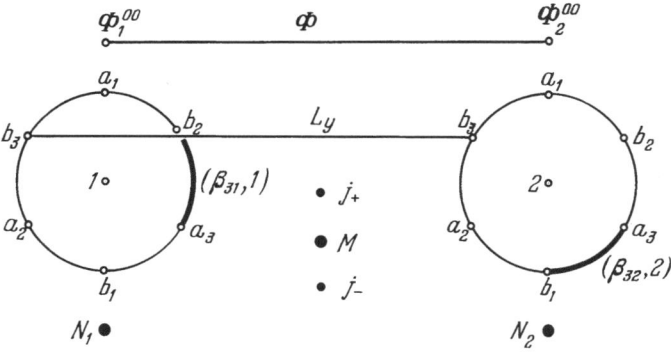

Fig. 18. The critical points of the dynamical system (8.10) on the manifold S for the homogeneous cosmological models of types III and VI

riant manifold Γ_w $(w = 0)$ and therefore do not correspond to any physical solutions. The three-dimensional separatrices approaching the critical points $j_\pm$ for $\tau_1 \to \infty$ from the physical region of the manifold S determine the following (unstable) asymptotic forms of the metric for the contraction of space $(t \to 0)$:

$$q_1 \approx q_1^0 t^{2(4+\bar{x})/(12-\bar{x})}, \quad q_2 \approx q_1^0 t^{2(4+\bar{x})/(12-\bar{x})},$$

$$q_3 \approx q_3^0 t^{(4-3x)/(12-x)}, \tag{10.12}$$

$$\varphi \cong -\frac{2a\bar{x}}{12-\bar{x}} \ln t + \varphi_0, \quad \bar{x} = 4\frac{-1 \pm 4\sqrt{1+3a^2}}{16a^2+5}.$$

Note that for the critical point j_- all powers of t in (10.12) are positive for all a, i.e. $q_i \to 0$. For j_+ we get $(4+\bar{x})/(12-\bar{x}) > 0$ for all a, whereas $(4-3\bar{x})/(12-\bar{x}) \leqslant 0$ for $a \geqslant 1$, i.e. in the asymptotic form (10.12) for $a > 1$ we have $q_1 \to 0, q_2 \to 0$ and $q_3 \to \infty \, (t \to 0)$. For $a = 1$ the asymptotic form (10.12) at the critical point j_+ takes the following form:

$$q_1 \cong q_2 \cong q_1^0 t, \quad q_3 \cong \text{const.}$$

The critical points j_+ and j_- belong to the corner $\Gamma_w \cap \Gamma_0 \, (w = 0, H_2 = 0)$ of the boundary Γ. This corner is a two-dimensional cylinder $S^1 \times I$. The entire cylinder is filled by the separatrices of the circles $(\psi, 1)$ and $(\psi, 2)$. The critical points j_+ and j_- are unstable in the cylinder and their separatrices approach the boundaries of the segments I_1 and I_2 in the circles $(\psi, 1)$ and $(\psi, 2)$ and the images of these boundaries under the reflection θ. These separatrices divide the cylinder into six parts. In four of these the trajectories have their beginning and end on one circle and in the other two the beginning and end of each trajectory belong to

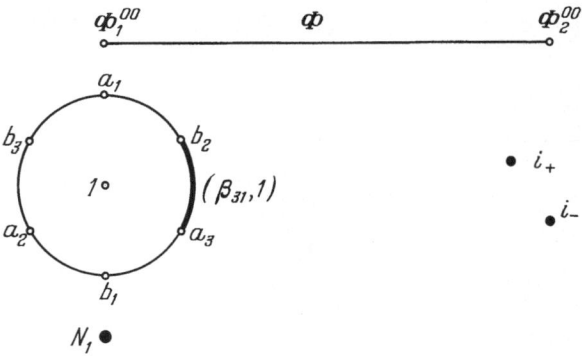

Fig. 19. The critical points of the dynamical system (8.10)–(8.18) on the manifold S for the homogeneous cosmological model of type IV

different circles. The critical points of the dynamical system (8.10) on the compact manifold S for the models of types III and VI are shown in Fig. 18.

Type IV. For $y_2 \neq 1$ the critical points of the dynamical system (8.10)–(8.18) on the compact manifold S for the model of type IV $\left(n_1 = 1, n_2 = n_3 = 0, a > 0\right)$ and the asymptotic properties of the corresponding solutions were analyzed in Sect. 9. On the invariant manifold $y_2 = 1$, $H_2 \geqslant 0$ the system (8.18) has four critical points. The first critical point

$$\Phi_2^{00}\left(s_1 = s_2 = -\frac{1}{\sqrt{3}}, z_1 = 0, w_1 = 0\right)$$

is the endpoint of the segment Φ. The second critical point

$$L_0\left(s_1 = s_2 = -\frac{1}{\sqrt{3}}, z_1 = 0, w_1 = \frac{1}{3a^2}\right)$$

is the endpoint of the segment L_y (9.6), $H_2(L_0) = 0$. The last two critical points are

$$i_+\left(s_1 = \frac{-2 + \sqrt{3}}{3\sqrt{2}}, s_2 = \frac{-2 - \sqrt{3}}{3\sqrt{2}}, z_1 = 0, w_1 = 0\right),$$

$$i_-\left(s_1 = \frac{-2 - \sqrt{3}}{3\sqrt{2}}, s_2 = \frac{-2 + \sqrt{3}}{3\sqrt{2}}, z_1 = 0, w_1 = 0\right),$$

$$H_2(i_\pm) = 0.$$

The entire collection of critical points of the dynamical system on the manifold S for the model of type IV is shown in Fig. 19. The eigenvalues of the system (8.18) at the critical points Φ_2^{00}, L_0, $i_\pm$ have the following form:

$$U_2^{00}: \lambda_{z_1} = \sqrt{3}(1 - k), \lambda_{s_i} = \sqrt{3}(1 - k), \lambda_{w_1} = -2\frac{1 + 3k}{\sqrt{3}}, \lambda_{\bar{y}_1} = 0;$$

$$L_0: \lambda_{z_1} = \frac{4}{\sqrt{3}}, \lambda_{s_i} = \frac{4}{\sqrt{3}}, \lambda_{w_1} = 2\frac{1 + 3k}{\sqrt{3}}, \lambda_{\bar{y}_1} = 0, \tag{10.13}$$

$$i_{\pm}: \lambda_{z_1} = 4(s_1 - s_2), \lambda_{s_i} = 3(1 - k)(s_1 + s_2) < 0,$$

$$\lambda_{w_1} = 4(s_1 + s_2) < 0, \lambda_{\bar{y}_1} = 4(s_2 - s_1).$$

According to (10.13) the critical points Φ_2^{00}, i_+ and i_- are unstable. The critical point L_0 belongs to the segment of repelling critical points L_y (9.6). The separatrices of the critical points Φ_2^{00}, i_-, L_0 lie in the components of the boundary Γ_w $(w = 0)$ and Y_2 $(\bar{y}_1 = 0, y_2 = 1)$ and therefore do not correspond to any physical solutions. The unstable critical point i_+ is approached by a three-dimensional separatrix from the physical region of S. It corresponds to a Kasner asymptotic form of the metric for the contraction of space $(t \to 0)$:

$$q_1 \approx q_1^0 t^{2/3(1+\sqrt{3})}, \quad q_2 \cong q_2^0 t^{2/3(1-\sqrt{3})}, \quad q_3 \cong q_3^0 t^{2/3}, \quad \varphi \to \text{const.} \tag{10.14}$$

Thus in the homogeneous model of type IV along with the stable (towards the contraction of space) Kasner asymptotic form of the metric (9.3) (where $q_1, q_2 \to 0$, $q_3 \to \infty$ for $t \to 0$) corresponding to the arc of attracting critical points $(\beta_{31}, 1)$, there is also one exceptional unstable Kasner asymptotic form of the metric (10.14) for which $q_2 \to \infty$ and $q_1, q_3 \to 0$.

Chapter III
Qualitative Theory of Homogeneous Cosmological Models with the Motion of Matter and Electromagnetic Fields

1. Einstein's System of Equations for the Homogeneous Cosmological Model of Type IX with the Motion of Matter

Homogeneous cosmological models without the motion of matter, studied in Chapt. II, are a special case of the solutions of Einstein's equations with a hydrodynamical stress-energy tensor of matter:

$$T_{ij} = (p + \varepsilon)u_i u_j - pg_{ij},$$

where u^i is the 4-velocity of the matter, ε is the energy density and p is the pressure $p = k\varepsilon$, $0 \leqslant k < 1$. More general homogeneous solutions are the ones for which the co-moving system of coordinates is not synchronous or conversely in the synchronous system of coordinates the matter moves (i.e. $u^i \neq 0$ $(i = 1, 2, 3)$).

Suppose that X^0, X^1, X^2, X^3 is a basis of right-invariant vector fields in the space-time manifold $M^4 = \mathbb{R}^1 \times G$ relative to the action of a three-dimensional Lie group G such that the fields X^1, X^2, X^3 are tangent to the group G and the commutation relations have the standard form (see Chapt. II, Table 1):

$$[X^0, X^i] = 0, \quad [X^i, X^j] = C_{ij}^k X^k,$$
$$C_{ij}^k = \varepsilon_{ijk} n^k + \delta_j^k \delta_i^3 a - \delta_i^k \delta_j^3 a, \quad an^3 = 0. \tag{1.1}$$

For homogeneous cosmological models the components of the metric g_{ij} and of the matter velocity u^i (and also the energy density ε and pressure p) in the basis of right-invariant vector fields X^0, X^1, X^2, X^3 depend only on time. From now on we will study only the (most important from the physical point of view) part of the space-time manifold $M^4 = \mathbb{R}^1 \times G$ in which the restriction of the metric g_{ij} to the group G is negative definite. In this case the basis of vector fields $X^0, X^i = \{e_\alpha^i\}$ can be chosen in such a way that the metric ds^2 has the following form:

$$ds^2 = g_{00}(\tau) d\tau^2 - g_{ij}(\tau)e_\alpha^i e_\beta^j dx^\alpha dx^\beta, \tag{1.2}$$

where the matrix $g_{ij}(\tau)$ is positive definite and $g_{00}(\tau)$ will be determined below by a special choice of the time scale.

In the basis X^0, X^1, X^2, X^3, Einstein's equations

$$R_{ij} - \tfrac{1}{2}g_{ij}R = T_{ij} \qquad (1.3)$$

determine a system of differential equations of second order in the matrix $g_{ij}(\tau)$. In Sects. 1–4 below we conduct a detailed qualitative analysis of Einstein's system of equations (1.3) for the homogeneous model of type IX. In Sect. 5 we show how the methods so developed are used to study the rest of the homogeneous cosmological models with the motion of matter.

To construct a dynamical system equivalent to the system (1.3) it is convenient to use the following identity [84]:

$$\int \delta R \sqrt{-g}\, dt\, dG = -\int (R^{ij} - \tfrac{1}{2}g^{ij}R)\sqrt{-g}\delta g_{ij}\, dt\, dG + \int g^{ij}\sqrt{-g}\delta R_{ij}\, dt\, dG.$$
$$(1.4)$$

As shown in [84] the second summand in the right-hand side of (1.4) is a full divergence and therefore becomes zero, if the spatial variations of the metric δg_{ij} are finite. Since the group SO(3) is compact, for the model of type IX the homogeneous variations of the metric are finite. Therefore by (1.4), Einstein's equations (1.3) (except the $(0, i)$ equations) follow from the variational principle

$$\delta \int R\sqrt{-g}\, dt = -\int T^{ij}\delta g_{ij}\sqrt{-g}\, dt, \qquad (1.5)$$

where the metric g_{ij} and its variations are homogeneous (i.e. depend only on t). For all other homogeneous models, the homogeneous variations of the metric are not finite. Nonetheless for all models of class A, the variational principle (1.5) holds, because for the homogeneous variations of the metric for the class A models we have $g^{ij}\delta R_{ij} = 0$ (see formulas (3.5) and (3.28) in Chapt. II). For the homogeneous models of class B the second summand in the right-hand side of (1.4) gives a non-zero contribution to the variational principle (1.5) (for a more detailed account see Sect. 5).

For the metric (1.2) the expression $T^{ij}\delta g_{ij}$ (see (1.5)) has the following form:

$$T^{ij}\delta g_{ij} = \varepsilon\delta\ln\frac{g_{00}}{|g|^k} + (1 + k)\varepsilon(g_{ab}u^a u^b g_{00}\delta g_{00} - u^i u^j \delta g_{ij}). \qquad (1.6)$$

From now on $|g| = \det \| g_{ab} \|$.

To make (1.5) look simpler, choose a time τ in such a way that

$$g_{00}(\tau) = |g|^k. \qquad (1.7)$$

Then the first summand in (1.6) is equal to zero. The time τ is related to the synchronous time t $(g_{00}(t) \equiv 1)$ by $dt = |g|^{k/2}\, d\tau$.

According to (1.6) and the variational principle (1.5), the "tensor components" $(i, j = 1, 2, 3)$ of Einstein's equations (1.3) for the homogeneous model of type IX are linear combinations of the equations

$$\frac{\partial L}{\partial g_{ij}} - \frac{d}{d\tau}\frac{\partial L}{\partial g_{ij}} = \frac{1 + k}{2}\varepsilon(u^i u^j - kg_{ab}u^a u^b g^{ij})|g|^{(1+k)/2} \qquad (1.8)$$

and

$$R_0^0 - \tfrac{1}{2}R = T_0^0$$

(because the variations δg_{00} are related by the condition (1.7)). The function $L(g_{ij}, \dot{g}_{ij})$ is obtained from $\tfrac{1}{2}R\sqrt{-g}$ by dropping the full derivative with respect to time and has the following form:

$$L = \tfrac{1}{8}|g|^{(1-k)/2}\left(\varkappa_\alpha^\alpha \varkappa_\beta^\beta - \varkappa_\alpha^\beta \varkappa_\beta^\alpha\right) - \tfrac{1}{4}|g|^{-(1-k)/2}\left(2|g|g^{\alpha\alpha} - g_{\alpha\beta}g_{\alpha\beta}\right), \qquad (1.9)$$

where $\varkappa_\alpha^\beta = g_{\alpha\gamma}g^{\gamma\beta}$.

Einstein's equations $R_0^0 - \tfrac{1}{2}R = T_0^0$ and $R_{0\alpha} = T_{0\alpha}$ are relations which allow us to express the energy density ε and the velocity of matter u^i in terms of the components of the metric g_{ij} and its first time derivatives $\dot{g}_{ij}$. Indeed Einstein's equation $R_0^0 - \tfrac{1}{2}R = T_0^0$ has the following form:

$$H = \varepsilon\left((1 + k)u_0^2 g^{00} - k)\right)|g|^{(1+k)/2},$$

$$H = g_{ij}\frac{\partial L}{\partial g_{ij}} - L \qquad (1.10)$$

$$= \frac{1}{8}|g|^{(1-k)/2}\left(\varkappa_\alpha^\alpha \varkappa_\beta^\beta - \varkappa_\alpha^\beta \varkappa_\beta^\alpha\right) + \frac{1}{4}|g|^{-(1-k)/2}\left(2|g|g^{\alpha\alpha} - g_{\alpha\beta}g_{\alpha\beta}\right).$$

Einstein's equations $R_{0\alpha} = T_{0\alpha}$ have the following form:

$$-\tfrac{1}{2}\varkappa_\beta^\gamma C_{\alpha\gamma}^\beta = (1 + k)\varepsilon u_0 u_\alpha, \qquad (1.11)$$

where $C_{\alpha\gamma}^\beta$ are the structure constants of the group of type IX $(SO(3))$ in the standard form (1.1): $n^i = 1$, $a = 0$. From the equations (1.10), (1.11) and the condition

$$u_0^2 g^{00} - u_a u_b g^{ab} = 1, \qquad (1.12)$$

the following equations are easily obtained:

$$\frac{\varepsilon u_0^2 |g|^{(1+k)/2}}{g_{00}} = \frac{1}{2}\left[H + \left(H^2 - \frac{16k}{(1+k)^2}X_a X_b g^{ab}|g|^k\right)^{1/2}\right], \qquad (1.13)$$

$$\frac{1+k}{2}\varepsilon u^i u^j |g|^{(1+k)/2} = \frac{4X^i X^j |g|^k}{(1+k)[H + (H^2 - 16k(1+k)^{-2}X_a X_b g^{ab}|g|^k)^{1/2}]}, \qquad (1.14)$$

where $X_a = -\tfrac{1}{4}\varkappa_\beta^\gamma C_{\alpha\gamma}^\beta |g|^{(1-k)/2}$.

Substituting (1.14) into the equations (1.8) we obtain a closed system of differential equations of second order in the components of the matrix $g_{ij}(\tau)$:

$$\frac{\partial L}{\partial g_{ij}} - \frac{d}{d\tau}\frac{\partial L}{\partial \dot{g}_{ij}} = h^{ij},$$

$$h^{ij} = \frac{4(X^i X^j - kg_{ab}X^a X^b g^{ij})|g|^k}{(1+k)[H + (H^2 - 16k(1+k)^{-2}X_a X_b g^{ab}|g|^k)^{1/2}]}. \qquad (1.15)$$

Let us transform this system (1.15) into a system of equations of first order defined in the phase space p^{ij}, g_{ij}. The momenta p^{ij} are determined by the following expressions:

$$p^{ij} = \frac{\partial L}{\partial \dot{g}_{ij}} = \frac{|g|^{(1-k)/2}}{4}\left(g^{ij}(\ln|g|)^{\cdot} - \dot{g}_{kl}g^{ik}g^{jl}\right). \tag{1.16}$$

In the phase space the system (1.15) becomes

$$\dot{p}^{ij} = -\frac{\partial H}{\partial g_{ij}} - h^{ij}, \quad \dot{g}_{ij} = \frac{\partial H}{\partial p^{ij}}. \tag{1.17}$$

In the coordinates p^{ij}, g_{ij} the function

$$H = \dot{g}_{ij}\frac{\partial L}{\partial \dot{g}_{ij}} - L$$

has the following form:

$$H = \frac{1}{|g|^{(1-k)/2}}\left[(\mathrm{Tr}(p \circ g))^2 - 2\,\mathrm{Tr}(p \circ g \circ p \circ g) + \frac{1}{4}(2|g|g^{\alpha\alpha} - \mathrm{Tr}(g^2))\right], \tag{1.18}$$

where $\mathrm{Tr}(Y)$ is the trace of the matrix Y and $p \circ g$ is the product of matrices $p = \|p^{ij}\|$ and $g = \|g_{jk}\|$.

According to this derivation the dynamical system (1.7) is equivalent to the complete system of Einstein's equations. Time dependence of the velocities u_α and the energy density ε is determined by the equations (1.10)–(1.12).

2. Transformation of the Dynamical System

The dynamical system (1.17) is defined in a 12-dimensional phase space $(p^{ij}$ and g_{ij} are symmetric matrices$)$. This system is invariant relative to the scale transformations

$$p^{ij} \to p^{ij}, \quad g_{ij} \to \lambda g_{ij}, \quad \tau \to \lambda^{(1+3k)/2}\tau$$

and therefore admits a reduction of order by 1. To study the system (1.17) using the methods of qualitative theory of dynamical systems we will transform it (using its scale-invariance) into a dynamical system defined in some 11-dimensional compact manifold S which has sufficiently simple (non-degenerate) critical points. It is convenient to break this transformation into three stages.

1. Let us introduce new coordinates $S_k^j = g_{ki}p^{ij}$. According to (1.16) we have

$$S_k^j = \frac{|g|^{(1-k)/2}}{4}\left(\varkappa_\alpha^\alpha\delta_k^j - \varkappa_k^j\right), \tag{2.1}$$

where $\varkappa_k^j = \dot{g}_{ki}g^{ij}$. The eigenvectors of the matrix S_k^j coincide with the eigenvectors of the matrix $\varkappa_k^j$ which are called the Kasner axes in [51]. If the velocities u_α are non-zero, then the matrix $S = \|S_k^j\|$ is not symmetric (as opposed to the matrices $g = \|g_{ij}\|$ and $p = \|p^{ij}\|$). Indeed using (1.11) and (1.16) we obtain

$$X_\alpha = -\tfrac{1}{4}\varkappa_\beta^\gamma C_{\alpha\gamma}^\beta |g|^{(1-k)/2} = \tfrac{1}{2}(1 + k)\varepsilon u_0 u_\alpha |g|^{(1-k)/2} = S_\beta^\gamma C_{\alpha\gamma}^\beta. \tag{2.2}$$

Obviously the matrices S and g satisfy

$$g \circ S^t = S \circ g, \tag{2.3}$$

where S^t is the transpose of S.

In the coordinates S_k^j, g_{ij} the system (1.17) takes the following form:

$$S_k^j = -\frac{1}{2|g|^{(1-k)/2}}[|g|(\delta_k^j g^{\alpha\alpha} - g^{kj}) - g_{ki}g_{ij}] + \delta_k^j\left(\frac{1-k}{2}\right)H - g_{ki}h^{ij}, \tag{2.4}$$

$$g_{ij} = \frac{2}{|g|^{(1-k)/2}}(g_{ij}S_k^k - 2g_{ik}S_j^k).$$

This system (2.4) has two first integrals L and K:

$$L = X_1^2 + X_2^2 + X_3^2, \tag{2.5}$$

$$K = \left[\left(H^2 - \frac{16k}{(1+k)^2}Z\right)^{1/2} + H\right]\left[\left(H^2 - \frac{16k}{(1+k)^2}Z\right)^{1/2} - \frac{1-k}{1+k}H\right]^{(1-k)/(1+k)}, \tag{2.6}$$

where $Z = X_a X_b g^{ab}|g|^k$. Using the formulas (2.2), (1.10) and (1.13) we can express these integrals in terms of the velocities u_α and the energy desity ε:

$$L = \left(\frac{1+k}{2}\right)^2 \varepsilon^2 \bar{u}_0^2 |g|(u_1^2 + u_2^2 + u_3^2), \tag{2.7}$$

$$K = 2\left(\frac{2k}{1+k}\right)^{(1-k)/(1+k)} \varepsilon^{2/(1+k)}\bar{u}_0^2 |g|,$$

where $\bar{u}_0$ is the component of velocity in the synchronous system of coordinates. If $k \to 0$, then instead of the integral K (2.6) we should use dK/dk (formally if $k = 0$, then (2.6) implies that $K = 0$). For $k = 1/3$ these integrals (2.7) are listed in [68]. Later, using the expressions (2.5) and (2.6) we will point out some important applications of these integrals to the question of typical states of the metric in the early stages of the expansion of space.

Just as for the general system of Einstein's equations in a synchronous system of coordinates (see Sect. 3 of Chapt. II), for the system (2.4) there is a monotone function

$$F = \frac{d}{dt}|g|^{1/6} = \frac{1}{3}(S_i^i)|g|^{-1/3}, \quad \frac{dF}{d\tau} \leqslant 0. \tag{2.8}$$

For the contraction of space we have $F < 0$ and $S_i^i < 0$. From the monotonicity of the function F it follows that the trajectories of the system (2.4) never leave the region $S_i^i \leqslant 0$. The process of expansion of space is described by the system (2.4) in the same region $(S_i^i \leqslant 0)$ but with the opposite direction of time.

2. Let us introduce new coordinates

$$\bar{s}_k^j = \frac{S_k^j}{G}, \quad y_{ij} = \frac{g_{ij}}{G}, \quad G, \tag{2.9}$$

where

$$G = \left(\sum_{\alpha, \beta}^{3} g_{\alpha\beta}^2 \right)^{1/2}.$$

These coordinates $y_{ij}, \bar{s}_k^j$ satisfy

$$\sum_{i,j=1}^{3} y_{ij}^2 = 1, \quad y \circ \bar{s}^t = \bar{s} \circ y.$$

The coordinates (2.9) are convenient in the analysis of the behavior of the system (2.4) near the state of maximal expansion, where $\det \| g_{ij} \|$ attains its maximum value and $\bar{s}_k^k = 0$.

3. To analyze the behavior of solutions (trajectories of the system (2.4)) near the cosmological singularity $G = 0$, it is necessary to complete the system of coordinates (2.9) by a boundary at infinity with respect to the coordinates $\bar{s}_k^j$. Such a completion is accomplished by introducing the coordinates

$$s_k^j = \frac{\bar{s}_k^j}{\bar{s}} = \frac{S_k^j}{\mathfrak{S}}, \quad y_{ij}, \quad w = \frac{G^2}{\mathfrak{S}^2}, \quad G, \tag{2.10}$$

where

$$\bar{s} = \left(\sum_{\alpha, \beta=1}^{3} (\bar{s}_\alpha^\beta)^2 \right)^{1/2}, \quad \mathfrak{S} = \left(\sum_{\alpha, \beta=1}^{3} (S_\alpha^\beta)^2 \right)^{1/2}.$$

The coordinates s_k^j satisfy

$$\sum_{k,j=1}^{3} (s_k^j)^2 = 1.$$

The two systems of coordinates $\bar{s}_k^j$, w and s_k^j, w together cover a nine-dimensional ball D^9. Near the center of the ball, the coordinates $\bar{s}_k^j$ are used. Near the boundary sphere, the coordinates s_k^j are used. The boundary sphere

$$\sum_{k,j=1}^{3} (s_k^j)^2 = 1$$

is determined by the condition $w = 0$ and corresponds to infinity with respect to the coordinates $\bar{s}_k^j$.

In the coordinates (2.10) and after a time substitution

$$\frac{d\tau_1}{d\tau} = \frac{\mathfrak{S}^2}{2|g|^{(1-k)/2}},$$ (2.11)

the dynamical system (2.4) takes the following form:

$$
\begin{aligned}
\dot{s}_k^j &= w\big[-|y|\big(\delta_k^j y^{\alpha\alpha} - y^{jk}\big) \\
&\quad + y_{ki} y_{ij} + s_k^j\big(|y|\big(s_\alpha^\beta y^{\beta\beta} - s_\alpha^\beta y^{\alpha\beta}\big) - y_{\alpha\beta} s_\gamma^\beta y_{\gamma\alpha}\big)\big] + \big(\delta_k^j - s_k^j s_\alpha^\alpha\big) H_1(1-k) \\
&\quad - \frac{8|y|w}{(1+k)\big[H_1 + \big(H_1^2 - 16k(1+k)^{-2} w x_a x_b |y| y^{ab}\big)^{1/2}\big]} \\
&\quad \times \big[x_k x_a y^{aj} - k x_a x_b y^{ab} \delta_k^j - s_k^j\big(x_a x_y y^{\beta\gamma} s_\alpha^\beta - k x_a x_b y^{ab} s_\gamma^\gamma\big)\big],
\end{aligned}
$$

$$\dot{y}_{ij} = 8\big(-y_{ij} s_j^k + y_{ij} y_{\alpha\beta} s_\gamma^\beta y_{\gamma\alpha}\big),$$ (2.12)

$$
\begin{aligned}
\dot{w} &= 2w\Big[s_\alpha^\alpha\big(4 - (1-k)H_1\big) - 8 y_{\alpha\beta} s_\gamma^\beta y_{\gamma\alpha} \\
&\quad + w\big(|y|\big(s_\alpha^\beta y^{\beta\beta} - s_\alpha^\beta y^{\alpha\beta}\big) - y_{\alpha\beta} s_\gamma^\beta y_{\gamma\alpha} \\
&\quad + \frac{8}{(1+k)} \frac{\big(x_a x_y |y| y^{\gamma\beta} s_\alpha^\beta - k x_a x_b |y| y^{ab} s_\gamma^\gamma\big)}{\big(H_1 + \big(H_1^2 - 16k(1+k)^{-2} w x_a x_b |y| y^{ab}\big)^{1/2}\big)}\big)\Big],
\end{aligned}
$$

$$\dot{G} = 4G\big(s_\alpha^\alpha - 2 y_{\alpha\beta} s_\gamma^\beta y_{\gamma\alpha}\big).$$

where

$$H_1 = H/\mathfrak{S}^2 = \big(s_\alpha^\alpha\big)^2 - 2 s_\alpha^\beta s_\beta^\alpha + \tfrac{1}{4} w\big(2|y| y^{\alpha\alpha} - 1\big), \quad x_\alpha = s_\beta^\gamma C_{\alpha\gamma}^\beta.$$

The system (2.12) contains a closed subsystem in the coordinates s_k^j, y_{ij}, w. This is a consequence of scale-invariance of the initial dynamical system (1.17).

The compact 11-dimensional manifold S, on which the dynamical system (2.12) is considered, is contained in a 17-dimensional space $D^9 \times S^8$, where S^8 is the unit sphere

$$\sum_{i,j}^{3} y_{ij}^2 = 1.$$

The manifold S is determined in $D^9 \times S^8$ by the following natural conditions:

$$y \circ s^t = s \circ y, \quad y_{ij} = y_{ji}, \quad \sum_{i,j=1}^{3} y_{ij}^2 = 1, \quad \sum_{k,j=1}^{3} \big(s_k^j\big)^2 = 1,$$ (2.13)

$$|y| = \det \|y_{ij}\| \geqslant 0, \quad 0 \leqslant w \leqslant \infty, \quad K \geqslant 0, \quad \bar{s}_k^k \leqslant 0.$$

From the condition $K \geqslant 0$ it follows in particular that

$$H_1 \geqslant 0, \quad H_1^2 \geqslant 16k(1+k)^{-2} w x_a x_b |y| y^{ab}.$$

The boundary Γ of the manifold S consists of four components Γ_0, Γ_1, Γ_w and Γ_m. These components are determined by the following conditions:

$$\Gamma_0: \det \|y_{ij}\| = 0; \quad \Gamma_1: K = 0; \quad \Gamma_w: w = 0; \quad \Gamma_m: s_k^k = 0.$$

Obviously the system (2.4)–(2.12) can be continuously extended to the components of the boundary Γ_0, Γ_1 and Γ_m. It can be extended to Γ_1 if $H_1 \neq 0$. Since

$$H_1^2 \geqslant 16k(1 + k)^{-2} w x_a x_b |y| y^{ab}, \tag{2.14}$$

the expressions containing H_1 in the denominator (see (2.12)) are bounded above as $H_1 \rightarrow 0$ by

$$cH_1 \frac{|x_k x_j| y | y^{ij}|}{x_a x_b |y| y^{ab}}, \quad c |x_k| (|y| y^{ii} w)^{1/2}. \tag{2.15}$$

Therefore such expressions tend to zero as $H_1 \rightarrow 0$ except for the points where the matrix y_{ij} is singly degenerate, where

$$x_a x_b |y| y^{ab} = 0, \quad w \neq 0, \quad (x_1, x_2, x_3) \neq 0.$$

Thus everywhere except for these points determined by the two conditions

$$\det \| y_{ij} \| = 0, \quad x_a x_b |y| y^{ab} = 0,$$

the system can be continuously extended to the boundary Γ as $H_1 \rightarrow 0$.

The components of the boundary Γ_0, Γ_1 and Γ_w are invariant manifolds of the system (2.12). Therefore a trajectory which starts on the boundary will remain there for all time.

3. Power Asymptotics. Typical States of the Metric in the Early Stages of the Expansion of Space

Let us find power (with respect to t) asymptotic forms of the metric for the model of type IX with the motion of matter for the contraction of space. In the coordinates (2.10) a metric having such an asymptotic form is represented by a trajectory of the system (2.12) approaching one of the critical points of this system. The existence of the monotone function $F = s_k^k / w^{1/2} |y|^{1/3}$ implies that all critical points of the system (2.12) belong to the components of the boundary Γ_0 ($|y| = 0$) or Γ_w ($w = 0$). The critical points fall into six sets: Φ, N, T, A, B and K. Analysis of the behavior of the dynamical system (2.12) near these critical sets shows that they are all unstable just as for the diagonal model of type IX (without the motion of matter).

1. The set Φ has dimension 5 and is determined by the conditions $s_k^j = -1/\sqrt{3}\, \delta_k^j$, $w = 0$, y_{ij} are arbitrary and $H_1(\Phi) = 1$. The system (2.12) on the manifold S has the following eigenvalues at the critical point Φ:

$$\lambda_1 = \lambda_2 = (1 - k)/\sqrt{3},$$
$$\lambda_3 = \lambda_4 = \lambda_5 = \sqrt{3}(1 - k) \quad \text{(variables } s_k^j\text{)},$$
$$\lambda_6 = -2(1 + 3k)/\sqrt{3} \quad \text{(variable } w\text{)},$$
$$\lambda_7 = \cdots = \lambda_{11} = 0 \quad \text{(variables } y_{ij}\text{)},$$

The five-dimensional set of critical points Φ is approached by a six-dimensional separatrix (corresponding to the negative eigenvalue λ_6) from the physical region of the manifold S. This separatrix corresponds to diagonalizable metrics (with stationary matter) which have the quasi-isotropic asymptotic form [48]:

$$g_{ij}(t) \approx t^{4/3(1+k)} g_{ij}^0 \tag{3.1}$$

(as the synchronous time t tends to 0).

2. The set N of dimension 2 is determined by the conditions

$$\|y_{ij}\| = Q_1 \begin{pmatrix} 1 & 0 & 0 \\ 0 & 0 & 0 \\ 0 & 0 & 0 \end{pmatrix} Q_1^t, \quad \|s_k^j\| = Q_1 \begin{pmatrix} s_1 & 0 & 0 \\ 0 & s_2 & 0 \\ 0 & 0 & s_2 \end{pmatrix} Q_1^t.$$

where from now on Q_1 is an arbitrary orthogonal matrix. We also have

$$s_1 = -2(3 + k)(43 + 2k + 3k^2)^{-1/2},$$

$$s_2 = -(5 - k) \times (2(43 + 2k + 3k^2))^{-1/2},$$

$$w = \frac{8(1 + 3k)(1 - k)}{43 + 2k + 3k^2}, \quad H_1(N) = \frac{8(5 - k)}{43 + 2k + 3k^2}.$$

The eigenvalues of the system (2.12) at these critical points are

$$\lambda_1 = \lambda_2 = \lambda_3$$
$$= -4(1 + 3k)(2/(43 + 2k + 3k^2))^{1/2} \qquad \text{(variables } y_{ij}),$$
$$\lambda_4 = \lambda_5 = \lambda_6 = \lambda_7$$
$$= 8(1 - k)(2/(43 + 2k + 3k^2))^{1/2} \qquad \text{(variables } s_k^j),$$
$$\lambda_{8,9} = 4\{1 - k \pm i(\tfrac{1}{2}(1 - k)(3 + 16k - 3k^2))^{1/2}\}$$
$$\times (2/(43 + 2k + 3k^2))^{1/2} \qquad \text{(variables } w \text{ and } s_k^j),$$
$$\lambda_{10} = \lambda_{11} = 0 \qquad \text{(variables } y_{ij} \text{ and } s_k^j).$$

The critical points N are approached from the physical region of S by a five-dimensional separatrix representing diagonalizable metrics with the asymptotic form

$$g_{ij}(t) \approx Q_1 \begin{pmatrix} C_1 t^{(1-k)/(1+k)} & 0 & 0 \\ 0 & C_2 t^{(3+k)/2(1+k)} & 0 \\ 0 & 0 & C_3 t^{(3+k)/2(1+k)} \end{pmatrix} Q_1^t. \tag{3.2}$$

The critical points Φ and N are non-degenerate and unstable.

3. The set T of dimension 5 is determined by the conditions

$$\|y_{ij}\| = Q_1 \begin{pmatrix} 2^{-1/2} & 0 & 0 \\ 0 & 2^{-1/2} & 0 \\ 0 & 0 & 0 \end{pmatrix} Q_1^t, \quad \|s_k^j\| = Q_1 \begin{pmatrix} s & 0 & 0 \\ 0 & s & x \\ 0 & 0 & 0 \end{pmatrix} Q_1^t.$$

where the coordinates s and x satisfy the conditions $s < 0$, $2s^2 + x^2 = 1$, $w > 0$ is an arbitrary number, and $H_1(T) = 0$. The set T is approached from the physical region of S by a seven-dimensional separatrix representing metrics (with rotation of axes for $x \neq 0$) having the asymptotic form generalizing the asymptotic form found by Taub:

$$g_{ij}(t) \approx Q_1 \begin{pmatrix} C_1 & 0 & 0 \\ 0 & C_1 + C_2 x^2 t^2 & -C_2 xst^2 \\ 0 & -C_2 xst^2 & C_2 s^2 t^2 \end{pmatrix} Q_1^t. \tag{3.3}$$

The boundary of the set T for $s = 0$ is a set of degenerate critical points T^0.

4. The sets A and B of dimension 6 are determined by the conditions

$$\|y_{ij}\| = Q_1 \begin{pmatrix} y_1 & 0 & 0 \\ 0 & y_2 & 0 \\ 0 & 0 & 0 \end{pmatrix} Q_1^t, \quad \|y_k^j\| = Q_1 \begin{pmatrix} s_1 & 0 & s_1^3 \\ 0 & s_1 & s_2^3 \\ 0 & 0 & s_2 \end{pmatrix} Q_1^t, \quad w = 0.$$

In the set A we have $s_1 = \frac{1}{4}s_2$. In the set B we have $s_2 = 0$. In both sets we have $H_1(A) = H_1(B) = 0$. There are no separatrices approaching the critical points of A and B. Therefore these critical points do not correspond to any asymptotics.

5. The set K has dimension 7 and is determined by the conditions

$$\|y_{ij}\| = Q_1 \begin{pmatrix} 1 & 0 & 0 \\ 0 & 0 & 0 \\ 0 & 0 & 0 \end{pmatrix} Q_1^t, \quad \|s_k^j\| = Q_1 \begin{pmatrix} s_1^1 & s_1^2 & s_1^2 \\ 0 & s_2^2 & s_2^3 \\ 0 & 0 & s_3^3 \end{pmatrix} Q_1^t, \tag{3.4}$$

$$w = 0, \quad H_1(K) = 0.$$

The set K belongs to the intersection of the components of the boundary $\Gamma_0 \cap \Gamma_1 \cap \Gamma_w$. These critical points are non-degenerate (for $s_1^1 \neq s_2^2 \neq s_3^3$) and unstable (a more detailed analysis of the critical points of K is done later in Sect. 4). The separatrices of critical points of K belong to the boundary Γ and go (for the direction of time towards the contraction of space) from one critical point of K to another. The critical points of K do not correspond to any power asymptotics. These critical points and their separatrices are an approximation (see Sect. 4) of the most general mode of behavior of the metric for the contraction of space namely the oscillatory mode [50, 51].

Let us show that if the matter moves, then the power asymptotic forms (3.1) and (3.2) are not applicable. From the integrals L (2.5) and K (2.6) we can form an integral M which is invariant relative to the following transformations:

$$g_{ij} \to \lambda^2 g_{ij}, \quad S_k^j \to \lambda^2 S_k^j, \quad t \to \lambda t. \tag{3.5}$$

In the coordinates s_k^j, y_{ij} and w the integral M has the following form:

$$M = L \cdot K^{-2(1+k)/(1+3k)}$$

$$= \left(x_1^2 + x_2^2 + x_3^2\right)|y|^{2(1-k)/(1+3k)}w^{3(1-k)/(1+3k)}$$

$$\times \left(Z_1^{1/2} - \frac{1-k}{1+k}H_1\right)^{-2(1-k)/(1+3k)} \left(Z_1^{1/2} + H_1\right)^{-2(1+k)/(1+3k)} \qquad (3.6)$$

$$= C(k)\varepsilon^{2(3k-1)/(1+3k)}\bar{u}_0^{2(k-1)/(1+3k)}|g|^{(k-1)/(1+3k)}$$

$$\times \left(u_1^2 + u_2^2 + u_3^2\right),$$

where

$$Z_1 = H_1^2 - 16k(1+k)^{-2}wx_\alpha x_\beta|y|y^{\alpha\beta}.$$

If all the way to the singularity the metric has the asymptotic form (3.1) or (3.2), then the corresponding trajectory of the system (2.12) approaches the critical points Φ or N. At the critical points Φ and N the integral M is equal to zero. Therefore a trajectory approaching these critical points corresponds to a metric without the motion of matter, i.e. in the presence of moving matter and for the contraction of space all the way to the singularity the asymptotic forms (3.1) and (3.2) are not applicable.

Note that the value of the parameter $k = 1/3$ is special because for $k = 1/3$ the integral M does not depend on the energy density ε:

$$M\left(\frac{1}{3}\right) = \frac{2^{4/3}}{9}\bar{u}_0^{-2/3}|g|^{-1/3}\left(u_1^2 + u_2^2 + u_3^2\right).$$

It seems natural to call the motion of matter fast if $M \gg 1$ and slow if $M \lesssim 1$.

From the existence of the monotone function F (see (2.8)), for the direction of time towards the contraction of space, all trajectories of the system (2.12) approach the boundary Γ. At the same time along each trajectory we have $F \to -\infty$ because on the boundary we have $F = -\infty$. After reaching a small neighborhood of the boundary Γ determined by the condition $|F| \gg 1$, a trajectory of the system (2.12) begins moving along the trajectories of this system which belong to the boundary Γ. All trajectories of the system (2.12) in the boundary Γ are separatrices of critical points and go from one critical point to another (we do not provide a separatrix diagram here, because basically it coincides with the separatrix diagram for the diagonal model of type IX given in Sect. 6 of Chapt. II). After a finite number of steps (which is never more than three) along the separatrices of critical points Φ, N, T, A and B, a trajectory ends up in the neighborhood of critical points of K and starts moving along their separatrices. During such movement of the trajectory, the metric varies in an oscillatory mode (see Sect. 4). Thus all metrics of the model of type IX with the motion of matter (except for the metrics which have the Taub asymptotic form (3.3)) undergo the oscillatory mode which therefore is the typical state of the metric for the contraction of space.

The definition of the concept of typical states of the metric was given in Sect. 6

of Chapt. II. For the homogeneous model of type IX with the motion of matter, the typical states of the metric in the early stage of the expansion of space (defined by the condition $|F| = |d(|g|^{1/6})/dt| \gg 1$) are the power modes (3.1)–(3.3) corresponding to the separatrices of critical points Φ, N and T. Along these separatrices the trajectories of the dynamical system (2.12) on the manifolds S can leave the boundary Γ.

In the presence of the motion of matter the typical states of the metric in the early stage of expansion substantially depend on the value of the integral M. If for some solution we have $M \gg 1$ (fast motion of matter), then the corresponding trajectory of the system (2.12) can never appear in the neighborhood of sets Φ and N because in these sets $M = 0$. Consequently for $M \gg 1$ the only typical state of the metric in the early stage of expansion is the power mode (3.3) which generalizes the Taub mode. If however $M \lesssim 1$ (slow motion of matter), then the typical states of the metric in the early stage of expansion are the power modes (3.1), (3.2) and (3.3) just as for the diagonal metric.

4. Combinatorial Model of the Oscillatory Mode

Let us fully integrate the separatrices of critical points and derive the combinatorial model of the oscillatory mode by means of an approximation of the trajectories of the system (2.12) by a sequence of separatrices of critical points of K along which this trajectory moves (for the contraction of space). For further analysis it is convenient to use the following invariant description of the set K. A point P of this set is determined by two matrices:

$$P = \left(y_{ij}, s_j^k\right) \left(\text{where } w = 0,\ H_1(P) = (\mathrm{Tr}(s))^2 - 2\,\mathrm{Tr}(s^2) = 0,\ s = \|s_j^k\|\right).$$

According to (3.4) the matrix y_{ij} has rank 1. Let e_y be the eigenvector of the matrix y_{ij} corresponding to its unit eigenvalue. According to (3.4) the vector e_y is also an eigenvector of the matrix s_j^k. Let s be the eigenvalue of this matrix which corresponds to e_y. Equivalently we can say that the matrix y_{ij} is a projection onto some eigenvector of the matrix s_j^k.

Suppose that $s_1 \geqslant s_2 \geqslant s_3$ are the three eigenvalues of the matrix s_j^k. It is convenient to divide the set K into three subsets K_1, K_2, K_3 determined by the following condition: in K_1 we have $s = s_1$. The eigenvalues of the system (2.12) and their eigenvectors at the critical points of K_1 are

$$\lambda_1 = 2(1 - k)(s_1 + s_2 + s_3) \qquad \text{(variables } s_k^j),$$

$$\lambda_2 = 8(s_n + s_m - s_1) \qquad \text{(variable } w), \qquad (4.1)$$

$$\lambda_3 = 8(s_1 - s_n),\ \lambda_4 = 8(s_1 - s_m) \qquad \text{(variables } y_{ij}),$$

where $(s_1, s_m, s_n) = (s_1, s_2, s_3)$. The remaining seven eigenvalues $\lambda_5, \ldots, \lambda_{11}$ are equal to zero and correspond to eigenvectors tangent to the set K. The conditions

$$H_1(K) = (s_1 + s_2 + s_3) - 2(s_1^2 + s_2^2 + s_3^2) = 0 \text{ and } s_1 + s_2 + s_3 \leqslant 0$$

imply that $s_i \leqslant 0$. The signs of the eigenvalues λ_1, λ_2, λ_3 and λ_4 at the critical points of K_1, K_2 and K_3 are shown in the following diagram

$$
\begin{array}{cccc}
 & K_1 & K_2 & K_3 \\
\lambda_1 & - & - & - \\
\lambda_2 & - & - & + \\
\lambda_3 & + & - & - \\
\lambda_4 & + & + & -
\end{array}
\qquad (4.2)
$$

Thus at each critical point of the sets K_1, K_2 and K_3 there are four non-zero eigenvalues with opposite signs. Consequently these critical points are non-degenerate and unstable.

Let us integrate the separatrices of the critical points of K. From each point $\bar{P}$ $(\bar{y}_{ij}, \bar{s}_k^j)$ contained in K_1 a two-dimensional separatrix leaves and goes along the component of the boundary Γ_w $(w = 0)$ with $H_1 = 0$ (see (4.1)). This separatrix has the following form:

$$w = 0, \quad s_k^j(\tau_1) = \bar{s}_k^j, \quad y_{ij}(\tau_1) = \frac{\exp(-8\tau_1 \bar{s}_k^j) \circ g_0}{[\mathrm{Tr}((\exp(-8\tau_1 \bar{s}_k^j)) \circ g_0)^2]^{1/2}}, \qquad (4.3)$$

where g_0 is such a symmetric matrix that $g_0 \bar{s}^t = \bar{s} g_0$ (in which case for all τ_1 we have $y(\tau_1) \bar{s}^t = \bar{s} y(\tau_1)$ (see (2.13)) and $y_{ij}(-\infty) = \bar{y}_{ij}$. Let $y_{ij}^1 = y_{ij}(+\infty)$. Obviously the matrix y_{ij}^1 has rank 1 and $y^1 \bar{s}^t = \bar{s} y^1$. It is not difficult to verify that the equality $y^1 \bar{s}^t = \bar{s} y^1$ implies that the matrix y^1 is a projection onto some eigenvector of $\bar{s}$, i.e. the point $P_1 = (y_{ij}^1, \bar{s}_k^j)$ belongs either to the set K_2 or the set K_3. From (4.1) it follows that a point of K_2 is approached only by a one-dimensional separatrix which goes along the manifold $w = 0$, $H_1 = 0$. Therefore almost all of the two-dimensional separatrix leaving the point $\bar{P}$ goes to the point $P^1 = (y_{ij}^1, \bar{s}_k^i)$ which belongs to K_3. From there a one-dimensional separatrix breaks off going to a point $(y_{ij}^2, \bar{s}_k^j)$ which belongs to K_2. The matrices y_{ij}^1 and y_{ij}^2 are projections onto the eigenvectors of the matrix $\bar{s}_k^j$ corresponding to the eigenvalues s_3 and s_2 respectively.

From each point $\bar{P} = (\bar{y}_{ij}, \bar{s}_k^j)$ contained in K_2 a one-dimensional separatrix of the form (4.3) leaves and goes to the point $P^1 = (y_{ij}^1, \bar{s}_k^j)$ contained in K_3. From each point $\bar{P} = (\bar{y}_{ij}, \bar{s}_k^j)$ contained in K_3 a one-dimensional separatrix leaves. It is of the form

$$y_{ij}(t) = \bar{y}_{ij},$$

$$s_k^j(t) = \bar{s}_k^j(\cosh t_0/\cosh t) + ((\sinh t - \sinh t_0)/\cosh t)\bar{y}_{ij},$$

$$w(t) = -4(\sinh t - \sinh t_0)(\sinh t + \sinh t_0 - 2(s_1 + s_2) \cdot \cosh t_0)\cosh^{-2} t,$$
$$\qquad (4.4)$$

where the time t is related to τ_1 by the relation $dt = w(t) d\tau_1$ and the constant t_0 is determined by the condition $\tanh t_0 = s_3$, where $0 \geqslant s_1 \geqslant s_2 \geqslant s_3$ are the eigenvalues of the matrix $\bar{s}_k^j$. The separatrix (4.4) is defined for $t_0 \leqslant t \leqslant t_1 < 0$. For

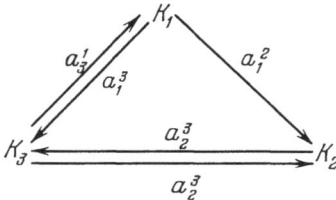

Fig. 20. Separatrix steps between the critical sets K_1, K_2 and K_3

$t = t_0$ we obtain the initial point $\bar{P} = (\bar{y}_{ij}, \bar{s}_k^j)$. For $t = t_1$, where

$$\sinh t_1 = \frac{2(s_1 + s_2) - s_3}{(1 - (s_3)^2)^{1/2}},$$

we obtain the final point $P^1 = (\bar{y}_{ij}, s_k^j(t_1)), w(t_1) = 0, H_1(t) \equiv 0$. From (4.4) it follows that the final matrix $s_k^j(t_1)$ is obtained as the first point of intersection of the shortest arc of a great circle passing (on the sphere

$$\sum_{k,j=1}^{3} (s_k^j)^2 = 1)$$

through the two matrices $\bar{s}_k^j$ and $\bar{y}_{ij}$ and the surface

$$H_1(s_k^j) = (\mathrm{Tr}(s))^2 - 2\,\mathrm{Tr}(s^2) = 0.$$

The matrix $s_k^j(t_1)$ has the following form:

$$s_k^j(t_1) = \frac{s_k^j + 2(s_1 + s_2 - s_3)y_{ij}}{\sqrt{1 + 4(s_1 + s_2)^2 - 4s_3(s_1 + s_2)}}. \tag{4.5}$$

The common eigenvector e_y of the two matrices $\bar{s}_k^j$ and y_{ij} is also an eigenvector of the matrix $s_k^j(t_1)$. Therefore the final point of the separatrix (4.4) $P^1 = (\bar{y}_{ij}, s_k^j(t_1))$ belongs to the set K. At this critical point the eigenvalue λ_2 corresponding to the change of coordinate w (see (4.1)) is negative. Therefore the point P_1 belongs to either the set K_1 or the set K_2 (see (4.2)).

The result of this integration of the separatrices is reflected in the separatrix diagram (Fig. 20). In the diagram an arrow and a symbol a_i^j denote a step along a separatrix going from the set K_i to the set K_j. Note that from each point of K_1 there are two possible steps: a_1^3 and a_1^2. However it is not difficult to verify that $a_2^3 \circ a_1^2 = a_1^3$, i.e. after these two steps we get the same point in the set K_3.

As noted earlier, in general a trajectory of the system (2.12), from some moment of time onwards starts moving along the separatrices of critical points of K_1. Therefore it can be approximated (according to (4.4)) by an infinite sequence of these separatrices and critical points. While a trajectory is in the neighborhood of the critical points of K_1, one of the eigenvalues of the metric $g_{ij}(t)$ is much larger than the other two, because in the set K the matrix

$$y_{ij} = g_{ij} \bigg/ \left(\sum_{\alpha,\beta=1}^{3} g_{\alpha\beta}^2 \right)^{1/2} \tag{4.6}$$

has rank 1. This maximal eigenvalue corresponds to an eigenvector of the matrix g_{ij} which is asymptotically close to the common eigenvector of the matrices y_{ij} and s_k^j.

Note that for almost all trajectories of the system (2.12) the order relations of the eigenvalues $q_1 > q_2 > q_3$ of the metric g_{ij} are preserved for all values of time t. This property is of purely geometric nature. It stems from the fact that in the six-dimensional space of symmetric matrices g_{ij} the manifold W of matrices with two equal eigenvalues has dimension 4, which means that almost all trajectories of the system (2.12) never intersect the manifold W. An analogous property of dynamical systems in the space of two-dimensional matrices is used in Sect. 3 of Chapt. VII.

Returning to the subject at hand we see that the infinite sequence of separatrices defined by the diagram (Fig. 20) is an approximation of the oscillatory mode of behavior of the metric $g_{ij}(t)$. Such an approximation can be mapped to the piecewise approximation of the oscillatory mode by Kasner solutions found in [51] as follows.

Suppose that e_i^1, e_i^2 and e_i^3 are the eigenvectors of the matrix s_k^j (as pointed out earlier they coincide with the eigenvectors of the matrix $\varkappa_k^j$, which in [51] are called the "Kasner axes") and s_1, s_2 and s_3 are the corresponding eigenvalues. As a trajectory moves along the separatrices a_1^3, a_1^2 and a_2^3, the metric $g_{ij}(t)$ is approximated by the following Kasner solution:

$$g_{ij}(t) = C_1 t^{2p_1} e_i^1 e_j^1 + C_2 t^{2p_2} e_i^2 e_j^2 + C_3 t^{2p_3} e_i^3 e_j^3, \tag{4.7}$$

where the Kasner exponents p_i are determined by the formulas

$$p_i = 1 - \frac{2s_i}{s_1 + s_2 + s_3}, \tag{4.8}$$

$$p_1 = \frac{-u}{1 + u + u^2}, \quad p_2 = \frac{1 + u}{1 + u + u^2}, \quad p_3 = \frac{u + u^2}{1 + u + u^2}.$$

The motion of a trajectory along the separatrices a_3^1 and a_3^2 in the model of the work [51] corresponds to "a change of Kasner exponents and a rotation of Kasner axes." Suppose that the matrix $\bar{s}_k^j$ is reduced to a triangular form. According to (4.5) the eigenvalues $s_i = s_i^i$ of the matrix $\bar{s}_k^j$ (here we have $s_1 < s_2 < s_3 < 0$, so in (4.5) we need to transpose s_1 and s_3) are transformed just like in the diagonal model of type IX. Consequently (see Sect. 6 of Chapt. II) the transformation (4.5) leads to the law of change of Kasner exponents given in [51]:

$$p_1' = p_2(u - 1), \quad p_2' = p_1(u - 1), \quad p_3' = p_3(u - 1).$$

The eigenvalues s_i correspond to the following eigenvectors of the matrices $\bar{s}_k^j$ and $s_k^j(t_1)$:

$$s_1 \rightarrow (1,0,0), \quad s_2 \rightarrow (s_1^2, s_2 - s_1, 0),$$

$$s_3 \rightarrow (s_1^2 s_2^3 + (s_3 - s_1)s_1^3, (s_3 - s_1)s_2^3, (s_3 - s_1)(s_3 - s_2)).$$

(4.9)

We will write the coordinates of these vectors in terms of the angles of rotation:

$$L\,(1,0,0), \quad M\,(\cos\theta_m, \sin\theta_m, 0),$$

$$N\,(\cos\theta_n, \sin\theta_n \cos\varphi_n, \sin\theta_n \sin\varphi_n).$$

(4.10)

From the formulas (4.5) and (4.8) we obtain relations for the angles of rotation of the eigenvectors of the matrices $\bar{s}_k^j$ and $s_k^j(t_1)$ after the separatrix steps a_3^1 and a_3^2:

$$(\tan\theta_n'/\tan\theta_n) = (u-2)/(u+2),$$

$$(\tan\theta_m'/\tan\theta_m) = -(2u-1)/(2u+1),$$

(4.11)

$$\varphi_n' = \varphi_n.$$

The obtained formulas coincide with the "law of rotation of Kasner axes" derived in [51]. Consequently the separatrix approximation of the metric $g_{ij}(t)$ defined by the diagram (Fig. 20) is isomorphic to the approximation in [51] under the condition that from the two possible steps a_1^3 and a_1^2 we choose the main step along the two-dimensional separatrix a_1^3.

Let us briefly describe the combinatorial model of the oscillatory mode obtained here. The main feature of this model is that a trajectory of the system (2.12) periodically returns to the neighborhood of critical points of K, which are obtained one from another by a successive action of some mapping T. A point of the set K is a pair of matrices (y_{ij}, s_k^j) satisfying the conditions

$$s_j^j \leqslant 0, \quad H_1(s) = (\mathrm{Tr}\, s)^2 - 2\,\mathrm{Tr}(s^2) = 0,$$

$$\sum_{i,j=1}^{3} y_{ij}^2 = 1, \quad \sum_{k,j=1}^{3} (s_k^j)^2 = 1,$$

(4.12)

where the matrix y_{ij} has rank 1 and is a projection onto some (real) eigenvector of the matrix s_k^j. Suppose that s_y is the eigenvalue of the matrix s_k^j corresponding to this eigenvector (by (4.12) we have $s_y \leqslant 0$).

There is a mapping T acting on the set K. It is defined as follows: if s_y is not the minimal eigenvalue of the matrix s_k^j, then the mapping T can be two-valued. It moves a point (y_{ij}, s_k^j) to a point $(\bar{y}_{ij}, s_k^j)$, where $\bar{y}_{ij}$ is a projection onto another eigenvector of the matrix s_k^j corresponding to an eigenvalue smaller than s_y. If s_y is the minimal eigenvalue of the matrix s_k^j, then the mapping T is single-valued. It moves a point (y_{ij}, s_k^j) to the point $(y_{ij}, \bar{s}_k^j)$, where the matrix $\bar{s}_k^j$ is the first point of intersection of the shortest arc of a great circle (on the sphere $\sum_{k,j=1}^{3}(s_k^j)^2 = 1$) passing through the points s_k^j and y_{ij} with the surface $H_1(s) = 0$.

The mapping of this combinatorial model to the model of the work [51] is determined by the fact that the eigenvectors of the matrix s_k^j coincide with the "Kasner axes" and the Kasner exponents are defined by the formulas

$$p_i = 1 - \frac{2s_i}{s_1 + s_2 + s_3},$$

where s_1, s_2 and s_3 are the eigenvalues of the matrix s_k^j.

5. Several Common Properties of the Dynamics of Homogeneous Cosmological Models with the Motion of Matter

I. Construction of the Dynamical System. In Sects. 1 and 2 we constructed a dynamical system on a compact manifold S, which was equivalent to Einstein's system of equations for the model of type IX. This can be generalized to all other homogeneous models with the motion of matter. This generalization was conducted in the work of Peresetsky [83] to whom the results of this section belong. The existence of a general oscillatory mode of behavior of the metric of the homogeneous cosmological model of type VII with the motion of matter was first discovered in [92]. As in Sect. 1 the vector fields X^0, X^1, X^2 and X^3 which are right-invariant under the action of the Lie group G are assumed to be chosen in such a way that the structure constants C_{ij}^k have the standard form (1.1) and the metric ds^2 has the form (1.2).

Let us substitute the tensor T^{ij} into the identity (1.4) instead of $R^{ij} - \frac{1}{2}g^{ij}R$ (see Einstein's equations (1.3)) and replace the variations δR_{ij} by their expressions in terms of the variations of the metric δg_{ij} using the formulas (3.5) and (3.28) of chapter II. After these substitutions and a choice of time τ according to (1.7), because of the independence of the homogeneous variations of the metric $\delta g_{ij}(i, j = 1, 2, 3)$ the identity (1.4) leads to the following system of equations:

$$\frac{\partial L}{\partial g_{ij}} - \frac{d}{d\tau}\frac{\partial L}{\partial \dot g_{ij}} = \Pi^{ij} + M^{ij}, \qquad (5.1)$$

where the functions Π^{ij} and M^{ij} have the form ($|g| = \det(g_{ij})$):

$$\Pi^{ij} = \frac{1}{2}|g|^{-(1-k)/2}g^{lj}[6a^2|g|g^{3i}\delta_l^3 - 2a^2|g|g^{33}\delta_l^i$$
$$+ a|g|g^{3m}\varepsilon_{lmi}n^i + ag_{lk}g_{km}\varepsilon_{ikm}n^k], \qquad (5.2)$$

$$M^{ij} = \frac{1+k}{2}\varepsilon(u^iu^j - kg_{ab}u^au^bg^{ij})|g|^{(1+k)/2}.$$

The function $L(g_{ij}, \dot g_{ij})$ in (5.1) is obtained from $\frac{1}{2}R\sqrt{|g|}$ by dropping the full derivative with respect to time and has the following form:

$$L = \frac{1}{8}|g|^{(1-k)/2}(\varkappa_\alpha^\alpha\varkappa_\beta^\beta - \varkappa_\alpha^\beta\varkappa_\beta^\alpha)$$
$$+ \frac{1}{4}|g|^{-(1+k)/2}[12a^2|g|g^{33} - |g|g^{ii}|\varepsilon_{ijk}|n^jn^k + g_{ij}^2n^in^j], \qquad (5.3)$$

where $\varkappa_\alpha^\beta = \dot g_{\alpha\gamma}g^{\gamma\beta}$.

Just as in Sect. 1, we introduce momenta $p^{ij} = \partial L/\partial \dot{g}_{ij}$ and coordinates $s_k^j = g_{ki}p^{ij}$. Obviously the form of the momenta p^{ij} and coordinates s_k^j is the same for all homogeneous models (because the kinetic energy in (5.3) does not depend on the type of the model) and is given by the formulas (1.16) and (2.1).

Einstein's equation $R_0^0 - \frac{1}{2}R = T_0^0$ has the following form:

$$H = \varepsilon |g|^{(1+k)/2}[(1+k)u_0^2(-g)^{-k} - k], \tag{5.4}$$

where the function

$$H = \dot{g}_{ij}\frac{\partial L}{\partial \dot{g}_{ij}} - L$$

is expressed in terms of the phase coordinates p^{ij} and g_{ij} according to the formula

$$H = \frac{1}{|g|^{(1-k)/2}}\left\{(\text{Tr}(p \circ g))^2 - 2\,\text{Tr}(p \circ g \circ p \circ g) \right.$$
$$\left. - \frac{1}{4}[12a^2|g|g^{33} - |g|g^{ii}|\varepsilon_{ijk}|n^j n^k + g_{ij}^2 n^i n^j]\right\}. \tag{5.5}$$

The system of equations (5.1) and (5.4) is equivalent to the system $(0,0)$ and (i,j) of Einstein's equations (1.3).

Einstein's equations $R_{0\alpha} = T_{0\alpha}$ have the following form:

$$s_\beta^\gamma C_{\alpha\gamma}^\beta - s_\alpha^\beta C_{\alpha\beta}^\alpha = \frac{1+k}{2}|g|^{(1-k)/2}\varepsilon u_0 u_\alpha. \tag{5.6}$$

From the equations (5.6) it follows that in the homogeneous model of type I $(C_{\alpha\gamma}^\beta \equiv 0)$ the inclusion of the motion of matter is impossible (because (5.6) implies that $u_\alpha \equiv 0$). For the homogeneous model of type II $(n^1 = n^2 = a = 0, n^3 = 1)$ from (5.6) we obtain $u_3 = 0$. Meanwhile u_1 and u_2 can be non-zero. For the remaining models (types III–VIII) it is possible to introduce all three components of the velocity of matter.

Just as for the homogeneous model of type IX, the equations (5.4) and (5.6) are relations which allow us to express the components of the velocity of matter u_i and the energy density ε in terms of the components of the metric g_{ij} and momenta p^{ij}. Taking the conditions (1.12) and (5.4) into account, we arrive at the following expressions:

$$2\varepsilon u_0^2|g|^{(1-k)/2} = H + \left(H^2 - \frac{16k}{(1+k)^2}X_a X_b g^{ab}|g|^k\right)^{1/2} = Z,$$

$$\frac{1+k}{2}\varepsilon u^i u^j|g|^{(1+k)/2} = \frac{4|g|^k X^i X^j}{(1+k)Z}, \tag{5.7}$$

$$X_i = s_k^j C_{ji}^k - s_i^j C_{kj}^k = s_k^j \varepsilon_{jik} n^k - as_k^k \delta_i^3 + 3as_i^3.$$

After substituting the expressions (5.6) and transforming into the phase coordinates p^{ij} and g_{ij} the system (5.1) takes the following form:

$$\dot{p}^{ij} = -\frac{\partial H}{\partial g^{ij}} - \Pi^{ij} - M^{ij}, \quad g_{ij} = \frac{\partial H}{\partial p^{ij}}, \tag{5.8}$$

where the functions Π^{ij} are defined by the formulas (5.2) and the functions M^{ij} by virtue of (5.7) have the following form:

$$M^{ij} = 4|g|^k [X^i X^j - kg^{ij}(g_{ab}X^a X^b)]/(1 + k)Z.$$

After a transformation into the coordinates (2.10) and a time change (2.11), the dynamical system (5.8) takes the following form:

$$\frac{ds_j^i}{d\tau_1} = (1 - k)H_1(\delta_j^i - s_j^i s_\alpha^\alpha)$$

$$+ \frac{8yw}{(1+k)Z_1}[B_j^i - s_j^i \operatorname{Tr}(s^t \circ B)] + w(Q_j^i - s_j^i \operatorname{Tr}(s^t \circ Q)), \tag{5.9}$$

$$\frac{dw}{d\tau_1} = 8w(\operatorname{Tr} s - 2\operatorname{Tr}(y^2 \circ s))$$

$$- 2w\left[(1 - k)H_1 \operatorname{Tr} s + \frac{8yw}{(1+k)Z_1}\operatorname{Tr}(s^t \circ B) + w\operatorname{Tr}(s^t \circ Q)\right],$$

$$\frac{dy_{ij}}{d\tau_1} = 8(-y_{ik}s_j^k + y_{ij}\operatorname{Tr}(y^2 \circ s)),$$

$$\frac{dG}{d\tau_1} = 4G(\operatorname{Tr} s - 2\operatorname{Tr}(y^2 \circ s)),$$

where we use the following notation:

$$B_j^i = -y^{il}x_l x_j + k\delta_j^i y^{ab}x_a x_b,$$

$$x_i = s_l^j \varepsilon_{jil} n^l - as_\alpha^\alpha \delta_i^3 + 3as_i^3,$$

$$Q_j^i = \delta_j^i(4a^2 yy^{33} - n^2 n^3 yy^{11} - n^1 n^3 yy^{22} - n^1 n^2 yy^{33})$$

$$+ a\varepsilon_{jli}yy^{3l}n^i + a\varepsilon_{i3l}y_{lm}y_{mj}n^m + \tfrac{1}{2}yy^{ij}n^k n^l|\varepsilon_{jkl}| + y_{ii}y_{ij}n^i n^l, \tag{5.10}$$

$$H_1 = (\operatorname{Tr} s)^2 - 2\operatorname{Tr}(s^2) - \tfrac{1}{4}w\operatorname{Tr} Q,$$

$$Z_1 = H_1 + \left(H_1 - \frac{16k}{(1+k)^2}|y|wy^{ab}x_a x_b\right)^{1/2}.$$

Just as for the homogeneous model of type IX, the dynamical system (5.9) is defined on a compact 11-dimensional manifold S determined by the natural conditions (2.13) in the coordinates s_k^j, y_{ij} and w. By (2.14) the system (5.9) can be continuously extended to the components of the boundary Γ_w and Γ_0 apart from the exceptional points (see the end of Sect. 2).

II. Separatrix Approximation of the Oscillatory Mode. On the component of the boundary Γ_w ($w = 0$) the dynamical system (5.9) does not depend on the type of homogeneous model. Therefore the critical points of the system (5.9) and their separatrices lying in the component of the boundary Γ_w for all homogeneous models are the same as for the model of type IX (see Sects. 3, 4). The eigenvalues of the critical points in Γ_w remain unchanged as well.

Let us analyze the separatrices of critical points of the set K which is broken up into three subsets K_1, K_2 and K_3 (see Sect. 4). The separatrices leaving the critical points of the sets K_1 and K_2 lie in the component of the boundary Γ_w and, as shown in Sect. 4, approach the critical points of the set K_3.

Let $P = (\bar{y}_{ij}, \bar{s}_j^k, w = 0)$ be a point contained in the set K_3. Recall that this means that the matrix $\bar{y}_{ij}$ has rank 1 and therefore can be written as $\bar{y}_{ij} = y_i y_j$, where (y_1, y_2, y_3) is the unit eigenvector e_y of the matrix $\bar{y}_{ij}$. The vector e_y is also an eigenvector of the matrix $\bar{s}_j^k$. It corresponds to the smallest eigenvalue and

$$H_1(P) = (\mathrm{Tr}(s))^2 - 2\,\mathrm{Tr}(s^2) = 0.$$

According to the formulas for the eigenvalues (4.1)–(4.2) a one-dimensional separatrix leaves the critical point P. Along this separatrix we have

$$y_{ij} = \mathrm{const}, \quad H_1 = 0$$

and the equations (5.9) take the following form:

$$\frac{ds_k^j}{d\tau_1} = w(Q_k^j - s_k^j\,\mathrm{Tr}(s \circ Q)),$$

$$\frac{dw}{d\tau_1} = 8w(\mathrm{Tr}\,s - 2\,\mathrm{Tr}(y^2 \circ s)) - 2w^2\,\mathrm{Tr}(s \circ Q). \tag{5.11}$$

The matrix $Q(\bar{y}_{ij})$ does not depend on time and, according to (5.10), has the following form:

$$Q(\bar{y}_{ij}) = Q = a \cdot y \circ N \circ y \circ K + y \circ N \circ y \circ N,$$

$$K = \begin{pmatrix} 0 & +1 & 0 \\ -1 & 0 & 0 \\ 0 & 0 & 0 \end{pmatrix}, \quad N = \begin{pmatrix} n_1 & 0 & 0 \\ 0 & n_2 & 0 \\ 0 & 0 & n_3 \end{pmatrix}. \tag{5.12}$$

Let $\|Q\| = \mathrm{Tr}^{1/2}(Q^t \circ Q) \neq 0$. After a time change $dt = w(\tau_1)\|Q\|\,d\tau_1$ the system (5.11) can be integrated:

$$s_k^j(t) = (\cosh t_0/\cosh t)\bar{s}_k^j + ((\sinh t - \sinh t_0)/\cosh t)\bar{Q}_k^j,$$

$$\bar{Q}_k^j = Q_k^j/\|Q\|, \tag{5.13}$$

$$w(t) = 4(2\,\mathrm{Tr}(y^2\S Q) - \mathrm{Tr}\,Q)(\sinh t - \sinh t_0)(\sinh t_1 - \sinh t),$$

where t_1 is defined by the condition

$$(\sinh t_1 - \sinh t_0)(2\,\mathrm{Tr}(\bar{y}^2\S\bar{Q}) - \mathrm{Tr}\,\bar{Q}) = 2\cosh t(\mathrm{Tr}\,\bar{s} - 2\,\mathrm{Tr}(\bar{y}^2\S\bar{s})).$$

At the time $t = t_1$ the separatrix (5.13) leaving the critical point $P(\bar{y}, \bar{s})$

contained in the set K_3 approaches the critical point $P'(y', s')$, where

$$w = 0, \quad y' = \bar{y}, \quad s' = s(t_1), \quad H_1(P') = 0.$$

Since at the critical point P' the eigenvalue λ_w is less than zero, the point P' belongs either to the set K_1 or the set K_2 (see (4.2)). As a consequence of the formulas (5.13) the matrix $s' = s(t_1)$ is the second point of intersection of the great circle on the sphere $\Sigma (s_k^j)^2 = 1$ passing through the two points $(\bar{s}_k^j)$ and $(\bar{Q}_k^j)$ with the surface $H_1(s_k^j) = 0$. Note that for the homogeneous model of type IX we have $Q = \bar{y}^2 = \bar{y}$ and the formulas (5.13) are equivalent to the formulas (4.4).

Suppose now that $Q(\bar{y}_{ij}) = 0$. Then the separatrix leaving the critical point $P = (\bar{y}_{ij}, \bar{s}_k^j)$ contained in K_3 leaves the set K. Indeed for $Q = 0$, the system (5.11) implies that

$$s_k^j = \text{const}, \quad \bar{y}_{ij} = \text{const}, \quad w = w_0 \exp[8\tau_1(\text{Tr}\, s - 2\,\text{Tr}(y^2 s))]. \quad (5.14)$$

The condition $Q(y_{ij}) = 0$ means that

$$\|Q\|^2 = \text{Tr}\, Q^t \circ Q$$
$$= (n_1 y_1^2 + n_2 y_2^2 + n_3 y_3^2)^2 \times [(ay_1 + n_2 y_2)^2 + (ay_2 - n_1 y_1)^2 + n_3^2 y_3^2]$$
$$= 0.$$

Since $a \cdot n_3 = 0$, the last equality holds simultaneously with

$$n_1 y_1^2 + n_2 y_2^2 + n_3 y_3^2 = 0.$$

Thus for models of types I and V the condition $Q(y_{ij}) = 0$ holds identically. For models of types II, III, IV, VI and VIII this condition $Q(y_{ij}) = 0$ determines a curve in the two-dimensional manifold RP^2 of matrices: $y_{ij} = y_i y_j$, $y_1^2 + y_2^2 + y_3^2 = 1$. For the model of type VII it determines a point and for the model of type IX it determines the empty set.

Thus for all homogeneous models, except for models of types I and V, at almost all points of the set K_3 we have $Q(y_{ij}) \neq 0$. Consequently the separatrix leaving these critical points is of the form (5.13) and approaches critical points of the sets K_1 and K_2. Therefore for all homogeneous models, except for types I and V, almost all critical points of the set K possess the following property: separatrix steps starting from a given critical point always return to a point of K and form infinite sequences.

By the existence of the monotone function

$$F = \frac{d(\det g_{ij})^{1/6}}{dt}$$

for the contraction of space the trajectories of the system (5.9) in a fashion completely analogous to the behavior of the trajectories of the system (2.12) approach the boundary Γ of the manifold S and start moving along the sequences of separatrices of the critical points of K (except for the separatrices of critical points of the types Φ, N, M and T). At the same time the quotients of the eigenvalues q_i/q_j of the metric g_{ij} oscillate, i.e. the metric varies in an oscillatory mode. Thus for all homogeneous models with the motion of matter, except for the models of types I and V, the general mode of behavior of the metric near a singularity is the oscillatory mode.

As a trajectory moves along the separatrix steps $K_1 \to K_3$ and $K_2 \to K_3$, the metric is approximated by the Kasner solution (4.7). As a trajectory moves along the separatrices (5.13) $K_3 \to K_1, K_2$ there is a change of Kasner exponents p_i and a rotation of Kasner axes (eigenvectors of the matrix s_k^j).

Let us show that during the step (5.13) from the critical point $P(\bar{y}_{ij}, \bar{s}_k^j)$ to the critical point $P'(\bar{y}_{ij}, (s_k^j)')$ the transformation of the eigenvalues of the matrix s_k^j (and the Kasner exponents p_i related to them (see (4.8)) does not depend on the type of the homogeneous model. Indeed the dynamical system (5.9)–(5.11) is invariant relative to the action of matrices A of the group of automorphisms Aut($\mathfrak{G}$) of the Lie algebra $\mathfrak{G}$ of the group G:

$$s \to P_1(AsA^{-1}), \quad y \to P_1(AyA^t), \tag{5.15}$$

where the operator P_1 denotes the projection of the matrix x_j^i to the unit sphere

$$P_1(x_j^i) = x_j^i \left(\sum_{\alpha, \beta}^3 (x_\beta^\alpha)^2 \right)^{-1/2}.$$

By means of transformations (5.15), the matrices $\bar{y}_{ij}$ and $\bar{s}_j^i$ can be written in the following form:

$$\bar{y}_{ij} = \begin{pmatrix} 1 & 0 & 0 \\ 0 & 0 & 0 \\ 0 & 0 & 0 \end{pmatrix}, \quad \bar{s}_k^j = \begin{pmatrix} s_1 & s_1^2 & s_1^3 \\ 0 & s_2 & s_2^3 \\ 0 & 0 & s_3 \end{pmatrix}, \quad s_1 < s_2 < s_3 < 0. \tag{5.16}$$

For the model of type VIII it is also possible to write the matrix $\bar{y}_{ij}$ as $\bar{y}_{ij} = \delta_i^3 \delta_j^3$ (see [83]).

By (5.13) we have

$$(s_k^j)' = s_k^j(t_1) = \text{const} \left[\|\bar{s}_k^j\| + 2(s_2 + s_3 - s_1) \begin{pmatrix} 1 & a & 0 \\ 0 & 0 & 0 \\ 0 & 0 & 0 \end{pmatrix} \right]. \tag{5.17}$$

Therefore during the step (5.13) the transformation of Kasner exponents $p_i = 1 - 2s_i(s_1 + s_2 + s_3)^{-1}$ does not depend on the type of the model (and is given in Sect. 4). The relations for the angles of rotation of the Kasner axes (which for the matrix s_k^j of the form (5.16) can be written in the form (4.10)) differ from (4.11) and are of the following form:

$$\tan \theta_n'/\tan \theta_n = (u - 2)/(u + 2 + 4 \tan \theta_n \cos \varphi_n),$$

$$\varphi_n' = \varphi_n, \tag{5.18}$$

$$(a + \cot \theta_m)/(a + \cot \theta_m') = -(2u - 1)/(2u + 1).$$

Note that when the matrix s_k^j is reduced to a triangular form (5.16) by the transformations (5.15), the eigenvectors of the matrix s_k^j are transformed by the matrices A. For all homogeneous models, except for the model of type IX, the matrices A are not orthogonal and therefore change the angles between the vectors. As a consequence of this, in general the law of rotation of Kasner axes depends not only on their relative position (the initial values of θ_m, θ_n and φ_n) but also on their position relative to the basis vectors X^1, X^2 and X^3.

Just as for the model of type IX (see Sect. 3), for all homogeneous models with the motion of matter all critical points of the dynamical system (5.9) and the corresponding asymptotics can be completely analyzed. This research can be found in [83]. It was shown there that all existing critical points and asymptotics are generalizations of critical points and asymptotics present in the homogeneous models without the motion of matter (see Chapt. II) and that there do not appear qualitatively new asymptotics and critical points in the models with the motion of matter.

6. Homogeneous Cosmological Model of Type IX with Electromagnetic Fields

I. Einstein-Maxwell Equations in Lagrangian Form. Just like the metric, an electromagnetic field in the homogeneous cosmological model of type IX is assumed invariant relative to the right translations in the group SO(3). The metric is of the form (1.2), where $g_{00}(t) \equiv 1$. In the basis X^0, X^1, X^2, X^3 the vector potential of the electromagnetic field has coordinates

$$A_i = \left(A_0(t), A_1(t), A_2(t), A_3(t)\right).$$

In the above basis the electromagnetic tensor $F_{ik} = A_{k;i} - A_{i;k}$ is of the following form:

$$F_{ik} = C_{ki}^s A_s + V_{X^i} A_k - V_{X^k} A_i, \tag{6.1}$$

where the symbol V_{X^i} denotes the Lie derivative with respect to the field X^i and C_{ki}^s are the structure constants:

$$[X_i, X_k] = C_{ik}^s X_s, \quad C_{ik}^0 = C_{0i}^s \equiv 0.$$

Choosing the structure constants of the group SO(3) in the standard form (1.1) we obtain from (6.1) that

$$F_{ik} = \begin{pmatrix} 0 & \dot{A}_1 & \dot{A}_2 & \dot{A}_3 \\ -\dot{A}_1 & 0 & -A_3 & A_2 \\ -\dot{A}_2 & A_3 & 0 & -A_1 \\ -\dot{A}_3 & -A_2 & A_1 & 0 \end{pmatrix} \quad \left(\dot{A}_i = \frac{dA_i}{dt}\right).$$

As a consequence of gradient invariance, the tensor F_{ik} does not depend on the components $A_0(t)$. From now on we assume that $A_0(t) \equiv 0$.

In a homogeneous cosmological model the metric g_{ij} and the vector potential A_i satisfy the Einstein-Maxwell equations (see [84, 93–99]):

$$R_{ik} - \frac{1}{2} g_{ik} R = \frac{8\pi k}{c^4} T_{ik} = \frac{2k}{c^4} \left(-F_{il} F_k^l + \frac{1}{4} F_{lm} F^{lm} g_{ik} \right),$$

$$F_{;k}^{ik} = 0, \quad i, k, l, m = 0, 1, 2, 3. \tag{6.2}$$

It is well known that these equations follow from the variational principle

$$\delta \int R \sqrt{-g}\, d\Omega + \frac{k}{c^4} \delta \int F_{ik} F^{ik} \sqrt{-g}\, d\Omega = 0 \tag{6.3}$$

through the independent variation of all components of the metric g_{ij} and the vector potential A_i. From now on we use a system of units in which $2k/c^4 = 1$.

Under the conditions of homogeneity of the metric and the electromagnetic field, the variational principle (6.3) becomes a Lagrangian principle

$$\delta \int (L_1 + L_2)\, dt = 0, \tag{6.4}$$

where

$$L_1 = \frac{1}{8} \sqrt{|g|} (\varkappa_\alpha^\alpha \varkappa_\beta^\beta - \varkappa_\alpha^\beta \varkappa_\beta^\alpha) - \frac{1}{4\sqrt{|g|}} (2|g| g^{\alpha\alpha} - g_{\alpha\beta} g_{\alpha\beta}),$$

$$\varkappa_\alpha^\beta = \dot{g}_{\alpha\gamma} g^{\gamma\beta}, \quad |g| = \det \| g_{\alpha\beta} \|, \tag{6.5}$$

$$L_2 = -\frac{1}{2} \dot{A}_\alpha \dot{A}_\beta g^{\alpha\beta} \sqrt{|g|} + \frac{1}{2} A_\alpha A_\beta g_{\alpha\beta} \frac{1}{\sqrt{|g|}}.$$

The Lagrange equations

$$\partial(L_1 + L_2)/\partial q = (d/dt)(\partial(L_1 + L_2)/\partial \dot{q})$$

corresponding to (6.4) are equivalent to the "tensor" components of Einstein's equations for $i, k = 1, 2, 3$ and Maxwell's equations for $i = 1, 2, 3$ (see (6.2)). Maxwell's equation $F_{;k}^{0k} = 0$ is satisfied identically, because of the homogeneity of the metric and the electromagnetic field. The Einstein equation

$$R_{00} - \tfrac{1}{2} g_{00} R = -F_{0l} F_0^l + \tfrac{1}{4} F_{lm} F^{lm} g_{00}$$

means that $H = 0$, where H is the energy corresponding to the Lagrangian $L_1 + L_2$. Einstein's equations

$$R_{0\alpha} = -F_{0\beta} F_{\alpha\gamma} g^{\gamma\beta}$$

have the following form:

$$-\tfrac{1}{2}\varkappa_\beta^\gamma C_{\gamma\alpha}^\beta = \dot{A}_\beta C_{\gamma\alpha}^\delta A_\delta g^{\gamma\beta}, \tag{6.6}$$

where $C_{\gamma\alpha}^\beta$ are the structure constants of SO(3) in their standard form (1.1).

Thus for the homogeneous cosmological model of type IX with an electromagnetic field the Einstein-Maxwell equations (6.2) are equivalent to a Lagrangian system with the Lagrangian $L_1 + L_2$ subject to (6.6) and $H = 0$. (Later we will show that the relations (6.6) are preserved by the Lagrangian system).

II. First Integrals of the Hamiltonian System. Let us transform the newly obtained Lagrangian system into a Hamiltonian system equivalent to it by means of the Legendre transformation. First we introduce momenta

$$p^{ij} = \frac{\partial L_1}{\partial \dot{g}_{ij}} = \frac{\sqrt{|g|}}{4}\left(g^{ij}(\ln|g|)^{\cdot} - \dot{g}_{kl}g^{ik}g^{ji}\right),$$

$$P_i = \frac{\partial L_2}{\partial \dot{A}_i} = -\dot{A}_j g^{ij}\sqrt{|g|} \quad i,j,k,l = 1,2,3. \tag{6.7}$$

In the phase space (p,q) with the coordinates p^{ij}, P_i, g_{ij} and A_i we obtain a Hamiltonian system

$$\dot{p} = -\frac{\partial H}{\partial q}, \quad \dot{q} = \frac{\partial H}{\partial p} \tag{6.8}$$

with the Hamiltonian $H = H_1 - H_2$, where

$$H_1 = \frac{1}{\sqrt{|g|}}[(\mathrm{Tr}(p\circ g))^2 - 2\,\mathrm{Tr}(p\circ g\circ p\circ g)$$

$$+ \frac{1}{4}(2|g|\,\mathrm{Tr}(g^{-1}) - \mathrm{Tr}(g^2))], \tag{6.9}$$

$$H_2 = \frac{1}{2\sqrt{|g|}}(P_i P_j + A_i A_j)g_{ij}.$$

where p and g are matrices

$$\|p^{ij}\|, \|g_{ij}\|; \quad |g| = \det\|g_{ij}\|.$$

Note that the Hamiltonian H_2 describing the change of the electromagnetic field is the Hamiltonian of a three-dimensional oscillator (with a changing metric).

The Lagrangian L and therefore the Hamiltonian system (6.8) has SO(3) as its group of symmetries. This group acts according to the law

$$g \to Q^t \circ g \circ Q, \quad p \to Q^t \circ p \circ Q, \quad A \to Q^t \circ A, \quad P \to Q^t \circ P, \tag{6.10}$$

where Q is an orthogonal matrix. According to Noether's theorem (see [100]), the existence of a three-dimensional group of symmetries (6.10) implies that the Hamiltonian system (6.8) has three first integrals ("momenta"):

$$M_\alpha = g_{\beta\gamma}P^{\gamma\delta}C_{\delta\alpha}^\beta + \tfrac{1}{2}A_\beta P_\gamma C_{\gamma\alpha}^\beta. \tag{6.11}$$

With the help of the integrals M_α it is easy to prove that the relations (6.6) are preserved by the Hamiltonian system (6.8). Indeed using the formulas (6.7) it is easy to verify that the relations (6.6) imply that $M_\alpha = 0$. Thus the Hamiltonian system (6.8) is considered subject to $H = 0$ and $M_\alpha = 0$.

The system (6.8) has one other first integral

$$W = \frac{1}{2}\left(\sum_{i=1}^{3} (P_i^2 + A_i^2) \right). \tag{6.12}$$

Indeed it is easy to verify that the Poisson bracket $[H, W]$ is zero.

III. Oscillatory Mode of Behavior of Solutions with a Diagonal Metric Near a Cosmological Singularity.
For the homogeneous cosmological model of type IX, the Einstein-Maxwell equations have partial solutions with a diagonal metric and a single-component vector potential A_i:

$$g_{ij} = \begin{pmatrix} q_1 & 0 & 0 \\ 0 & q_2 & 0 \\ 0 & 0 & q_3 \end{pmatrix}, \quad A_i = (0, A_1, 0, 0). \tag{6.13}$$

Such solutions identically satisfy the relations $M_\alpha = 0$. Under the conditions (6.13), the Hamiltonian $H = H_1 - H_2$ (6.9) has the following form:

$$H = \frac{1}{\sqrt{q_1 q_2 q_3}}\left[\left(\sum_{i=1}^{3} p_i q_i \right)^2 - 2\sum_{i=1}^{3} p_i^2 q_i^2 \right.$$
$$\left. + \frac{1}{4}\left(\left(\sum_{i=1}^{3} q_i \right)^2 - 2\sum_{i=1}^{3} q_i^2 \right) - \frac{(P_1^2 + A_1^2)}{2} q_1 \right]. \tag{6.14}$$

After a time change

$$d\tau_1/dt = 1/(2\sqrt{q_1 q_2 q_3})$$

and a change of coordinates $\bar{s}_i = p_i q_i$ $(i = 1, 2, 3)$ the corresponding Hamiltonian system subject to $H = 0$ becomes the following system:

$$\dot{\bar{s}}_i = -q_i\left(\sum_{k=1}^{3} q_k - 2q_i \right) + 2H_0\delta_i^1, \quad \dot{q}_i = 4q_i\left(\sum_{k=1}^{3} \bar{s}_k - 2\bar{s}_i \right), \tag{6.15}$$

where

$$H_0 = H_1\sqrt{q_1 q_2 q_3} = \left(\sum_{i=1}^{3} \bar{s}_i \right)^2 - 2\sum_{i=1}^{3} \bar{s}_i^2 + \frac{1}{4}\left(\sum_{i=1}^{3} q_i \right)^2 - \frac{1}{2}\sum_{i=1}^{3} q_i^2.$$

At the same time the vector potential satisfies the following equations:

$$P_1 = \sqrt{W} \sin \varphi(\tau_1), \quad A_1 = \sqrt{W} \cos \varphi(\tau_1), \quad \dot{\varphi} = 2q_1. \tag{6.16}$$

where $W = 2H_0/q_1$ is a first integral of the system (6.15).

Consider the behavior of the metric (6.13) during the contraction of space

$((q_1 q_2 q_3)^{\cdot} < 0)$ in a neighborhood of the cosmological singularity $q_1 q_2 q_3 = 0$. Contraction of space corresponds to the region $\bar{s}_1 + \bar{s}_2 + \bar{s}_3 \leqslant 0$. According to the meaning of the problem we have

$$q_i \geqslant 0, \quad H_0 = H_1 \sqrt{q_1 q_2 q_3} = H_2 \sqrt{q_1 q_2 q_3} \geqslant 0.$$

The system (6.15) has a two-dimensional set of non-degenerate critical points. This set is a cone L:

$$\left(\sum_{i=1}^{3} \bar{s}_i\right)^2 - 2 \sum_{i=1}^{3} \bar{s}_i^2 = 0, \quad q_1 = q_2 = q_3 = 0.$$

The cone L lies in the negative quadrant $\bar{s}_1, \bar{s}_2, \bar{s}_3 < 0$ and is tangent to the bisectors of the coordinate angles $\bar{s}_i = \bar{s}_j, \bar{s}_k = 0$. The bisectors divide the cone into three parts. Let L_i denote the part of the cone closest to the axis $\bar{s}_i$.

The eigenvalues of the critical points of L are

$$\lambda_{q_1} = \lambda_{s_1} = 4(\bar{s}_2 + \bar{s}_3 - \bar{s}_1), \quad \lambda_{q_2} = 4(\bar{s}_3 + \bar{s}_1 - \bar{s}_2),$$

$$\lambda_{q_3} = 4(\bar{s}_1 + \bar{s}_2 - \bar{s}_3), \quad \lambda_{s_2} = \lambda_{s_3} = 0.$$

Obviously in the set L_i we have

$$\lambda_{q_j} > 0, \quad \lambda_{q_j} < 0, \quad \lambda_{q_k} < 0 \quad (i, j, k = 1, 2, 3).$$

Consequently a separatrix leaves the set L_i along which the coordinates q_i and $\bar{s}_i$ change and the coordinates $q_j = q_k = 0$, $\bar{s}_j$ and $\bar{s}_k$ remain constant.

After the time change $d\tau/d\tau_1 = q_1$, a separatrix leaving a point $(\bar{s}_1^0, \bar{s}_2^0, \bar{s}_3^0)$ in the set $L_1(\bar{s}_2^0 + \bar{s}_3^0 - \bar{s}_1^0 > 0)$ has the following form:

$$q_1(\tau) = W \cos 2\tau + 2(\bar{s}_2^0 + \bar{s}_3^0 - \bar{s}_1^0)\sin 2\tau - W,$$

$$\bar{s}_1(\tau) = (\bar{s}_1^0 - \bar{s}_2^0 - \bar{s}_3^0)\cos 2\tau + \frac{W}{2}\sin 2\tau + \bar{s}_2^0 + \bar{s}_3^0, \qquad (6.17)$$

$$q_2 = q_3 = 0, \quad \bar{s}_2 = \bar{s}_2^0, \quad \bar{s}_3 = \bar{s}_3^0, \quad 0 \leqslant \tau \leqslant \tau_1.$$

The final point of this separatrix is reached for $\tau = \tau_1$, where $\tan \tau_1 = 2(\bar{s}_2^0 + \bar{s}_3^0 - \bar{s}_1^0)/W$, $q_1(\tau_1) = 0, \bar{s}_1(\tau_1) = 2(\bar{s}_2^0 + \bar{s}_3^0) - \bar{s}_1^0$. This point belongs either to the set L_2 or L_3. In the coordinates $\bar{s}_1, \bar{s}_2, \bar{s}_3$ the separatrix (6.17) moves from the point $(\bar{s}_1^0, \bar{s}_2^0, \bar{s}_3^0)$ to the point $(2(\bar{s}_2^0 + \bar{s}_3^0) - \bar{s}_1^0, \bar{s}_2^0, \bar{s}_3^0)$ along a straight line parallel to the $\bar{s}_1$ axis. Along (6.17) the maximum value of q_1 is attained when $\tau = \tau_m$, where $\tan(2\tau_m) = 2(\bar{s}_2^0 + \bar{s}_3^0 - \bar{s}_1^0)/W$ and is equal to

$$q_{1\,max} = \left(W^2 + 4(\bar{s}_2^0 + \bar{s}_3^0 - \bar{s}_1^0)^2\right)^{1/2} - W. \qquad (6.18)$$

Analogous formulas hold for the separatrices of critical points of L_2 and L_3. It is only necessary to permute the indices and set $W = 0$.

The critical points of L_1 and their separatrices are an approximation of the oscillatory mode. A trajectory moving in the neighborhood of a separatrix (6.17) reaches a neighborhood of some unstable critical point in L_i and then moves along a separatrix of type (6.17) leaving this critical point until it reaches a neighborhood of a new unstable critical point in L_j and so on. Thus along a trajectory, one of the three quantities q_1, q_2, q_3 in succession is much larger than the other two. The formulas (6.17) show that the succession of the maximum values of q_1, q_2, q_3 does not depend on the value of W (presence of an electromagnetic field) and is the same as in empty space $(W = 0)$. Consequently the law of change of "Kasner exponents" [50] also holds. However according to (6.18) the maximum values of q_1 are smaller in the presence of an electromagnetic field. For a strong field $(W \gg q_1)$ we have

$$q_{1\,max}(W) \cong (q_{1\,max}(0))^2/2W,$$

where $q_{1\,max}(0)$ is the appropriate maximum value of q_1 in vacuum $(W = 0)$.

A complete qualitative analysis of the dynamical system (6.15) can be done by means of the methods used in Sect. 6 of Chapt. II. After a transformation into the coordinates

$$s_i = \bar{s}_i/P, \quad y_i = q_i/G, \quad w = G^2/P^2,$$
$$G = (q_1^2 + q_2^2 + q_3^2)^{1/2}, \quad P = (\bar{s}_1^2 + \bar{s}_2^2 + \bar{s}_3^2)^{1/2} \tag{6.19}$$

and a time change $d\tau_2/d\tau_1 = P$, the system (6.15) takes the following form:

$$\dot{s}_i = w[-y_i(y_j + y_k - y_i) + s_i(y_\alpha s_\alpha(y_\beta + y_\gamma - y_\alpha))]$$
$$+ 2(\delta_i^1 - s_i s_1)H_2,$$

$$\dot{w} = 2w\left[w(y_\alpha s_\alpha(y_\beta + y_\gamma - y_\alpha)) - 2s_1 H_2 + 4\left(\sum_{\alpha=1}^{3} s_\alpha - 2\sum_{\beta=1}^{3} s_\beta y_\beta^2\right)\right], \tag{6.20}$$

$$\dot{y}_i = 8y_i(s_\alpha y_\alpha^2 - s_i),$$

$$\dot{G} = 4G\left(\sum_{\alpha=1}^{3} s_\alpha - 2\sum_{\beta=1}^{3} s_\beta y_\beta^2\right).$$

where

$$H_2 = \frac{H_0}{P^2} = 2\sum_{i<j}^{3} s_i s_j - 1 + \frac{1}{4}w\left(2\sum_{i<j}^{3} y_i y_j - 1\right).$$

Just like the system (4.9) in Chapt. II, this system (6.20) is defined on a compact manifold S determined by the conditions

$$H_2 \geqslant 0, \quad y_i \geqslant 0, \quad s_1 + s_2 + s_3 \leqslant 0.$$

On the surface $H_2 = 0$ these two systems obviously coincide. Therefore, according to the results of Sect. 6 of Chapt. II, the system (6.20) has the following sets of critical points lying on the surface $H_2 = 0$: three circles (ψ, i) and nine segments A_i, B_i and T_i. It is not difficult to verify that the system (6.20) has no other critical points (in the region $H_2 > 0$). Therefore for the contraction of space all power asymptotic forms of the metric correspond to separatrices of the critical points T_i.

The critical points in (ψ, i), A_i and B_i have no separatrices approaching them from the physical region.

Let us show without calculating the eigenvalues at the critical points T_i that the only possible asymptotic form in the considered problem is the Taub asymptotic form

$$q_1 \cong C_1 t^2, \quad q_2 \equiv q_3 \to C_2 > 0 \text{ for } t \to 0$$

which corresponds to the critical points T_i. Indeed, from the condition $H = 0$ (see (6.14)) it follows that

$$H_0/q_1 = (P_1^2 + A_1^2)/2 = W = \text{const.}$$

From this we see that $y_1 = H_2 G/wW$. For the Taub asymptotic form we have $G \to \text{const}$, $w \to \text{const}$ and $H_2 \to 0$, so $y_1 \to 0$, i.e. only the asymptotic form T_1 is possible. Such an asymptotic form holds for the exact solutions $(q_2 \equiv q_1)$ found in [93]. All other solutions in the considered problem enter the oscillatory mode as $t \to 0$ and the corresponding trajectories of the system (6.20) move along the separatrices of the three circles of critical points (ψ, i).

Note that in the Lorentz system of coordinates the electromagnetic field components are

$$E_1 = -P_1/\sqrt{q_2 q_3}, \quad H_1 = A_1/\sqrt{q_2 q_3}.$$

By the existence of the integral $P_1^2 + A_2^2 = 2W$ this implies that for a general solution near the cosmological singularity we have $E_1^2 + H_1^2 \to \infty$, whereas for partial solutions $q_2 \equiv q_3$, E_1 and H_1 are bounded.

Chapter IV
Self-Similar Spherically Symmetric Solutions for the General Theory of Relativity

Along with the homogeneous cosmological models there is another important class of solutions of Einstein's equations for which the equations of the general theory of relativity are reduced to a system of ordinary differential equations. This is the class of self-similar solutions. Self-similar spherically symmetric solutions in the general theory of relativity were first considered in [101]. Einstein's equations for these solutions were transformed in various ways in [102, 103] and were considered in the co-moving system of coordinates in [104, 105]. Self-similar generalizations of homogeneous cosmological models were pointed out in [106]. A classification of partially invariant solutions of Einstein's equations including the self-similar solutions was given in [107]. In connection with the problem of formation of non-stationary black holes, self-similar solutions were studied in [108]. Self-similar solutions which are non-singular inside the event horizon were studied in [109]. The methods used in the listed works permitted an analysis of several properties of self-similar solutions. However they did not provide a full picture of the dynamics of these solutions.

In this chapter we conduct a qualitative analysis of the three-dimensional dynamical system describing the self-similar solutions for the general theory of relativity. We have discovered for the first time the existence of radial oscillations of gas in self-similar solutions (this effect has an analogue in the classical theory as well (see Chapt. V)). We found self-similar solutions for which the three-dimensional spatial sections have the same topology as in the Schwarzschild-Kruskal solution.

1. Einstein's System of Equations for Spherically Symmetric Self-Similar Solutions

I. Spherically symmetric self-similar solutions of Einstein's equations are four-dimensional manifolds $\mathcal{M}^4$ with an Einsteinian metric $ds^2 = g_{ij} dx^i dx^j$ which have a three-dimensional group of isometries $SO(3)$ acting with two-dimensional orbits (spheres S^2) and a one-dimensional group of transformations

$$\varphi_\lambda: \ \mathcal{M}^4 \to \mathcal{M}^4,$$

under whose action the metric is stretched:

$$\varphi_\lambda: \ ds^2 \to c(\lambda) ds^2, \ c(\lambda_1 + \lambda_2) = c(\lambda_1)c(\lambda_2).$$

Self-similar solutions can be written in various equivalent forms (see Sect. 3 below). From now on we will study the metric of self-similar solutions in their conformally static form:

$$ds^2 = l^2 e^{2\tau}\left(\sigma e^{\nu(r)} d\tau^2 - \sigma e^{\lambda(r)} dr^2 - r^2 d\Omega^2\right), \tag{1.1}$$

where $d\Omega^2 = d\theta^2 + \sin^2\theta \, d\varphi$ is the standard metric with a constant positive curvature $\kappa = +1$ on S^2, $\sigma = \pm 1$, l is a constant with the dimensions of length and variables τ and r are dimensionless. In the self-similar solution (1.1) the group φ_μ acts by translations with respect to τ. Along with (1.1) we can consider self-similar solutions with the metric $d\Omega^2 = d\theta^2 + \sinh^2\theta \, d\varphi^2$ of constant negative curvature $\kappa = -1$ and with the metric $d\Omega^2 = d\theta^2 + d\varphi^2$ of zero curvature $\kappa = 0$.

Suppose that in the self-similar solution (1.1) the stress-energy tensor is hydrodynamical:

$$T_{ij} = (p + \varepsilon)u_i u_j - p g_{ij}, \tag{1.2}$$

where the energy density ε, pressure p and the 4-velocity of matter u^i are of the form

$$\varepsilon = \varepsilon_0 \bar{\varepsilon}(r)e^{-2\tau}, \quad p = k\varepsilon, \quad 0 \leqslant k < 1,$$

$$(u^0, u^1, u^2, u^3) = e^{-\tau}\left[\left(\frac{\sigma e^{-\nu}}{1-u^2}\right)^{1/2}, u\left(\frac{\sigma e^{-\lambda}}{1-u^2}\right)^{1/2}, 0, 0\right], \tag{1.3}$$

where ε_0 is a constant with the dimensions of energy density. For $\sigma = \pm 1$ we have $u(r) = v(r)/c$ and $|u| < 1$, whereas for $\sigma = -1$ we have $u(r) = c/v(r)$ and $|u| > 1$, where $v(r)$ is the three-dimensional radial velocity of matter.

Einstein's equations for self-similar solutions are of the following form (where $\kappa_0 = 8\pi k/c^4$):

$$R_0^0 - \frac{1}{2}R = \kappa_0 T_0^0:$$

$$e^{-\lambda}\left(\frac{\lambda'}{r} - \frac{1}{r^2}\right) + 3e^{-\nu} + \frac{\kappa}{r^2} = \kappa_0 \varepsilon_0 \bar{\varepsilon}\frac{1 + ku}{1 - u^2}, \tag{1.4}$$

$$R_1^1 - \frac{1}{2}R = \kappa_0 T_1^1:$$

$$-e^{-\lambda}\left(\frac{\nu'}{r} + \frac{1}{r^2}\right) + e^{-\nu} + \frac{\kappa}{r^2} = -\kappa_0 \varepsilon_0 \bar{\varepsilon}\frac{u^2 + k}{1 - u^2}, \tag{1.5}$$

$$R_2^2 - \frac{1}{2}R = \kappa_0 T_2^2:$$

$$-\frac{1}{4}e^{-\lambda}\left(2\nu'' + \nu'^2 - \frac{2}{r}\lambda' + \frac{2}{r}\nu' - \nu'\lambda'\right) + e^{-\nu} = -\kappa_0 k\varepsilon_0 \bar{\varepsilon}, \tag{1.6}$$

$$R_{01} = \kappa_0 T_{01}: \quad \nu' = -(1 + k)\varepsilon_0 \bar{\varepsilon}\frac{\sigma u}{1 - u^2}e^{(\nu + \lambda)/2}. \tag{1.7}$$

II. Consider the system of two Einstein's equations

$$R_0^0 - R_1^1 = \kappa_0(T_0^0 - T_1^1), \quad R_2^2 - \tfrac{1}{2}R = \kappa_0 T_2^2 \tag{1.8}$$

and one hydrodynamical equation $T_{1;k}^k = 0$:

$$\frac{1}{\sqrt{g}}\frac{\partial\sqrt{g}T_1^1}{\partial r} + 2T_1^0 - \frac{1}{2}v'T_0^0 - \frac{1}{2}\lambda'T_1^1 - \frac{2}{r}T_2^2 = 0, \tag{1.9}$$

where $\sqrt{g} = r^2 e^{(v+\lambda)/2}$. After eliminating $\varepsilon_0\bar\varepsilon$ from the equation (1.7) and a transformation into new variables

$$\zeta = \ln r, \quad Q = re^{(\lambda-v)/2}, \quad w = \frac{dv}{d\zeta} = v'r \tag{1.10}$$

the equations (1.8) and (1.9) determine a self-contained (closed) three-dimensional dynamical system:

$$\frac{dQ}{d\zeta} = \dot{Q} = Q\left(1 - w - Q^2 - wQ\frac{1+u^2}{2u}\right),$$

$$\dot{w} = w\left\{-1 - w - Q^2 - Q\left[\frac{w(1+u^2)}{2u} + \frac{2k(1-u^2)}{(1-k)u} + \frac{1+u^2}{u}\right]\right\},$$

$$\dot{u} = -\frac{1-u^2}{u^2-k}\left[\frac{wu(1-k)}{2} - 2ku + Q\frac{u^2(1-k)-k(1+3k)}{(1+k)}\right]. \tag{1.11}$$

The trajectories of the dynamical system (1.11) completely determine the self-similar solution (1.1). After eliminating $\varepsilon_0\bar\varepsilon$ from equation (1.7), Einstein's equation (1.5) determines a relation which is preserved by (1.11):

$$V = w + 1 - Q^2 + wQ\frac{k+u^2}{(1+k)u} = \sigma\kappa e^{\lambda} \tag{1.12}$$

where $\kappa = \pm 1, 0$ is equal to the constant curvature of the two-dimensional metric $d\Omega^2$ (see (1.1)).

The dynamical system (1.11) is defined in the region

$$Q > 0, \quad \text{sign } w = \text{-sign } u,$$

where the last condition follows from (1.7) because ε is positive. According to (1.12) the system (1.11) defines spherically symmetric solutions ($\kappa = \pm 1$) in the region $\sigma V > 0$. In the region $\sigma V < 0$ it describes solutions with a two-dimensional metric $d\Omega^2$ of negative curvature $\kappa = -1$ (the group of symmetries is $SL(2,\mathbb{R})$). On the invariant manifold $V = 0$ it describes solutions with $\kappa = 0$ (flat symmetry). The two types of metric (1.1) ($\sigma = \pm 1$) are described by the system (1.1) in various regions: for $\sigma = +1$ we have $|u| < 1$ and for $\sigma = -1$ we have $|u| > 1$.

In the invariant manifolds $u = \delta = \pm 1$ the system (1.11) has a first integral

$$K = \frac{wQ}{(1 + w - \delta Q)^2} \qquad (1.13)$$

and can be integrated explicitly. This system describes directed streams of neutrinos (for a classical definition of the stress-energy tensor

$$T_0^0 = -T_1^1 = -\delta T_{01} e^{-(\nu + \lambda)/2}$$

see [71]).

2. Analysis of the Dynamical System

I. Conditions at the Discontinuities. Just as in classical gas dynamics an important property of self-similar solutions in the general theory of relativity is the existence of shock waves. In the language of the dynamical system (1.11) the appearance of shock waves is determined by the existence of a surface of non-extendability of the solutions $L_{\pm}$: $u = \pm k^{1/2}$. On the two sides of this surface $(u^2 - k < 0$ and $u^2 - k > 0)$ the vector field of the system (1.11) points in opposite directions and in the limit as $|u| \to k^{1/2}$ is perpendicular to the surface $L_{\pm}$. As a consequence of such non-extendability some solutions can be defined for all r only by introducing a discontinuity, i.e. a shock wave.

The position of the front of the shock wave is determined by some constant value of the coordinate r. Therefore the system of coordinates (1.1) is a system which moves with the shock wave. At the front of the shock wave we glue solutions of the system (1.11) located on the opposite sides of the surface of non-extendability of solutions $L_{\pm}$. In other words the subsonic solution $(|u|^2 < k = dp/d\varepsilon)$ is glued to the supersonic solution $(|u| > k^{1/2})$. It is possible to consider a more general situation where in the region behind the shock wave the equation of state of the matter is $p = k_1 \varepsilon$ and in the region in front of the shock wave the equation of state is $p = k_2 \varepsilon$ $(0 \leqslant k_2 \leqslant k_1)$. The most interesting cases are $k_2 = k_1$ and $k_2 = 0$. The limit values of the parameters on both sides of the discontinuity are related by the following natural conditions:
1) the coefficients of the metric ν, λ and r are always continuous
2) at the discontinuity conservation laws are satisfied (see [110]): $[T_i^k n_k] = 0$, where $n_k = (0, 1, 0, 0)$ is a vector perpendicular to the wave front.
These conditions lead to the following equalities (the indices 1 and 2 refer to the two sides of the discontinuity, the subsonic and the supersonic regions respectively):

$$[T_1^1] = 0: \quad \frac{\bar{\varepsilon}_1(u_1^2 + k_1)}{1 - u_1^2} = \frac{\bar{\varepsilon}_2(u_2^2 + k_2)}{1 - u_2^2},$$

$$[T_0^1] = 0: \quad \frac{(1 + k_1)\bar{\varepsilon}_1 u_1}{1 - u_1^2} = \frac{(1 + k_2)\bar{\varepsilon}_2 u_2}{1 - u_2^2}.$$

$$(2.1)$$

From this we obtain the relations

$$\frac{(1 + k_1)u_1}{u_1^2 + k_1} = \frac{(1 + k_2)u_2}{u_2^2 + k_2}, \quad \frac{\bar{\varepsilon}_1}{\bar{\varepsilon}_2} = \frac{1 + k_2}{1 + k_1} \cdot \frac{u_2}{u_1} \cdot \frac{1 - u_1^2}{1 - u_2^2}, \tag{2.2}$$

which determine u_2 and $\bar{\varepsilon}_2$ in terms of u_1 and $\bar{\varepsilon}_1$ which were found from the solution. Note that from the two values of u_2 $(k_2 \leqslant k_1)$ we choose the supersonic one $(u_2^2 > k_2)$. From the condition $|u_2| < 1$ and (2.2) it follows that $|u_1| > k_1$. If $k_2 = k_1$, then $u_1 u_2 = k_1$. If $k_2 = 0$, then $u_2 = (u_1^2 + k_1)/u_1(1 + k_1)$. From (1.12), the continuity of the functions v, λ, r and the first relation (2.2) it follows that at the front of the shock wave the function w (and thus $v' = wr$) is continuous. Thus the functions v, λ, r, v', w and Q are continuous and the jump in u and ε is determined by (2.2).

Let us turn to the analysis of the dynamical system (1.11) in the region $|u| < 1$. The behavior of solutions in this region is of the most interest, because it is precisely here that the discontinuities corresponding to shock waves appear.

II. Non-Extendability of Solutions for $0 < u < 1$. Spherically symmetric self-similar solutions in the region $0 \leqslant u \leqslant 1$ are described by trajectories of the dynamical system (1.11) contained in the region

$$S_0: Q \geqslant 0, w \leqslant 0, V \geqslant 0, 0 \leqslant u \leqslant 1.$$

From the form of the function V (1.12) it easily follows that this region is bounded, because $0 \leqslant Q \leqslant 1, -1 \leqslant w \leqslant 0$.

Let us show that for some finite $r = r_1$ all trajectories in the region S_0 leave the surface L_+ $(u = k^{1/2})$. In the region $0 < u < k^{1/2}$ this follows because by (1.11) we have $\dot{u} < 0$ and $|\dot{u}| \to \infty$ on L_+. By (1.11) in the region $k^{1/2} < u < 1$ we have $\dot{w} > -w > 0$. Therefore by the boundedness of the region S_0 all trajectories for some finite r leave the surface L_+. As a consequence of this, in the region S_0 we cannot obtain any solutions defined for $r \to 0$ even by introducing a discontinuity. Therefore, the solutions in the region S_0 $(0 < u < 1)$ have no physical meaning and will not be considered further.

III. Resolution of Singularities of the Dynamical System for $-1 \leqslant u \leqslant 0$. In the case under consideration the self-similar spherically symmetric solutions are defined in the region

$$w \geqslant 0, \quad V \geqslant 0, \quad -1 \leqslant u \leqslant 0.$$

The region S_1 is unbounded with respect to w. If $-k < u < 0$, then from the condition $V > 0$ we obtain $Q < 1$. If $-1 < u < -k$, then $Q < (1 + k)|u|(k + u^2)^{-1}$. On the surface $u = 0$ the condition $V > 0$ determines a segment l_1 $(u = w = 0, 0 \leqslant Q \leqslant 1)$ and a ray l_2 $(u = Q = 0, w \geqslant 0)$, where the dynamical system (1.11) is singular. To analyze the dynamical system (1.11) in the region S_1 we transform this system into a system defined on some closed three-dimensional manifold S with boundary Γ with only non-degenerate critical points. The construction of the manifold S is done by means of the following changes of coordinates:

1. Resolution of singular lines l_1 and l_2:

$$q = \frac{Q}{u}, \quad u, \quad w, \quad \zeta; \tag{2.3}$$

$$Q, \quad u_1 = \frac{u}{Q}, \quad w, \quad \frac{d\zeta_1}{d\zeta} = -\frac{Q}{u} > 0. \tag{2.4}$$

2. Completion of the manifold S_1 at infinity with respect to the coordinate w:

$$q = \frac{Q}{u}, \quad u, \quad v = \frac{1}{w}, \quad \frac{d\zeta_2}{d\zeta} = w. \tag{2.5}$$

The closed three-dimensional manifold S is covered by the systems of coordinates (1.10), (2.3)–(2.5) and is determined in them by the following conditions:

$$Q, w, v \geqslant 0, \quad u, u_1, q \leqslant 0,$$
$$V(Q, u, w) \geqslant 0, \quad -1 \leqslant u \leqslant 0. \tag{2.6}$$

The boundary Γ of the manifold S consists of six components $\Gamma_i (i = 1, \ldots, 6)$. The conditions which determine the components of the boundary and the systems of coordinates extended to Γ_i have the following form:

$$\Gamma_1: \ v = 0 \ (2.5); \qquad\qquad \Gamma_2: \ u = -1 \ (1.10);$$
$$\Gamma_3: \ u = 0 \ (2.3) - (2.4); \quad \Gamma_4: \ V = 0 \ (1.10);$$
$$\Gamma_5: \ Q = 0 \ (1.10), (2.3); \quad \Gamma_6: \ w = 0 \ (1.10), (2.4).$$

Now let us transform the system (1.11) into the coordinates (2.3)–(2.5). In the coordinates (2.3) and variable ζ the system (1.11) has the following form:

$$\dot{q} = q \left[1 - w - q^2 u^2 - wq \frac{1 + u^2}{2} \right.$$
$$\left. + \frac{1 - u^2}{u^2 - k} \left\{ \frac{w}{2}(1 - k) - 2k + \frac{1}{1 + k} q[u^2(1 - k) - k(1 + 3k)] \right\} \right],$$

$$\dot{w} = w \left[-1 - w - q^2 u^2 - wq \frac{1 + u^2}{2} - 2q \frac{k}{1 + k}(1 - u^2) \right.$$
$$\left. - q(1 + u^2) \right], \tag{2.7}$$

$$\dot{u} = -u \frac{1 - u^2}{u^2 - k} \left\{ \frac{w}{2}(1 - k) + \frac{1}{1 + k} q[u^2(1 - k) \right.$$
$$\left. - k(1 + 3k)] - 2k \right\}.$$

In the coordinates (2.4) and variable ζ_1 the system (1.11) has the following form:

$$\dot{Q} = Q\left(u_1(-1 + w + Q^2) + w\frac{1 + u_1^2 Q^2}{2}\right),$$

$$\dot{u}_1 = u_1\left[\frac{1 - u_1^2 Q^2}{u_1^2 Q^2 - k}\left(\frac{wu_1(1-k)}{2} - 2ku_1 + \frac{u_1^2 Q^2(1-k) - k(1 + 3k)}{1 + k}\right)\right.$$

$$\left. - u_1(-1 + w + Q^2) - w\frac{1 + u_1^2 Q^2}{2}\right], \tag{2.8}$$

$$\dot{w} = w\left[u_1(1 + w + Q^2) + \frac{w(1 + u_1^2 Q^2)}{2} + \frac{2k(1 - u_1^2 Q^2)}{1 + k} + 1 + u_1^2 Q^2\right].$$

In the coordinates (2.5) and variable ζ_2 the system (1.11) has the following form:

$$\dot{q} = q\left[v - 1 - vq^2 u^2 - q\frac{1 + u^2}{2}\right.$$

$$\left. + \frac{1 - u^2}{u^2 - k}\left\{\frac{1 - k}{2} + \frac{1}{1 + k}vq(u^2(1 - k) - k(1 + 3k)) - 2vk\right\}\right],$$

$$\dot{v} = v\left[1 + v + vq^2 u^2 + q\frac{1 + u^2}{2} + 2qv\frac{k}{1 + k}(1 - u^2)\right. \tag{2.9}$$

$$\left. + qv(1 + u^2)\right],$$

$$\dot{u} = -u\frac{1 - u^2}{u^2 - k}\left[\frac{1 - k}{2} + \frac{1}{1 + k}qv(u^2(1 - k) - k(1 + 3k)) - 2vk\right].$$

The dynamical systems (1.11), (2.7)–(2.9) in the coordinates (1.10), (2.3)–(2.5) define a dynamical system on the manifold S. This system is smoothly continued to the boundary Γ and has only non-degenerate critical points.

IV. Analysis of Critical Points of the Dynamical System on the Manifold S.
The dynamical system considered has eight isolated critical points:

$$A\,(v = 0, q = -1, u = 0), \quad B\,(v = q = u = 0), \quad C\,(w = q = u = 0),$$

$$G\,(v = q = 0, u = -1), \quad H\,(w = q = 0, u = -1),$$

$$Z_1\,(w = u = 0, q = -3(1 + k)/(1 + 3k)),$$

$$Z_2\,(Q = 1, w = 0, u = u_2$$

$$= [k(1 - k) - \{k^2(1 + k) + k(1 - k)(1 + 3k)\}^{1/2}]/(1 - k),$$

$$Z_3\,(u = 0, q = -1, w = 4k/(1 + k))$$

and three lines of critical points:

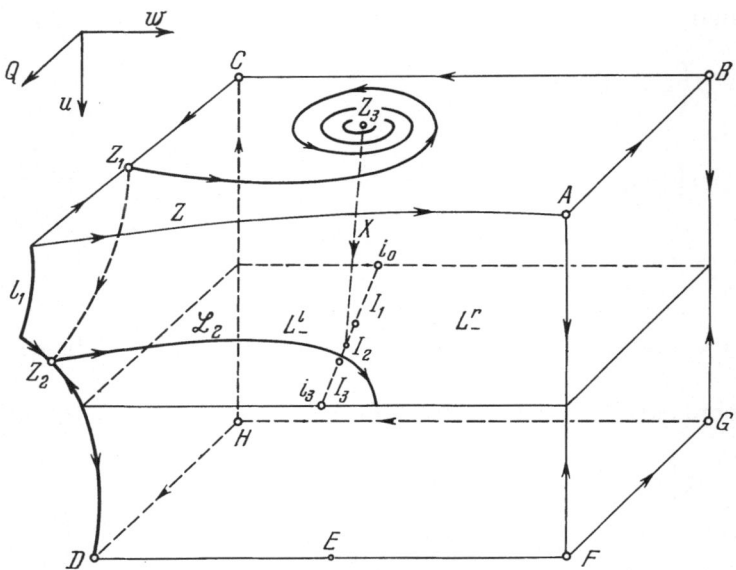

Fig. 21. The general distribution of critical points and separatrices of the dynamical system (1.11) in the closed manifold S

$$l_1 \left(u_1 = w = 0, 0 \leqslant Q \leqslant 1 \right), \quad DE \left(Q = 1, u = -1, 0 \leqslant w \leqslant \infty \right),$$

$$I \left(u = -k^{1/2}, w(1 - k) + 8Qk^{1/2}(1 + k)^{-1} = 4k \right).$$

All critical points belong to the various components of the boundary Γ except for the line I which belongs to the surface of non-extendability of solutions L_-. The manifold S along with the critical points of the dynamical system and their separatrices is pictured in Fig. 21.

The critical points A, B, C, G and H belong to the corners of the boundary Γ and are unstable (saddle points). All their separatrices belong to the various components of the boundary Γ and do not correspond to any exact physical solutions.

The critical points on the line l_1 are repellors. Their eigenvalues are

$$\lambda_{u_1} = \lambda_w = \frac{1 + 3k}{1 + k}, \quad \lambda_Q = 0.$$

Along each separatrix leaving the line of critical points l_1 the parameter r varies up to some finite final value $r = r_0$, where $u(r_0) = w(r_0) = 0$. For $r < r_0$ the corresponding solutions can be smoothly extended to the region $u > 0$. However, in this region the solutions have no physical meaning as a consequence of the existence of non-extendability. The solutions which correspond to the separatrices of critical points of l_1 are defined for $r \geqslant r_0$ and describe the formation of an expanding cavity inside the gas. Indeed, $r = r_0$ is the boundary of the gas preserved during the motion, because $u(r_0) = 0$. A solution for $r < r_0$ can be attached (while preserving the continuity of the metric) to the Minkowski space whose metric has the following form in the coordinates (1.1):

$$ds^2 = e^{2\tau}\left(C_1\, d\tau^2 - \frac{C_1}{C_1 + r^2}\, dr^2 - r^2\, d\Omega^2 \right).$$

Meanwhile the energy density ε and pressure p have a discontinuity at the boundary of the cavity just as in solutions with an expanding cavity in classical gas dynamics [7]. In order to supply physical meaning to these solutions, it is necessary to assume (just like in classical gas dynamics (see [7])) that the gas is being pushed out by some kind of a spherical piston equalizing the non-zero pressure at the internal boundary.

The critical points on the straight line DF have the following eigenvalues:

$$\lambda_1 = \lambda_2 = w - 2, \quad \lambda_3 = 0.$$

Thus the segment DE $(0 \leqslant w < 2)$ consists of attracting critical points, while the segment EF $(2 < w \leqslant \infty)$ consists of repelling ones. All trajectories of the system (1.11) in the component of the boundary Γ_2 $(u = -1)$ begin for some $w = w_0$ on the segment EF and end for $w = 4/w_0$ on the segment DE as follows from the existence of the first integral $K = wQ(1 + w + Q)^{-2}$ (see (1.13)). The trajectories of the system (1.11) which are near the component of the boundary Γ_2 $(u \approx -1)$ behave analogously. The metric corresponding to the trajectories approaching the critical points of DE (for $r \to \infty$) or EF (for $r \to 0$) is not complete and can be smoothly continued in the synchronous system of coordinates (see Sect. 3 below).

The line of critical points I divides the surface of non-extendability of solutions L_- $(u = -k^{1/2})$ into the left and right parts: L_-^l, L_-^r (see Fig. 21). Trajectories leave the left part L_-^l and go inside the manifold S (in the neighborhood of L_-^l we have $\dot{u} > 0$ if $u > -k^{1/2}$ and $\dot{u} < 0$ if $u < -k^{1/2}$). The right side L_-^r attracts trajectories from the manifold S. In both cases the trajectories intersect the surface L_- for a finite value of r.

To study the behavior of trajectories in a small neighborhood of the line I we perform the following change of coordinates and a (non-monotone) change of variable ζ:

$$\alpha = \frac{1}{Q}, \quad \beta = \frac{(1 - k)w}{2Q} - \frac{2k}{Q}, \quad \frac{d\tau_1}{d\zeta} = -\frac{Q}{u^2 - k}. \tag{2.10}$$

The line I $(\beta = -4k^{3/2}(1 + k)^{-1}, u = -k^{1/2}, \alpha > 0)$ consists of non-degenerate critical points of the system (1.11) in the coordinates (2.10). The eigenvalues of the obtained system at these critical points are

$$\lambda_\pm = -k^{1/2}(1 - k) \pm k^{1/2}(1 - k)(1 + 2Z_0(1 - k)^{-1})^{1/2}, \quad \lambda_3 = 0,$$

$$Z_0 = \frac{2k(1 + 3k)}{1 - k}\alpha^2 - \frac{2k}{1 - k^2}[(1 + k)^3 + 2k(3 + k^2)]\alpha$$

$$+ \frac{16k^2(1 + k^2)}{(1 + k)^2(1 - k)} + 6k. \tag{2.11}$$

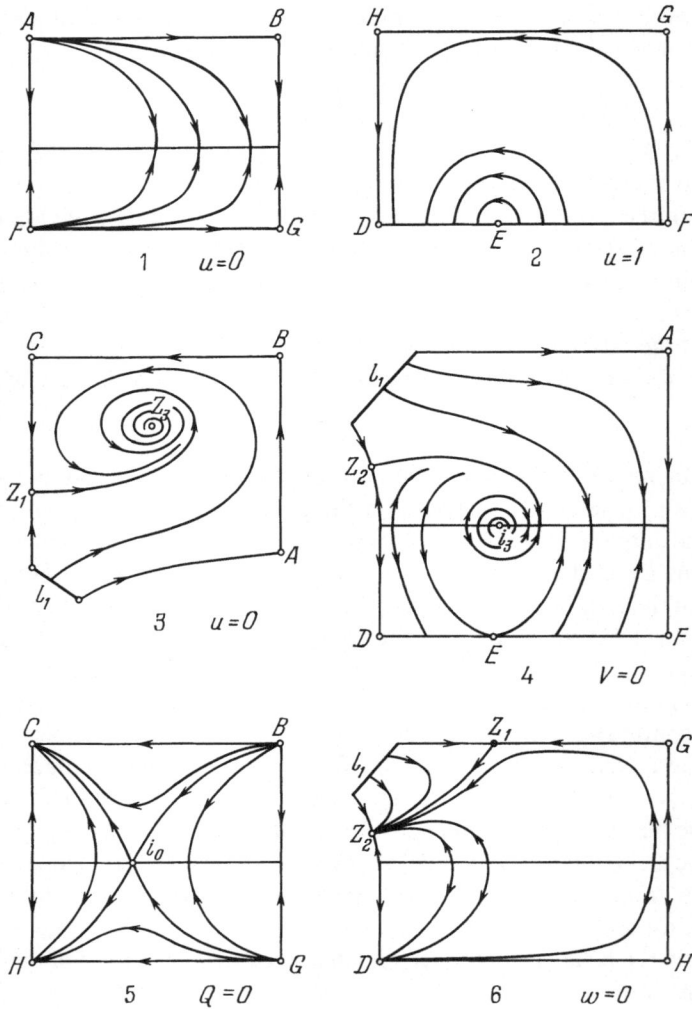

Fig. 22. The movement of the trajectories of the dynamical system in the components of the boundary Γ_k (the numbers below the pictures correspond to the values of $k = 1, \ldots, 6$)

The line of critical points I inside the manifold S is divided into three segments

$$I_1: \ \infty > \alpha > \alpha_1, \quad I_2: \ \alpha_1 > \alpha > \alpha_2, \quad I_3: \ \alpha_2 > \alpha > \alpha_3,$$

where α_1 and α_2 are the largest roots of quadratic equations $Z_0(\alpha_1) = 0$, $Z_0(\alpha_2) = -(1 - k)/2$ and α_3 is the coordinate of the point i_3 of intersection of the line I with the component of the boundary Γ_4 ($V = 0$, Fig. 22, $k = 4$).

From (2.11) it follows that the critical points of the segment I_1 are unstable saddles. The critical points of the segment I_2 are attracting nodes. The critical points of the segment I_3 are attracting foci. The segment I_1 intersects with the

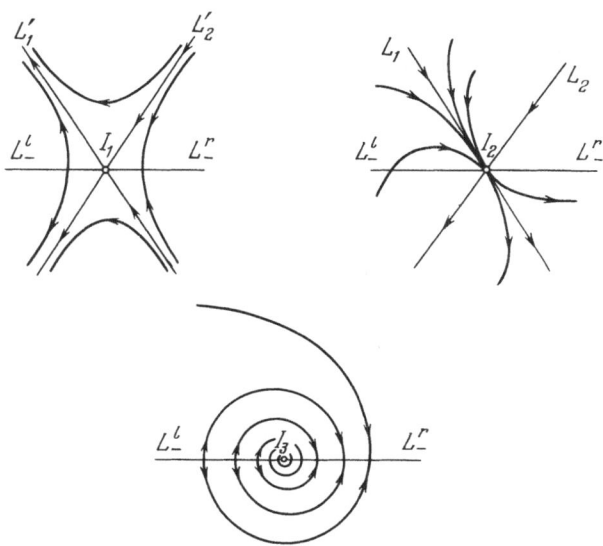

Fig. 23. The qualitative behavior of the trajectories of the dynamical system (1.11) in neighborhoods of critical points contained in the segments I_1, I_2 and I_3

component of the boundary Γ_5 $(Q = 0)$ at the point i_0 $(\alpha = \infty)$ (see Fig. 22, $k = 5$).

Returning to the initial coordinates w, Q and u we see that a two-dimensional separatrix L_2' approaches the segment I_1 from the subsonic region of the manifold S $(|u| < k^{1/2})$ and a two-dimensional separatrix L_1' leaves I_1 (Fig. 23). A three-dimensional separatrix L approaches the segment I_2. There is a two-dimensional separatrix L_2 which is a smooth continuation of L_2' corresponding to the eigenvalues $\lambda_-(\alpha)$. All other trajectories approaching this segment are tangent to a two-dimensional separatrix L_1 which corresponds to the eigenvalues $\lambda_+(\alpha)$. In all cases when a trajectory approaches the segment I the parameter $\zeta = \ln r$ remains finite.

The critical points Z_1, Z_2, Z_3 and their separatrices have the most physical meaning. In the manifold S the dynamical system has the following eigenvalues at these critical points (indices below indicate the corresponding eigenvectors):

$$Z_1: \quad \lambda_q = -3, \lambda_w = 2, \lambda_u = 1;$$

$$Z_2: \quad \lambda_Q = -2, \lambda_w = -\frac{(1-k)u_2}{k} > 0,$$

$$\lambda_u = \frac{2(1-u_2^2)(u_2(1-k) - k(1+k))}{(k - u_2^2)(1+k)} < 0; \tag{2.12}$$

$$Z_3: \quad \lambda_{q,w} = \frac{-(1+3k) + i(7 + 42k - k^2)^{1/2}}{2(1+k)},$$

$$\lambda_u = \frac{1-k}{1+k}.$$

Thus the critical points Z_1 and Z_2 are unstable saddles. The critical point Z_3 (also unstable) is an attracting focus in the component of the boundary Γ_3 ($u = 0$). It has a one-dimensional separatrix X leaving it and going inside the manifold S.

All trajectories in the components of the boundary Γ_6 ($w = 0$) (for $0 > u > -k^{1/2}$) and Γ_3 approach the attracting (in these components of the boundary) critical points Z_2 and Z_3 respectively. In the component of the boundary Γ_6 this follows from the fact that $\dot{Q} > 0$ in Γ_6. In the component of the boundary Γ_3 ($u = 0$) this follows from the fact that for $u = 0$ the dynamical system (2.7) has no limit cycles. Indeed at $u = 0$ on the line $q = -(1 + k)/k$, due to (2.7) we have $\dot{q} = (1 + k)/k^2$, so cycles can occur only in the region $q > -(1 + k)/k$ (the critical point Z_3 ($q = -1$) is situated in this region). After a transformation into the coordinates

$$q, \; V_1 = 1 + w(1 + qk/(1 + k))$$

(in the region $q > -(1 + k)/k$ we have $V_1 > 1$) the system (2.7) takes the following form for $u = 0$:

$$\dot{q} = q\left[3 + \frac{1 + k}{2k} - \frac{1 + k}{2k}V_1 + \frac{1 + 3k}{1 + k}q\right] = P,$$

$$\dot{V} = -\frac{V_1(V_1 - 1)}{1 + q(k/(1 + k))}(1 + q) = Q. \tag{2.13}$$

By the Dulac-Bendixon criterion the system (2.13) in the region $V_1 > 1$ (and therefore the dynamical system in the component of the boundary Γ_3) has no limit cycles, because the following identity holds:

$$\frac{\partial fP}{\partial q} + \frac{\partial fQ}{\partial V_1} = -\frac{1 + 3k}{(1 + k)V_1(V_1 - 1)} < 0,$$

where $f = (qV_1(1 - V_1))^{-1}$.

The qualitative behavior of the dynamical system in the components of the boundary Γ_k is shown in Fig. 22 ($k = 1, \ldots, 6$). In this construction we used the absence of limit cycles in the component of the boundary Γ_3 (proved above), the existence of simple monotone functions in the components $\Gamma_1, \Gamma_4, \Gamma_5, \Gamma_6$ and the existence of the integral (1.13) of the dynamical system in the component of the boundary Γ_2.

A calculation of the eigenvalues (2.12) shows that a two-dimensional separatrix Z leaves the critical point Z_1. One dimensional separatrices obtained by intersecting Z with the components of the boundary Γ_6 and Γ_3 connect the critical point Z_1 to Z_2 and Z_3. Therefore the one-dimensional separatrix $\mathscr{L}_2$ leaving the critical point Z_2 (see Fig. 22, $k = 4$) is the intersection of the separatrix Z with the component of the boundary Γ_4 ($V = 0$) (the separatrix $\mathscr{L}_2$ corresponds to some exact solution with flat symmetry). One-dimensional separatrix X leaving the critical point Z_3 is a limit line onto which the two-dimensional separatrix Z is wound (see Fig. 21).

The separatrix X leaving the critical point Z_3 can be integrated explicitly. On this trajectory we have

$$Q = -u = \frac{C_1 r^{1-\alpha}}{|C_1^2 r^{2(1-\alpha)} - 1|^{1/2}}, \quad w = 2\alpha, \quad \alpha = \frac{2k}{1+k}. \tag{2.14}$$

Trajectory X corresponds to the following exact static solution of Einstein's equations found by Oppenheimer and Volkoff (see [111, 112]):

$$ds^2 = R^{2\alpha} dt^2 - A \, dR^2 - R^2(d\theta^2 + \sin^2\theta \, d\varphi^2), \tag{2.15}$$

where

$$A = (1 + 6k + k^2)(1 + k)^{-2}, \quad \varepsilon = 4k[\kappa_0(1 + 6k + k^2)R^2]^{-1}, \quad p = k\varepsilon.$$

The self-similar variable in the solution (2.15) is $\mu = tR^{\alpha-1}$, The relationship between the self-similar variables μ and r is given by the expression

$$r(\mu) = \frac{1}{C_1} \left| 1 - \frac{(\alpha-1)^2}{A} \mu^2 \right|^{1/(2\alpha-2)}.$$

One separatrix leaving the critical point Z_1 (and contained in the two-dimensional separatrix Z) can be integrated explicitly too. On this trajectory we have

$$Q = \frac{u}{\beta - 1 - \beta u^2}, \quad w = -\frac{2\beta u^2}{\beta - 1 - \beta u^2}, \quad u = (\beta - 1)\mu,$$

$$r(\mu) = C_2 \mu(1 - (\beta - 1)^2 \mu^2)^{1/(2\beta-2)}, \quad \beta = \frac{2}{3(1+k)}. \tag{2.16}$$

Trajectory (2.16) corresponds to the flat Friedmann solution:

$$ds^2 = dt^2 - t^{2\beta}(dR^2 + R^2(d\theta^2 + \sin^2\theta \, d\varphi^2)), \quad \varepsilon = \frac{c_0}{t^2}, \tag{2.17}$$

which is self-similar with the self-similar variable $\mu = Rt^{\beta-1}$ for all k. The relationship between the self-similar variables μ and r is given by (2.16). All other separatrices emerging from the critical point Z_1 thus describe some perturbations of the flat Friedmann solution. All these solutions can be extended to the center of symmetry and have the following asymptotic form for $r \to 0$ (where $T = \exp\tau$):

$$l^{-2} ds^2 \approx \left(1 + \frac{c_1}{2}r^2\right) dT^2$$

$$- T^2\left[\left(1 + \left(\frac{c_1}{1+3k} - 1\right)r^2\right) dr^2 + r^2(d\theta^2 + \sin^2\theta \, d\varphi^2)\right], \tag{2.18}$$

$$u \approx -\frac{1+3k}{3(1+k)}r, \quad \varepsilon \approx \frac{3c_1}{1+3k}\frac{1}{T^2}.$$

Obviously the solutions with the asymptotic form (2.18) are regular at the center of symmetry $r = 0$ as well.

V. Gravitational Radius of Matter in Self-Similar Solutions. For spherically symmetric solutions of Einstein's equations the gravitational radius function r_g is defined by the following expression [73, 108]:

$$r_g = R\left(1 + g^{ij}\frac{\partial R}{\partial x^i}\frac{\partial R}{\partial x^j}\right), \tag{2.19}$$

where the function R is the radius of the two-dimensional sphere S^2. For the conformally static metric (1.1) we have $R = re^\tau$ and the gravitational radius r_g is of the following form:

$$r_g = R(1 + \sigma e^{-\lambda}(r^2 e^{\lambda - \nu} - 1)). \tag{2.20}$$

After substituting this expression into the formulas (1.10) and (1.12) we obtain

$$\frac{r_g}{R} = \frac{w(1 + Q(k + u^2)/(1 + k)u)}{w(1 + Q(k + u^2)/(1 + k)u) + 1 - Q^2}. \tag{2.21}$$

By definition, matter is outside the gravitational radius if $r_g/R < 1$ and inside the gravitational radius if $r_g/R > 1$. According to (2.21) $r_g/R > 1$ in two cases. The first case is in the region $-1 < u < 0$ ($\sigma = +1$ and $V > 0$ (see (1.12))) for $Q > 1$ (for this it is necessary that $-1 < u < -k$). The second case is in the region $|u| > 1$ ($\sigma = -1$ and $V < 0$) for $Q < 1$.

At the critical point Z_1 ($w = u = Q = 0$) we have $r_g/R = 0$. At the critical point Z_3 and on the whole separatrix X (2.14) (i.e. for the static solution (2.15)) we obtain $r_g/R = 4k(1 + 6k + k^2)^{-1} < 1$. At the critical point Z_2 the expression (2.21) is undefined. In the invariant plane $u = -1$ this expression (2.21) is of the form $r_g/R = w/(w + 1 + Q)$. Consequently for all trajectories of the system (1.11) moving in the neighborhood of the plane $u = -1$ we have $r_g/R < 1$, i.e. in the corresponding self-similar solutions the matter is distributed outside the gravitational radius.

3. Transformation of Self-Similar Solutions in Various Coordinates

I. Self-similar spherically symmetric metrics, except for the conformally static one (1.1), can also be written in the following form:

$$ds^2 = \exp\left(v_0\left(\frac{ct}{R}\right)\right)c^2\,dt^2 - \exp\left(\lambda_0\left(\frac{ct}{R}\right)\right)dR^2 - R^2\exp\left(\mu_0\left(\frac{ct}{R}\right)\right)d\Omega^2, \tag{3.1}$$

where the functions v_0, λ_0 and μ_0 satisfy an arbitrary (non-degenerate) relation

$F(v_0, \lambda_0, \mu_0, ct/R) = 0$. The energy density ε and the 4-velocity of matter u^i are defined by the formulas

$$\varepsilon = \varepsilon_0 \frac{\bar{\varepsilon}(ct/R)}{R^2},$$

$$(u^0, u^1, u^2, u^3) = \left(\frac{\exp(-v_0/2)}{(1-v^2)^{1/2}}, v \frac{\exp(-\lambda_0/2)}{(1-v^2)^{1/2}}, 0, 0 \right),$$

(3.2)

where $cv(ct/R)$ is the three-dimensional radial velocity of matter. Self-similar solutions can be written in the form (3.1) in the co-moving system of coordinates as well (see for example [104, 105]), where instead of the relation $F = 0$ we have the relation $v = 0$. Often the metrics of self-similar solutions are written in the following form:

$$ds^2 = X\left(\frac{ct}{R}\right) c^2 dt^2 - Y\left(\frac{ct}{R}\right) dR^2 - R^2(d\theta^2 + \sin^2\theta \, d\varphi^2)$$

(3.3)

(with the relation $F = \mu_0 = 0$) and in the synchronous system of coordinates (with the relation $F = v_0 = 0$):

$$ds^2 = c^2 dt^2 - U\left(\frac{ct}{R}\right) dR^2 - R^2 V\left(\frac{ct}{R}\right)(d\theta^2 + \sin^2\theta \, d\varphi^2).$$

(3.4)

II. The metric (1.1) is mapped to the synchronous metric of the form (3.4) as follows:

$$\frac{ct}{l} = f_1(\zeta)\exp\tau, \quad \frac{R}{l} = f_2(\zeta)\exp\tau, \quad \zeta = \ln r,$$

(3.5)

where the functions f_1 and f_2 satisfy the equations

$$\frac{df_1}{d\zeta} = Q(f_1^2 - \sigma \exp(v))^{1/2},$$

$$\frac{df_2}{d\zeta} = f_1 f_2 Q(f_1^2 - \sigma \exp(v))^{-1/2}.$$

(3.6)

The coefficients of the metric (3.4), the radial velocity v and the energy density ε are of the following form:

$$U = (f_1^2 - \sigma \exp(v))f_2^{-2}, \quad V = \exp(2\zeta)f_2^{-2},$$

$$v = \frac{(f_1^2 - \sigma \exp(v))^{1/2} + uf_1}{f_1 + u(f_1^2 - \sigma \exp(v))^{1/2}}, \quad \varepsilon = \frac{\varepsilon_0 \sigma(1 - u^2)wf_2^2 \exp(-v)}{\beta Q(1+k)uR^2}.$$

(3.7)

Using the formulas (3.5)–(3.7), let us write the asymptotic forms of solutions approaching (for $\zeta \to +\infty$) the attracting critical points on the segment DE (see Sect. 2) in the synchronous system of coordinates. We obtain the following asymptotic forms:

$$f_1 \approx \exp \zeta, \quad f_2 \approx C_2 \exp \zeta, \quad \frac{R}{ct} = \frac{f_2}{f_1} \approx C_2,$$

$$U \approx V \approx C_2^{-2}, \quad \varepsilon \to \text{const}, \quad v \to \text{const.}$$
(3.8)

Thus under the mapping (3.5), as $\zeta \to \infty$ the metric (1.1) becomes the metric (3.4) defined in the region $R/ct < C_2$. Note that for $R/ct \to C_2$ the metric (3.4) is non-degenerate. Therefore the solution can be smoothly extended to the region $ct/R < C_2^{-2}$. When this happens the self-similar variable ct/R becomes time-like and the straight line $ct/R = C_2^{-1}$ becomes a light line. Such continuation of the solution is possible up to $ct/R = 0$. All parameters of the solution remain regular and new qualitative singularities of the solution do not occur. In the region $ct/R < C_2^{-1}$ the metric (3.4) corresponds to the metric (1.1) with $\sigma = -1$. The trajectory of the system (1.11) corresponding to it smoothly continues the initial trajectory to the region $u < -1$.

In the synchronous system (3.4) the metrics (1.1) corresponding to the trajectories emerging (for $\zeta \to -\infty$) from the segment of repelling critical points EF are continued analogously.

III. The metric (1.1) can be written in the form (3.3) by means of the following transformation:

$$R = l \exp(\tau + \zeta), \quad \frac{ct}{l} = \exp(\tau + \varphi(\zeta)),$$
(3.9)

where $d\varphi/d\zeta = Q^2$. At the same time the components of the metric, the velocity of matter v and the energy density ε take the following form:

$$X\left(\frac{ct}{R}\right) = \frac{\sigma \exp(v - 2\varphi)}{1 - Q^2}, \quad Y\left(\frac{ct}{R}\right) = \frac{\sigma \exp(v - 2\zeta)Q^2}{1 - Q^2},$$

$$v = \frac{Q + u}{1 + uQ}, \quad \varepsilon = -\frac{\varepsilon_0 \sigma(1 - u^2)w \exp(2\zeta - v)}{\beta Q(1 + k)uR^2}.$$
(3.10)

Note that the velocity v of motion of the shock wave (on which we have $ct/R = \text{const}$) in the system of coordinates (3.3) is of the form $v/c = (Y/X)^{1/2}$ $(R/ct) = Q$.

After the transformation (3.9) the asymptotic forms of solutions corresponding to the separatrices Z for $\zeta \to -\infty$ have the following form (for $R/ct \to 0$):

$$ds^2 \approx \left(1 + \frac{d}{2}\left(\frac{R}{ct}\right)^2\right)c^2 dt^2 - \left(1 + \frac{d}{1 + 3k}\left(\frac{R}{ct}\right)^2\right)dR^2 - R^2 d\Omega^2,$$
(3.11)

$$vc \approx \frac{2}{3(1 + k)}\frac{R}{t}, \quad \varepsilon \approx \frac{3\varepsilon_0 d}{\beta(1 + 3k)}\frac{1}{t^2}, \quad d = \text{const.}$$

Just as for the mapping (3.5), we have $R/ct \to C_1$ if $\zeta \to +\infty$. The metric (3.3) can be smoothly continued to the region $ct/R < C_1^{-1}$. However generally speaking for

$ct/R \to 0$ this metric has a non-physical singularity with the asymptotic form $X \approx (ct/R)^{\gamma}$, $Y \to$ const. This singularity can be removed by means of the change (3.5) to synchronous coordinates mentioned above.

4. The Problem of Breakdown of Equilibrium of a Star in the General Theory of Relativity

I. The statement of the problem of a stellar explosion in classical gas dynamics [7] (see also Chapt. V) is naturally carried over to the general theory of relativity. Suppose that at the initial moment of time the space-time metric and the distribution of gas inside the star are static and have the form (2.15). Assume that as a result of some perturbation a shock wave emerges from the center of symmetry at the time $t = 0$. Also assume that the motion of gas and the space-time metric are self-similar beyond the front of the shock wave. Thus the solution of the problem of the breakdown of equilibrium of a star in the general theory of relativity consists of finding self-similar solutions of Einstein's equations which by virtue of the conditions (2.2) can be glued to the static solution (2.15) at the shock wave.

As shown in the well known paper by Oppenheimer and Volkoff [111], the equation of state of matter $p = k\varepsilon$ can be applied (for $k = 1/3$) in the neighborhood of the center of a neutron star. However the density of matter at the center of a stable neutron star cannot exceed $\rho_c^0 \approx 10^{15}$ g/cm^3. This value corresponds to the Oppenheimer-Volkoff limit of the mass of a neutron star $M_{OV} \approx 1.6$ M$_\odot$, where M$_\odot$ is the mass of the sun (see [111, 73]). All equilibrium configurations with the central density $\rho_c > \rho_c^0$ are unstable [73]. As shown in [113, 112], the mass of such equilibrium configurations (with the real equation of state of matter taken into account) is a periodic function of ρ_c for $\rho_c \to \infty$ (the graph of this function (dependence of M on ρ_c) obtained in [114] is shown in Fig. 31). In particular the equilibrium configuration with $\rho_c = \infty$ which coincides with the exact solution (2.15) in the neighborhood of the center $r = 0$ has finite mass and is unstable if we take into account (in the outer layers of the star) the real equation of state of the Fermi-gas of neutrons.

Self-similar solutions considered in this section in connection with the above describe a special explosive type of breakdown of unstable equilibrium of a star. These solutions can be considered as a model of evolution of a star with the mass $M < M_{OV}$ (such a star need not collapse into a black hole) and central density $\rho_c > \rho_c^0$.

Self-similar solutions of the form (1.1) glued at the shock wave to the static solution (2.15) (by virtue of the conditions (2.2) for $k_1 = k_2$) are described by trajectories of the dynamical system (1.11) approaching the line Y (in the subsonic region $|u| < k^{1/2}$) for some $r = r_1$:

$$u = -\frac{k}{Q}, \quad w = 2\alpha, \quad \alpha = \frac{2k}{1+k}. \tag{4.1}$$

After the transformation (2.2) $(w \to w, Q \to Q, u_1 \to k/u_2)$ the line Y becomes the trajectory X (2.14) corresponding to the static solution (2.15). The position of the front of the shock wave is determined by the condition $r = r_1$. The shock wave diverges. Indeed the radius of the shock wave $R = r_1 \exp(\tau)$ tends to infinity for $\tau \to \infty$. For $r > r_1$ the solution is described by a segment of the trajectory X and is static. A natural parameter on the line Y is the Mach number $M = Qk^{-1/2}$ of motion of the shock wave $(M = v/(ck^{1/2}))$ in the system of coordinates (3.3).

II. In the three-dimensional manifold S the trajectory X, the line Y and the line of critical points I all intersect in one point Y_1 $(Q = -u = k^{1/2}, w = 2, M = 1)$ (the intersection of these three lines in one point is not a degeneracy and in fact must occur, as follows from their definition). Let us analyze the behavior of trajectories of the dynamical system (1.11) in the neighborhood of the critical point Y_1. The eigenvalues (2.11) at the critical point Y_1 are

$$\lambda_+^0 = -\frac{2k^{1/2}(1-k)^2}{1+k}, \quad \lambda_-^0 = -\frac{4k^{3/2}(1-k)}{1+k}, \quad \lambda_3 = 0. \tag{4.2}$$

The corresponding eigenvectors l_+, l_-, l_3 (the vector l_3 is tangent to the line of critical points I) and l_4 (the vector l_4 is tangent to the line Y at the point Y_1) in the coordinates α, u, β $(\alpha = 1/Q, \beta = ((1-k)w/2 - 2k)/Q$ (see (2.10))) have the following coordinates:

$$l_+ : \left(\frac{1}{k}, 1, -\frac{4k}{1-k}\right),$$

$$l_- : \left(\frac{1-k}{2k^2}, 1, -2\frac{1-k}{1+k}\right), \tag{4.3}$$

$$l_3 : (1, 0, 0),$$

$$l_4 : \left(-\frac{1}{k}, 1, \frac{4k^2}{1+k}\right).$$

In the neighborhood of the critical point Y_1 there are two surfaces L_1 and L_2 invariant relative to the system (1.11), formed by the separatrices of critical points in I corresponding to the eigenvalues λ_+ and λ_-. At the point Y_1 the surface L_1 is tangent to the two-dimensional plane $\mathscr{L}_+$ generated by the vectors l_+ and l_3 and the surface L_2 is tangent to the two-dimensional plane $\mathscr{L}_-$ generated by vectors l_- and l_3.

The planes $\mathscr{L}_+$ and $\mathscr{L}_-$ intersect the u, β plane at right angles, because the vector l_3 is perpendicular to this plane. Therefore the relative position of the surfaces L_1, L_2 and the line Y is determined by the position of the projections of the vectors l_+, l_- and l_4 to the u, β plane. Their position is shown in Fig. 24. Here

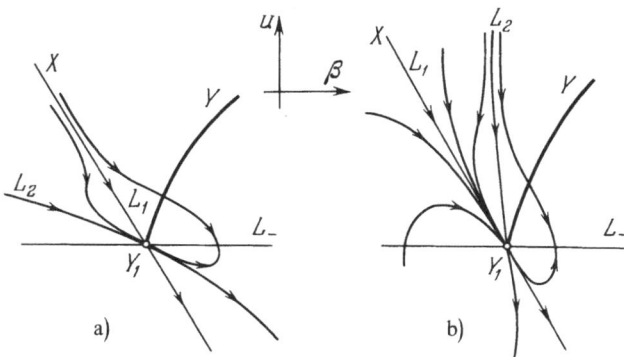

Fig. 24. Qualitative behavior of the trajectories of the dynamical system (1.11) in the neighborhood of the critical point Y_1 (projected to the u, β plane) for $0 \leqslant k \leqslant 1/3$ (a) and $1/3 < k < 1$ (b)

the cases $k < 1/3$ and $k > 1/3$ are qualitatively different, because for $k < 1/3$ we have $4k/(1 + k) < 2(1 - k)(1 + k)$ and vice versa (see (4.3)).

The trajectory X is a separatrix approaching the critical point Y_1 which corresponds to the eigenvalue λ_+^0. Therefore it belongs to the surface L_1. For $k < 1/3$ we have $\lambda_+^0 < \lambda_-^0$. Consequently all trajectories in the subsonic region $|u| < k^{1/2}$ (except for the trajectories which lie in the surface L_1) approach the attracting critical points of the line I, where they are tangent to the surface L_2. In particular all trajectories in the neighborhood of the trajectory X diverge from the surface L_1 as they approach the critical point Y_1 for $k < 1/3$ and end up in a small neighborhood of the surface L_2. Therefore due to the position of the line Y shown in Fig. 24, some segment $Y_1 Y(M_0)$ of the line Y $(M_0 > 1)$ is completely intersected by trajectories which earlier were in the neighborhood of the trajectory X.

As shown in Sect. 2, there is a two-dimensional surface invariant relative to the system (1.11) in the neighborhood of the trajectory X which is wound around it infinitely many times. This surface is the separatrix Z of the critical point Z_1. It intersects the line Y at an infinite number of points $Y(M_i)$, $M_i \to 1$ for $i \to \infty$. Thus for $k < 1/3$ there is an infinite number of solutions of the problem of breakdown of stellar equilibrium. These solutions are separatrices of the critical point Z_1 and therefore can be continued without singularities to the center of symmetry. Just like the analogous solutions in classical gas dynamics for $\gamma < \gamma_1 < 4/3$, $\omega < 5/2$ (see Chapt. V) they describe the breakdown of equilibrium of a star without the emission of energy.

All solutions corresponding to Mach numbers $M: M_i > M > M_{i+1}$ are described by trajectories emerging from the critical points of the line l_1. As shown in Sect. 2, these solutions have an expanding cavity $(r < r_1)$ inside the gas formed as a result of the expulsion of gas from the inside by a spherical piston.

Note that each solution of the problem of breakdown of equilibrium of a star corresponding to a Mach number M of motion of the shock wave (in the system of

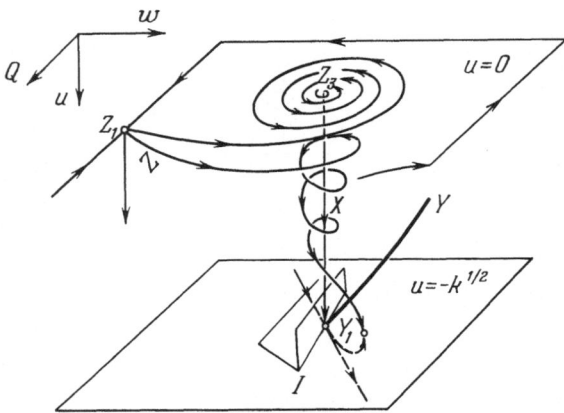

Fig. 25. To the problem of breakdown of equilibrium of a star in the general theory of relativity

coordinates (3.3)) makes some finite number $N(M)$ of turns around the trajectory X (see Fig. 25). If $M \to 1$, then $N(M) \to \infty$. The radial velocity of gas v in the system of coordinates (3.3) is of the form (3.10): $v = (Q + u)/(1 + uQ)$. Since we have $Q + u = 0$ on the trajectory X, in the considered solutions the velocity of gas v has $2N(M)$ zeroes. Consequently in these solutions the gas makes a finite number $N(M)$ of radial oscillations (with respect to the coordinate R (3.3) which has an obvious physical meaning), so that $N(M) \to \infty$ for $M \to 1$. In each interval $M_i > M \geqslant M_{i+1}$ the number of oscillations $N(M)$ is constant and in two adjacent intervals increases by 1. In particular $N(M_{i+1}) = N(M_i) + 1$. After the radial pulsations of gas stop there is a monotone motion of gas away from the center with the asymptotic form (2.18)–(3.11) for the solutions with $M = M_i$ and with some asymptotically (for $r \to r_1$) constant speed in the system of coordinates (3.3) for the solutions with $M \neq M_i$.

    III. For all k the critical points contained in the line I in the neighborhood of the point Y_1 are attracting (see (4.2)). If a trajectory not lying in the surface L_1 approaches a critical point in the line I for $k < 1/3$, then it is tangent to the surface L_2, because $\lambda_+^0 < \lambda_-^0$. For $k > 1/3$ all trajectories except those which are contained in the surface L_2 are tangent to the surface L_2, because $\lambda_+^0 > \lambda_-^0$ (see Fig. 24). For each critical point contained in the line I in the neighborhood of the point Y_1 there exists a two-dimensional invariant surface $\mathscr{P}$ formed by the separatrices approaching this critical point.

    The two-dimensional surface $\mathscr{P}(Y_1)$ intersects the two-dimensional separatrix Z wound around the trajectory X an infinite number of times in an infinite number of trajectories R_i. After entering the critical point Y_1 these trajectories can be continued by the trajectory X. The solutions thus obtained can be continued to the center of symmetry and describe the breakdown of equilibrium of a star without the emission of energy and without a shock wave. There is a weak discontinuity for these solutions at the surface where the self-similar solution is glued to the static one. All physical parameters remain continuous, but their derivatives do not. For $k \leqslant 1/3$ the first derivatives already have a discontinuity (see Fig. 24a). For $k > 1/3$ the first derivatives remain continuous (see Fig. 24b), but the second derivatives do not.

All other trajectories on the surface $\mathscr{P}(Y_1)$ emerge from the line of critical points l_1 and correspond to solutions with an expanding cavity inside the gas. They have an analogous weak discontinuity. All conclusions drawn above pertaining to the finite number of radial oscillations of the gas in the system of coordinates (3.3) also hold for the newly obtained solutions with weak discontinuities.

Note that for all $k \, (0 < k < 1)$ the two-dimensional separatrix Z intersects the surface L_1 at an infinite number of trajectories which can be smoothly extended through the line of critical points I. The corresponding solutions are some smooth perturbations of the static solution (2.15). They do not have a singularity at the center of symmetry and entail an arbitrary finite number of radial oscillations of the gas.

5. Self-Similar Solutions with Expanding and Converging Shock Waves

The previous analysis of the dynamical system (1.11) shows that in the subsonic region of the manifold $S \, (0 > u > -k^{1/2})$ the trajectories of this system fall into the following four categories:

A) trajectories emerging from the repelling critical points in l_1,
B) trajectories which form the two-dimensional separatrix Z of the critical point Z_1 (in particular the separatrices $\mathscr{L}_2$ and X of the critical points Z_2 and Z_3),
C) trajectories which form the two-dimensional separatrix L'_1 leaving the segment of critical points I_1,
D) trajectories leaving the surface V^1_-.

Apparently in the manifold S there are no trajectories winding onto some invariant subsets, so each trajectory in the subsonic region belongs to one of the above categories. Among the trajectories of the above four types we shall call those trajectories "of type E" which approach the segments of critical points I_1 and I_2 (and thus form the separatrices L'_2 and L (see Sect. 2)).

I. Trajectories of types A and B were analyzed in section 4, where we listed their applications in the solution of the problem of breakdown of equilibrium of a star. These trajectories can also be used to analyze the propagation of an expanding shock wave in a moving stream of gas (in the system of coordinates (3.3)). In this case the discontinuity on the trajectory is introduced for $r = r_0$ at some point in the subsonic region of the manifold S not contained in the line Y

(4.1). In the subsonic region $(|u| < k^{1/2})$ of the solution we have $r < r_0$. Consequently the velocity of gas u $(u < 0)$ is directed towards the interior of the subsonic region, where according to (2.2) we have $\bar{\varepsilon}_1 > \bar{\varepsilon}_2$. In other words the shock wave is a wave of compression. The radius of the shock wave is equal to $R_0 = 1\exp(\tau)r_0$. Therefore $R_0 \to \infty$ for $\tau \to \infty$, i.e. the shock wave is expanding.

In the case of a strong discontinuity (i.e. for $u_1 \approx -k$) we have $u_2 \approx -1$ for the supersonic solution glued to the given solution at the discontinuity. Therefore as observed in Sect. 2, the corresponding trajectory moves along the component of the boundary Γ_2 $(u = -1)$ until it approaches some attracting critical point in the segment DE as $r \to \infty$. Meanwhile the self-similar variable $\zeta = \ln r$ becomes time-like and as shown in Sect. 3 the solution continues in the synchronous system (3.4) in the region $ct/R \geq 0$. The discontinuity in the solution occurs for a constant value of $\zeta = \ln r_0$, i.e. the shock wave corresponds to $ct/R = \text{const}$. Consequently in the synchronous system (3.4) the shock wave moves with some constant velocity v_0.

The formation of the shock wave occurs at the center of symmetry $R = 0$ for $t = 0$. At this moment the three-dimensional metric (3.4) $U(0)\,dR^2 + R^2V(0)\,d\Omega^2$ has a conical singularity at zero if $U(0) \neq V(0)$. At infinity the metric is obviously flat. The velocity of gas $v(0)$ is constant in space and the energy density is $\varepsilon = c_0/R^2$ (such a distribution of energy density in a bounded region can occur for example as a result of constant leakage of matter out of the star).

Trajectories of type A describe solutions for which the gas is pushed out from the inside by a spherical piston (see Sect. 2). Solutions corresponding to trajectories of type B have the asymptotic form (3.11) for $R/ct \to 0$ in the system of coordinates (3.3). From this we obtain a corollary:
1) solutions of type B with an expanding shock wave can be continued to the center of symmetry $R = 0$ such that after the shock wave leaves the center of symmetry $(t > 0)$ the metric has no singularity there,
2) in the neighborhood of the center of symmetry the gas moves away from the center (with a linear asymptotic form for the velocity (3.11) for $R \to 0$),
3) the energy density near the center is approximately constant in R and decreases like c_0/t^2.

II. As the parameter ζ decreases, the trajectories of type C are continuously extended through the segment I_1 into the supersonic region (see Fig. 23). After such single-valued continuation the trajectories of type C (separatrices L'_1) passing through the segment I_1 in the neighborhood of the point i_0 are approximated by a sequence of separatrices:

$$\overrightarrow{FG},\ \overrightarrow{Gi_0},\ \overrightarrow{i_0C},\ \overrightarrow{CZ_1},\ Z \quad (\text{see Fig. 21 and 22, } k = 5).$$

Thus as ζ increases, the trajectories of type C move in the neighborhood of the separatrix Z and therefore (just like the trajectories of type B) correspond to solutions with an expanding shock wave. After the introduction of a discontinuity these trajectories continue in the synchronous system of coordinates (3.4) in the region $V_1: C_1 \geq ct_1/R_1 \geq 0$ (where the self-similar variable ζ is time-like).

However the behavior of solutions of type C in the region beyond the shock wave is substantially different from the behavior of solutions of type B. From the above separatrix approximation it follows that after a smooth continuation through the points of the segment I_1 into the supersonic region, the solutions of type C approach the attracting (for this direction of ζ) critical points of the segment EF for $\zeta \to -\infty$. The self-similar variable ζ becomes light-like again and (as shown in Sect. 3) the solution smoothly continues for the time-like variable ζ in the static system of coordinates (3.4) in the new region V_2: $C_2 \geqslant ct_2/R_2 \geqslant 0$.

An entire solution of type C is glued from the regions V_2, V_{2c}, V_g, V_{1c} and V_1 bordering with each other in the above order. The subsonic region V_g is described by the motion of the trajectory along the separatrices $i_0 \overrightarrow{C}$, $C\overrightarrow{Z_1}$ and Z to the point of discontinuity (see Fig. 21 and 22, $k = 5$). The region V_{2c} is described by the motion of the trajectory along the separatrices $\overrightarrow{FG}$ and $\overrightarrow{Gi_0}$. The region V_{1c} is glued to the region V_g at the shock wave and is described by the motion of the trajectory in the supersonic region of the manifold S from the point of discontinuity to the attracting segment DE.

Regions V_1 and V_2 are bounded by the light-like surfaces $ct_1/R_1 = C_1$ and $ct_2/R_2 = C_2$ and therefore cannot communicate by physical signals. The shock wave moves along the joint boundary of the regions V_g and V_{1c}. The formation of the shock wave occurs at the center of symmetry just as in solutions of type B. However for $t > 0$ a spatial section has a "throat" and topologically is the product of a two-dimensional sphere and a line just as in the well known Kruskal solution [89].

III. Trajectories of type D are defined for $r > r_1 > 0$. Therefore in order to continue the corresponding solutions for all $r > 0$ it is necessary to introduce a discontinuity which in this case for the positive direction of τ corresponds to the shock wave of evacuation. However the shock waves of evacuation in matter with normal properties are impossible. Therefore solutions of type D can have physical meaning only after a change of parameters $t \to -t$, $u \to -u$. As a result of this change the discontinuity corresponds to a shock wave of compression (see below). Such a solution will be defined for all r (with no additional discontinuities of evacuation) only if the corresponding trajectory continues through the line of critical points I, i.e. belongs to category E.

Some trajectories of type E were already used above (see Sect. 4) as solutions with a weak discontinuity for the problem of breakdown of equilibrium of a star. Let us point out some applications of other trajectories of type E which form the separatrix L_2' and pass through the segment of critical points I in the neighborhood of the point i_0. As the parameter ζ decreases, such trajectories are approximated by the stable sequence of separatrices $i_0\overrightarrow{B}$, $B\overrightarrow{A}$, $A\overrightarrow{l_1}$ and reach the attracting (for a decreasing ζ) critical points of the segment l_1. As ζ increases, the separatrices L_2' move along the separatrices $i_0\overrightarrow{H}$, $H\overrightarrow{D}$ and consequently reach the segment of attracting critical points DE. After the introduction of a discontinuity in the neighborhood of the separatrix $B\overrightarrow{i_0}$ the separatrices L_2' determine solutions with a collapsing shock wave.

Indeed the solutions of Einstein's equations of the type (1.1) are mapped to other solutions under the transformation $\tau_1 = -\tau, u_1(r) = -u(r)$. All other functions of r do not change. Under this transformation the metric (1.1) takes the following form:

$$ds^2 = l^2 \exp(-2\tau_1)[\exp(v(r))\, d\tau_1^2 - \exp(\lambda(r))\, dr^2 - r^2\, d\Omega^2].$$

Suppose that in the above trajectories a discontinuity is introduced for $r = r_0$ (a discontinuity is necessary because these trajectories emerge from l_1 and L_-^1). In the region beyond the shock wave, where the flow of gas is subsonic, we have $r > r_0$. The velocity of gas $u_1(r) = -u(r) > 0$ is directed towards the interior of this region. Therefore this shock wave is a wave of compression. The radius of the shock wave is $R_0 = 1\exp(-\tau_1)r_0$. Consequently $R_0 \to 0$ for $\tau_1 \to +\infty$, i.e. the shock wave is converging.

An entire solution of type E is glued from the regions V_2, V_{2c}, V_g, V_{1c} and V_1. The region V_g is described by the motion of the trajectory (as the variable ζ increases) in the subsonic part of the manifold S from the discontinuity to the arrival at the segment of critical points I_2. The region V_{1c} is described by the motion of the trajectory in the supersonic part of the manifold S from the segment I_2 to the segment of attracting critical points DE. Later (as shown in Sect. 3) the solution continues in the region V_1, where the variable ζ is time-like. In the synchronous system (3.4) the region V_1 is determined by the condition $-C_1 \leqslant ct_1/R_1 \leqslant 0$. The region $V_g + V_{1c}$ is determined by $-C_0 \leqslant ct_1/R_1 \leqslant -C_1$. The equation of the shock wave is $ct_1/R_1 = -C_0$. The shock wave collapses with some constant speed to the center of symmetry for $t_1 = 0$.

The region V_{2c} is glued to the region V_g at the shock wave and is described (as the variable ζ decreases) by the movement of the trajectory in the supersonic part of the manifold S from the discontinuity to the segment EF. Then the solution enters the region V_2, where the variable ζ is again time-like. In the static system of coordinates the region V_2 is determined by the condition $-C_2 \leqslant ct_2/R_2 \leqslant 0$. Since the surfaces $ct_1/R_1 = -C_1$ and $ct_2/R_2 = -C_2$ are light-like, the regions V_1 and V_2 do not communicate by physical signals. The spatial sections in these solutions topologically are products of a two-dimensional sphere and a line just like in the Kruskal solution (see Sect. 2 of Chapt. II) and have a "throat" which contracts to a point as the shock wave collapses to the center. However for the considered solutions by virtue of (2.21) we have $r_g/R < 1$, i.e. in solutions of type E the matter is located outside its gravitational radius.

Along the trajectories of motion of the gas the derivative of the function $R = re^{-\tau}$ (see (5.1)) has the following form:

$$R' = \frac{\partial R}{\partial x^i} u^i = -\frac{e^{-\lambda/2}}{(\sigma(1 - u^2))^{1/2}}(Q + u).$$

In the two outer regions V_1 and V_2 described by the movement of trajectories of the system (1.11) in the region $u < -1$, $Q + u > 0$, we have $R' < 0$. Therefore a solution of type E describes the collapse of matter (in the presence of pressure) accompanied by the formation of a converging shock wave.

Chapter V
Self-Similar Motion of Self-Gravitating Gas in Stars

In the Newtonian theory the self-similar spherically symmetric motion of an ideal self-gravitating gas with shock waves was first considered in [7, 116] as a model for supernova explosions. In particular, [7] contained an analysis of the conservation laws for self-similar solutions and some solutions were analyzed numerically. Numerical methods were the principal tool used in [116]. Subsequent efforts [117–123] made the results of [7, 116] more precise and further developed the methods used in these papers. Self-similar motion of an ideal gas in the gravitational field of an attracting center was considered in [124–127].

In this chapter the model of stellar explosions [7, 116] is analyzed by means of the methods of qualitative theory of multi-dimensional dynamical systems. We also qualitatively analyze the model of explosions in stellar envelopes. We obtain all possible asymptotic forms for self-similar accretion of self-gravitating gas to the center and construct several solutions with converging shock waves.

1. Resolution of Singularities of the Dynamical System

I. The Equations of Gas Dynamics for Self-Similar Solutions. The principal physical parameters for the spherically symmetric motion of an ideal gas in the Newtonian theory are the radial velocity $v(r, t)$, the density of gas $\rho(r, t)$ and the pressure of gas $p(r, t)$, where r is the radial coordinate and t is the time. The temperature of an ideal gas T is determined by the equation of state

$$p = \frac{R_0}{\mu} \rho T, \tag{1.1}$$

where R_0 is the gas constant and μ is the mass of one mole of gas. The entropy S of a unit mass of an ideal gas is determined by $S = C_V \ln(p/\rho^\gamma)$, where $\gamma = C_p/C_V > 1$ and C_p and C_V are the specific heats of the gas under constant pressure and volume. The motion of the gas with no infusion of energy from the outside is called adiabatic. In this case $dS/dt = 0$ and γ is called the adiabatic constant.

In the presence of Newtonian gravitation of the gas, particles having mass m located at a distance r from the center of symmetry are attracted by a gravitational force given by the following formula:

$$F = -G \frac{m \cdot \mathcal{M}(r, t)}{r^2}, \tag{1.2}$$

where $\mathcal{M}(r,t)$ is the entire mass of gas inside a sphere of radius r at the time t and G is the gravitational constant. The dimensions of this constant G are $[G] = M^3 L^{-1} T^{-2} = \mathrm{cm}^3 \mathrm{g}^{-1} \mathrm{sec}^{-2}$. By definition we have

$$\frac{\partial \mathcal{M}}{\partial r} = 4\pi r^2 \rho. \tag{1.3}$$

In the spherically symmetric case the equations for the adiabatic motion of an ideal self-gravitating gas expressing the laws of conservation of mass, momentum and energy have the following form:

$$\frac{\partial \rho}{\partial t} + \frac{\partial \rho v}{\partial r} + \frac{2\rho v}{r} = 0, \tag{1.4}$$

$$\frac{\partial v}{\partial t} + v\frac{\partial v}{\partial r} + \frac{1}{\rho}\frac{\partial p}{\partial r} + G\frac{\mathcal{M}}{r^2} = 0, \tag{1.5}$$

$$\frac{\partial (p/\rho^\gamma)}{\partial t} + v\frac{\partial (p/\rho^\gamma)}{\partial r} = 0. \tag{1.6}$$

The equations of gas dynamics (1.3)–(1.6) are invariant relative to the action of the three-dimensional group $\mathbb{R}^3$ of scale transformations which correspond to a stretching of the units of physical quantities:

$$r' = lr, \quad t' = \tau t, \quad \mathcal{M}'(r',t') = \mu \mathcal{M}(r,t),$$

$$v'(r',t') = \frac{l}{\tau} v(r,t), \quad \rho'(r',t') = \frac{\mu}{l^3} \rho(r,t), \tag{1.7}$$

$$p'(r',t') = \frac{\mu}{\tau^2 l} p(r,t), \quad G' = \frac{l^3}{\tau^2 \mu} G.$$

By definition the self-similar solutions of the equations of gas dynamics are those solutions which are invariant relative to some one-parameter subgroup $\mathbb{R}^1$ of the scale transformations which preserve the gravitational constant G. All such subgroups $\mathbb{R}^1$ are determined by one parameter ω such that

$$l = \tau^{2/\omega}, \quad \mu = \tau^{2(3/\omega - 1)}, \quad \tau \in \mathbb{R}^1_+.$$

The corresponding self-similar solutions have the following form [7]:

$$\rho = \frac{1}{Gt^2} R(\lambda), \quad p = \frac{r^2}{Gt^4} P(\lambda),$$

$$\mathcal{M} = \frac{r^3}{Gt^2} m(\lambda), \quad v = \frac{r}{t} V(\lambda), \tag{1.8}$$

where λ is a dimensionless self-similar variable:

$$\lambda = \frac{r}{(AGt^2)^{1/\omega}} \tag{1.9}$$

and the constant A has the dimensions of $M \cdot L^{\omega-3}$. Self-similar solutions depend on two dimensionless constants: the adiabatic constant $\gamma > 1$ and the self-similarity constant ω (since p and $\mathcal{M}$ are positive in an equilibrium distribution of gas, it follows that $1 < \omega < 3$ (see [7] and section 5 below)).

After substituting the expressions (1.8) and using the new variable $z(\lambda) = \gamma P/R$ instead of $P(\lambda)$, the equations of gas dynamics (1.3), (1.4), (1.5) and (1.6) are transformed into the following system of ordinary differential equations [7]:

$$\lambda\left[V' - \left(\frac{2}{\omega} - V\right)\frac{R'}{R}\right] = 2 - 3V,$$

$$\lambda m' = -3m + 4\pi R,$$

$$\lambda\left[\left(\frac{2}{\omega} - V\right)V' - \frac{z}{\gamma}\left(\frac{z'}{z} + \frac{R'}{R}\right)\right] = m + V^2 - V + \frac{2z}{\gamma}, \qquad (1.10)$$

$$\lambda\left(V - \frac{2}{\omega}\right)\left[\frac{z'}{z} - (\gamma - 1)\frac{R'}{R}\right] = -2(V + \gamma - 2),$$

where prime denotes derivatives with respect to λ. As shown in [7] the system of differential equations has two integrals whose existence can be verified by direct substitution. The first integral is the integral of mass:

$$\lambda^3[(1 - 3/\omega)m - 2\pi R(V - 2/\omega)] = C_1 \qquad (1.11)$$

The second integral is the adiabatic integral:

$$zR^{1-\gamma}m^{(\gamma+(2/\omega)-2)/(1-(3/\omega))}\lambda^{(3\gamma-4)/(1-(3/\omega))} = C_2. \qquad (1.12)$$

In this chapter self-similar solutions are considered subject to $C_1 = 0$. Physically this is the most important case of the absence of a source of matter at the center of symmetry [7]. In this case the integral (1.11) gives the following relation:

$$R = \frac{1 - 3/\omega}{2\pi} \cdot \frac{m}{V - 2/\omega}. \qquad (1.13)$$

After the substitution of this relation (1.13) and a change to a new variable $\tau = \ln \lambda$ the system of equations of gas dynamics for self-similar solutions (the system (1.10)) becomes the following three-dimensional dynamical system:

$$\frac{dz}{d\tau} = z' = \frac{z}{\gamma(V - 2/\omega)(z - (V - 2/\omega)^2)}[z(2\gamma - 4(1 - 1/\omega)(\gamma - 1) - 2\gamma V)$$

$$- \gamma(2 + (1 - 3\gamma)V)(V - 2/\omega)^2 - \gamma(\gamma - 1)(m + V^2 - V)(V - 2/\omega)],$$

$$V' = \frac{z(4 - 3\gamma V - 4/\omega) + \gamma(m + V^2 - V)(V - 2/\omega)}{\gamma(z - (V - 2/\omega)^2)}, \qquad (1.14)$$

$$m' = m\frac{2 - 3V}{V - 2/\omega}.$$

The first equation of the system (1.14) can also be written in the following form:

$$z' = \frac{z}{V - 2/\omega}[-(\gamma - 1)V' + 2 + (1 - 3\gamma)V]. \tag{1.15}$$

The dynamical system (1.14) is considered in the region S_1 determined by the following natural conditions:

$$z > 0, \quad m > 0, \quad V - 2/\omega < 0 \tag{1.16}$$

where by virtue of (1.13) the last condition implies that $\rho > 0$ if $\omega < 3$.

The surface $L = z - (V - 2/\omega)^2 = 0$ is a surface of non-extendability of the trajectories of the system (1.14), because on the two sides of this surface the vector field of the system (1.14) has opposite directions. As mentioned in [7], from a formal point of view the existence of such a surface leads to the generation of a shock wave in some self-similar solutions. The law of propagation of the shock wave is determined by the condition $\lambda = \lambda_*$.

With the help of self-similar solutions of the type (1.8) we can construct models of such important astrophysical phenomena as the expansion of self-gravitating gas of a star as a result of an explosion, the collapse of a self-gravitating gas to the center, the contraction of a self-gravitating gas with a converging shock wave and so on. In all these models the motion of gas is described by some trajectories of the three-dimensional dynamical system (1.14). The analysis of self-similar solutions by means of numerical methods and the methods of search of exact solutions based on finding the first integrals conducted in [7, 116] allowed us to obtain important information about the behavior of solutions in the models of stellar explosions only for special values of the parameters γ and ω. In order to study the self-similar motion of a self-gravitating gas in greater detail, it is necessary to conduct a complete analysis of the dynamical system (1.14) using the methods of qualitative theory of multi-dimensional dynamical systems outlined in Chap. I. Such analysis is done in this section and in Sects. 2 and 3.

II. Construction of the Closed Manifold S. To completely analyze the behavior of trajectories of the system (1.14) in the region S_1 (see (1.16)) we transform this system into a system defined on some closed three-dimensional manifold S with boundary Γ. Note that the boundary Γ is not a smooth manifold. It consists of several components which intersect along the corners of the boundary. The above transformation includes a resolution of complex singularities of the system (1.14) (e.g. the critical point $z = m = 0$, $V = 2/\omega$). The next task is to write down the successive changes of the coordinates which will aid us in the construction of the closed manifold S including changing the variable τ to variables τ_i for which the dynamical system (1.14) is extended to the corresponding components of the boundary Γ.

Obviously after a change of variable $d\tau_3/d\tau = -(V - 2/\omega)$ the dynamical system (1.14) in the region S_1 (1.16) is continued to the three coordinate planes which are the three components of the boundary Γ of the manifold S: Γ_6 $(V - 2/$

$\omega = 0$), Γ_7 ($m = 0$) and Γ_8 ($z = 0$). These three components of the boundary are invariant manifolds of the dynamical system (1.14). In the region S_1 it is convenient to use the following coordinates:

$$u = z^{1/2} > 0, \quad v_0 = V - 2/\omega < 0,$$

$$m_0 = -(m(V - 2/\omega))^{1/3} > 0. \tag{1.17}$$

In order to study the behavior of trajectories of the dynamical system (1.14) for large values of the coordinates u, v_0 and m_0 it is necessary to complete the region S_1 by a boundary at infinity with respect to the coordinates u, v_0 and m_0. Such a completion is achieved by passing over to projective coordinates:

$$v_1 = \frac{v_0}{u}, \quad \rho_0 = \frac{1}{u}, \quad m_1 = \frac{m_0}{u}, \quad \frac{d\tau_1}{d\tau} = -\frac{u}{v_0}, \tag{1.18}$$

$$v_0 = \frac{v_0}{m_0}, \quad u_1 = \frac{u}{m_0}, \quad \mu = \frac{1}{m_0}, \quad \frac{d\tau_2}{d\tau} = -\frac{m_0}{v_0}, \tag{1.19}$$

$$\eta = \frac{1}{v_0}, \quad u_2 = \frac{u}{v_0}, \quad m_4 = -\frac{m_0}{v_0}, \quad \tau. \tag{1.20}$$

The infinity with respect to the coordinates u, v_0 and m_0 in the coordinates (1.18) corresponds to the plane $\rho_0 = 0$ which is the component of the boundary Γ_1 of the manifold S. In the coordinates (1.18)–(1.20) and corresponding variables τ_i the dynamical system (1.14) is continued smoothly to the component of the boundary Γ_1 which is an invariant manifold of the dynamical system.

The dynamical system (1.14) in the region S_1 completed by the boundary Γ_1 has several degenerate critical points (a critical point is called degenerate if some eigenvalue $(\lambda_1, \lambda_2, \lambda_3)$ at the point is zero). The first point O_1 ($v_1 = m_1 = \rho_0 = 0$) belongs to the component of the boundary Γ_1. The second point is O_2 ($u = v_0 = m_0 = 0$). In fact there is a line of degenerate critical points:

$$l: u = v_0 = 0, \quad 0 \leqslant m_0 \leqslant \infty.$$

To resolve the critical point O_1 we make the following transformation of coordinates:

$$v_0 = \frac{v_1}{\rho_0}, \quad m_2 = \frac{m_1^3}{\rho_0} = -\frac{v_0 m}{u^2}, \quad \rho_0 = \frac{1}{u}, \quad \frac{d\tau_3}{d\tau} = -\frac{1}{v_0}, \tag{1.21}$$

$$v_1 = \frac{v_0}{u}, \quad m_3 = \frac{m_1^3}{v_1} = \frac{m}{u^2}, \quad \rho_2 = -\frac{\rho_0}{v_1} = -\frac{1}{v_0}, \quad \tau, \tag{1.22}$$

$$v_3 = \frac{v_1}{m_1^3} = -\frac{u^2}{m}, \quad M_1 = m_1^3, \quad \rho_3 = \frac{\rho}{m_1^3} = -\frac{u^2}{v_0 m}, \quad \frac{d\tau_4}{d\tau} = \frac{m}{u^2}. \tag{1.23}$$

In the coordinates (1.21)–(1.23) instead of the point O_1 we have a two-dimensional plane ($v_1 = 0$ and the coordinates $m_3 \geqslant 0$ and $\rho_2 \geqslant 0$ vary) which is the component of the boundary Γ_2 of the manifold S. In the coordinates (1.21)–(1.23) and corresponding variables τ_i the dynamical system (1.14) is smoothly continued to the component of the boundary Γ_2 which is an invariant manifold of the dynamical system. All critical points of the dynamical system which belong to the component of the boundary Γ_2 are non-degenerate. This means that the transformations (1.21)–(1.23) fully resolve the critical point O_1.

In order to resolve the degenerate critical point O_2 $(u = v_0 = m_0 = 0)$ we make the following transformations of the coordinates:

$$v_1 = \frac{v_0}{u}, \quad u, \quad M_2 = \frac{M_0}{u} = -\frac{v_0 m}{u}, \quad \frac{d\tau_5}{d\tau} = -\frac{1}{uv_0}, \tag{1.24}$$

$$v_0, \quad u_2 = -\frac{u}{v_0}, \quad m = -\frac{M_0}{v_0}, \quad \frac{d\tau_6}{d\tau} = \frac{1}{v_0^2}, \tag{1.25}$$

$$v_5 = \frac{v_0}{M_0} = -\frac{1}{m}, \quad u_3 = \frac{u}{M_0} = -\frac{u}{v_0 m}, \quad M_0, \quad \frac{d\tau_7}{d\tau} = \frac{1}{v_0^2 m}. \tag{1.26}$$

where

$$M_0 = m_0^3 = -m(V - 2/\omega).$$

In the coordinates (1.24)–(1.26) instead of the critical point O_2 we have a two-dimensional plane $(M_0 = 0$ and the coordinates $v_5 \leqslant 0$ and $u_3 \geqslant 0$ vary$)$ which is the component of the boundary Γ_3 $(M_0 = 0)$ of the manifold S. Just like Γ_1 and Γ_2, this component of the boundary Γ_3 is an invariant manifold of the dynamical system (1.14) in the manifold S. Along with non-degenerate critical points the dynamical system has one degenerate critical point O_3 $(v_1 = u = M_2 = 0)$. To resolve this critical point we make the following transformations of the coordinates:

$$v_4 = \frac{v_1}{u} = \frac{v_0}{u^2}, \quad u, \quad m_2 = \frac{M_2}{u} = -\frac{v_0 m}{u^2}, \quad \tau_3, \tag{1.27}$$

$$v_1 = \frac{v_0}{u}, \quad u_4 = -\frac{u}{v_1} = -\frac{u^2}{v_0}, \quad m = -\frac{M_2}{v_1}, \quad \frac{d\tau_8}{d\tau} = u^{-2}, \tag{1.28}$$

$$v_5 = \frac{v_1}{M_2} = -\frac{1}{m}, \quad u_5 = \frac{u}{M_2} = -\frac{u^2}{v_0 m}, \quad M_2 = -\frac{v_0 m}{u}, \quad \tau_4. \tag{1.29}$$

In the coordinates (1.27)–(1.29 instead of the critical point O_3 we have a two-dimensional plane $(M_2 = 0$ and the coordinates $v_5 \leqslant 0$ and $u_5 \geqslant 0$ vary$)$ which is the component of the boundary Γ_4 $(M_2 = 0)$ of the manifold S. Just like Γ_1, Γ_2 and Γ_3, this component of the boundary is an invariant manifold of the dynamical system in the manifold S. All critical points of the dynamical system contained in the component of the boundary Γ_4 are non-degenerate. This means that the transformations (1.24)–(1.29) fully resolve the degenerate critical point O_2.

In order to resolve the line of degenerate critical points

$$l(u = u_3 = v_5 = v_0 = 0, \quad 0 \leqslant M_0 \leqslant \infty)$$

in the neighborhood of $M_0 = 0$ we introduce the following coordinates (see (1.26)):

$$v_5 = -\frac{1}{m}, \quad u_2 = -\frac{u_3}{v_5} = -\frac{u}{v_0}, \quad M_0, \quad \frac{d\tau_0}{d\tau} = \frac{m}{v_0^2}, \tag{1.30}$$

$$v_6 = \frac{v_5}{u_3} = \frac{v_0}{u}, \quad u_3 = \frac{u}{M_0}, \quad M_0, \quad \tau_4. \tag{1.31}$$

In the neighborhood of $M_0 = \infty$ we introduce the following coordinates (see (1.19)):

$$v_2 = \frac{v_0}{m_0}, \quad u_2 = -\frac{u_1}{v_2} = -\frac{u}{v_0}, \quad \mu = \frac{1}{m_0}, \quad \tau_9, \tag{1.32}$$

$$v_1 = \frac{v_2}{u_1} = \frac{v_0}{u}, \quad u_1 = \frac{u}{m_0}, \quad \mu, \quad \tau_4. \tag{1.33}$$

In the coordinates (1.30)–(1.33) instead of the line l we have a two-dimensional plane ($v_5 = 0$ and the coordinates $u_2 \geqslant 0$ and $M_0 \geqslant 0$ vary) which is the component of the boundary Γ_5 of the manifold S. In the coordinates (1.30)–(1.33) and corresponding variables τ_i the dynamical system (1.14) is smoothly continued to the component of the boundary Γ_5 which is an invariant manifold of the dynamical system. All critical points of the dynamical system in the manifold S contained in the component of the boundary Γ_5 are non-degenerate. In other words the transformations (1.30)–(1.33) fully resolve the line of critical points l.

After the above transformations of the coordinates (1.17)–(1.33), it is easy to complete the construction of the closed three-dimensional manifold S. The closed three-dimensional manifold S is covered by systems of coordinates (1.17)–(1.33) and is determined in them by the following natural conditions:

$$\eta, \quad v_i \leqslant 0; \quad u_i, \quad m_i, \quad M_i, \quad \rho_i, \quad \mu \geqslant 0.$$

The boundary Γ of the manifold S consists of eight components $\Gamma_i \, (i = 1, \ldots, 8)$. The conditions which determine the components of the boundary Γ_i and the systems of coordinates which extend to Γ_i have the following form:

$$\Gamma_1: \mu = 0 \quad (1.18)–(1.20); \quad \Gamma_2: v_1 = 0 \quad (1.21)–(1.23);$$

$$\Gamma_3: M_0 = 0 \quad (1.24)–(1.26); \quad \Gamma_4: M_2 = 0 \quad (1.27)–(1.29);$$

$$\Gamma_5: u = 0 \quad (1.30)–(1.33); \quad \Gamma_6: v_0 = 0 \quad (1.17);$$

$$\Gamma_7: m = 0 \quad (1.16); \quad \Gamma_8: z = 0 \quad (1.16).$$

III. Transformation of the Dynamical System. The dynamical system in the manifold S is defined after transforming the system (1.14) into the local charts (1.17)–(1.33). Actually making these transformations easily convinces us that the system (1.14) is smoothly continued (in the appropriate variable τ_j) to the components of the boundary Γ_i. All components of the boundary Γ_i and their intersections are invariant submanifolds of the dynamical system in the manifold S.

In further analysis the transformations of the dynamical system (1.14) in the two local charts (1.21) and (1.27) are the ones most frequently used. We will now show these transformations explicitly. In the coordinates (1.21) and variable τ_3 the dynamical system (1.14) has the following form:

$$\dot{v}_0 = -\frac{v_0}{\gamma(1 - v_0^2 \rho_0^2)}\left[\left(4 - \frac{4}{\omega} - \frac{6\gamma}{\omega}\right) - 3\gamma v_0 - \gamma m_2 \right.$$

$$\left. + \gamma v_0 \left(v_0 + \frac{2}{\omega}\right)\left(v_0 + \frac{2}{\omega} - 1\right)\rho_0^2 \right],$$

$$\dot{m}_2 = -\frac{m_2}{2\gamma(1 - v_0^2 \rho_0^2)}\left[4\gamma - \frac{12\gamma}{\omega} - 2\gamma v_0 - 2\gamma^2 m_2 \right.$$

$$+ 2\gamma^2 v_0 \left(v_0 + \frac{2}{\omega}\right)\left(v_0 + \frac{2}{\omega} - 1\right)\rho_0^2 + 4\gamma\left(1 + \frac{1 - 3\gamma}{\omega} + \frac{1 - 3\gamma}{2} v_0\right)v_0^2 \rho_0^2$$

$$\left. + 2\gamma\left(2\frac{\omega - 3}{\omega} - 3v_0\right)(1 - v_0^2 \rho_0^2) \right],$$

$$\dot{\rho}_0 = \frac{\rho_0}{2\gamma(1 - v_0^2 \rho_0^2)}\left[4 - 2\gamma - \frac{4}{\omega} - 2\gamma v_0 + \gamma(\gamma - 1)m_2 \right.$$

$$- 2\gamma\left(1 + \frac{1 - 3\gamma}{\omega} + \frac{1 - 3\gamma}{2} v_0\right)v_0^2 \rho_0^2$$

$$\left. - \gamma(\gamma - 1)v_0 \left(v_0 + \frac{2}{\omega}\right)\left(v_0 + \frac{2}{\omega} - 1\right)\rho_0^2 \right]. \tag{1.34}$$

In the coordinates (1.27) and variable τ_3 the system (1.14) has the following form:

$$\dot{v}_4 = -\frac{v_4}{\gamma(1 - v_4^2 u^2)}\left[2\gamma - \frac{6\gamma}{\omega} - \gamma v_4 u^2 - \gamma^2 m_2 \right.$$

$$+ \gamma^2 v_4 \left(\left(v_4 u^2 + \frac{2}{\omega}\right)^2 - \left(v_4 u^2 + \frac{2}{\omega}\right)\right)$$

$$\left. + 2\gamma v_4^2 u^2 \left(1 + \frac{1 - 3\gamma}{\omega} + \frac{1 - 3\gamma}{2} v_4 u^2\right) \right],$$

$$\dot{m}_2 = -\frac{m_2}{\gamma(1 - v_4^2 u^2)}\left[2\gamma - \frac{6\gamma}{\omega} - \gamma v_4 u^2 - \gamma^2 m_2 \right.$$

$$+ \gamma\left(2 - \frac{6}{\omega} - 3v_3 u^2\right)(1 - v_4^2 u^2) + \gamma^2 v_4 \left(\left(v_4 u^2 + \frac{2}{\omega}\right)^2 - \left(v_4 u^2 + \frac{2}{\omega}\right)\right)$$

$$\left. + 2\gamma v_4^2 u^2 \left(1 + \frac{1 - 3\gamma}{\omega} + \frac{1 - 3\gamma}{2} v_4 u^2\right) \right],$$

$$\dot{u} = -\frac{u}{2\gamma(1 - v_4^2 u^2)}\left[4 - \frac{4}{\omega} - 2\gamma v_4 u^2 + \gamma(\gamma - 1)m_2 - 2\gamma\right.$$

$$- \gamma(\gamma - 1)v_4\left(\left(v_4 u^2 + \frac{2}{\omega}\right)^2 - \left(v_4 u^2 + \frac{2}{\omega}\right)\right)$$

$$\left. - 2\gamma v_4^2 u^2\left(1 + \frac{1 - 3\gamma}{\omega} + \frac{1 - 3\gamma}{2}v_4 u^2\right)\right]. \tag{1.35}$$

IV. Critical Points of the Dynamical System in the Manifold S. It is not difficult to verify that the following functions Φ_μ vary monotonically along the trajectories of the system (1.14):

$$\Phi_\mu = z\left(V - \frac{2}{\omega}\right)^{\gamma - 1} m^\mu,$$

$$\frac{d\Phi_\mu}{d\tau} = \Phi_\mu \frac{2(1 - 3\gamma + \omega + \mu(\omega - 3)) + \omega(1 - 3\gamma - 3\mu)(V - 2/\omega)}{\omega(V - 2/\omega)}, \tag{1.36}$$

where μ is bounded by two numbers: $\mu_1 = (1 - 3\gamma)/3$ and $\mu_2 = (1 - 3\gamma + \omega)/(3 - \omega)$. If $\gamma < 4/3$, then $\mu_1 < \mu_2$ and $\Phi_\mu' < 0$. If $\gamma > 4/3$, then $\mu_1 > \mu_2$ and $\Phi_\mu' > 0$. If $\gamma = 4/3$, then $\mu_1 = \mu_2 = -1$ and Φ_{-1} is a first integral of the system (1.14). Note that the monotone function Φ_{μ_2} can also be obtained from the adiabatic integral (1.12).

The existence of the monotone function (1.36) proves that for $\gamma \neq 4/3$ the system (1.14) has no critical points in the interior of the manifold S. All critical points of this system belong to the various components of the boundary Γ (and also to the surface of non-extendability $L = 0$). In the manifold S the dynamical system has sixteen isolated critical points:

$$A(v_1 = m_3 = \rho_2 = 0), \quad B(v_3 = \rho_3 = M_1 = 0), \quad C(v_1 = u_1 = \mu = 0),$$

$$C_1(v_2 = u_2 = \mu = 0), \quad D(v_6 = u_3 = M_0 = 0), \quad H(\eta = u_2 = m_4 = 0),$$

$$Z_1\left(v_0 = \frac{4(\omega - 1) - 6\gamma}{3\gamma\omega}, m_2 = \rho_0 = 0\right), \quad Z_2\left(m_3 = \frac{1}{\gamma - 1}, v_1 = \rho_2 = 0\right),$$

$$Z_3\left(v_0 = -\frac{2}{\omega}, m_2 = \frac{4(\omega - 1)}{\gamma\omega}, \rho_0 = 0\right), \quad Z_4(v_0 = m_2 = \rho_0 = 0),$$

$$Z_5(v_4 = m_2 = u = 0), \quad Z_6(\omega > 2)\left(v_4 = -\frac{\omega(3 - \omega)}{\gamma(\omega - 2)}, m_2 = u = 0\right),$$

$$Z_6(\omega < 2)(z = m = 0, V - 1), \quad Z_8(v = z = m = 0),$$

$$Z_7\left(V = \frac{2}{3\gamma - 1}, z = 6\gamma\frac{(3\gamma - (1 + \omega))(\gamma - 1)}{(3\gamma - 1)^2[2(\omega - 1) - 3\gamma(\omega - 2)]}, m = 0\right),$$

$$Z_9\left(V = \frac{2}{3}, z = 0, m = \frac{2}{9}\right), \quad Z_{10}(\eta = u_2 = 0, m_4 = 2^{-1/3})$$

and four lines of critical points:

$$EF\left(u_5 = M_2 = 0, -\infty \leqslant v_5 \leqslant 0\right), \quad D_1 G\left(M_0 = u_3 = 0, -\infty \leqslant v_5 \leqslant 0\right),$$

$$I_1 I_2, I_3 I_4\left(z = v_0^2, m = 2v_0^2 + v_0[\omega + 2 - 4(\omega - 1)/\gamma]\omega^{-1} + 2(\omega - 2)\omega^{-2}\right),$$

$$J_1 J_2\left(2 < \omega < 3\right)\left(v_5 = -\omega^2/2(\omega - 2), u_5 = 0, 0 \leqslant M_2 \leqslant \infty\right).$$

For $\gamma = 5/3$ there is a line of critical points $\mathscr{L}$ in the component of the boundary Γ_1 which passes through the critical points Z_2, I_4 and Z_{10}:

$$m_1^3 = -v_1(3 + v_1^2)/2, \quad \rho_0 = 0, \quad -\infty \leqslant v_1 \leqslant 0.$$

For $\gamma = 4/3$ there are two lines of critical points: $\mathscr{L}_1$ (which passes through the critical points Z_7, Z_9 and intersects the segment of critical points

$$I_3 I_4: \ 4/27 - 2m/3, V = 2/3, 0 \leqslant m \leqslant 2/9$$

(see [7])) and $\mathscr{L}_2$ (which passes in the component of the boundary Γ_2 through the critical points

$$Z_1, Z_2, Z_3: \ \rho_0 = 0, v_0 = -((3 - \omega)/\omega + m_2/3), 0 \leqslant m_2 \leqslant \infty.$$

For $\gamma = \gamma_1 = 2(\omega - 1)/\omega$ there is a line of critical points $\mathscr{L}_3$ which is the intersection of the components of the boundary Γ_6 and Γ_7 and passes through the critical points

$$Z_4 \text{ and } Z_5: v_4 = m_2 = 0, 0 \leqslant u \leqslant \infty.$$

In Fig. 26 we see the manifold S "through" the component of the boundary Γ_1. All critical points are shown here, except the segments $I_1 I_2$ and $I_3 I_4$ which belong to the surface of non-extendability of solutions $L = 0$ (dotted lines denote the intersection of the components of the boundary with the surface $L < 0$). Figure 26 shows the qualitative behavior of the dynamical system in the components of the boundary Γ for

$$2 < \omega < 3, \quad \gamma > \{\gamma_0 = 4(\omega - 1)(2 - (\omega - 2)^{1/2})^{-2}, 4/3\}.$$

V. Analysis of the Line of Critical Points $I_1 I_2, I_3 I_4$. The line of critical points $I_1 I_2, I_3 I_4$ belongs to the surface of non-extendability of solutions $L = 0$. For $\omega > 2$ this line consists of two segments: $I_1 I_2$ and $I_3 I_4$ if $\gamma > \gamma_0$. If however $\gamma < \gamma_0$, then there is only one segment $I_1 I_4$. The endpoints are determined by the following conditions:

$$I_1\left(v_0 = 0\right), \quad I_2, I_3\left(m(v_{0; 2, 3}) = 0\right), \quad I_4\left(v_0 = -\infty\right)$$

and belong to the components of the boundary Γ_3, Γ_7 and Γ_1 respectively. For $\omega < 2$ there is only one segment $I_3 I_4$.

To analyze the line of critical points $I\left(I_1 I_2, I_3 I_4\right)$ we make the following non-monotone change of variable:

$$\frac{d\tau_0}{d\tau} = -\left[\gamma\left(V - \frac{2}{\omega}\right)\left(z - \left(V - \frac{2}{\omega}\right)^2\right)\right]^{-1}.$$

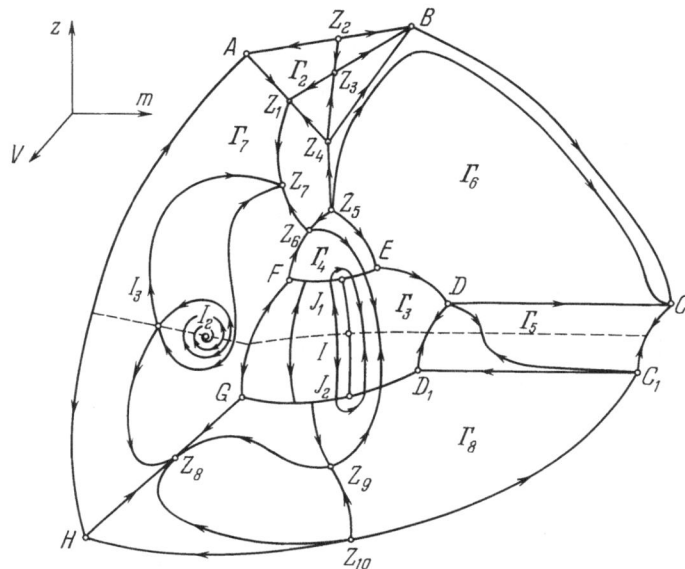

Fig. 26. The manifold S after the resolution of degenerate critical points of the dynamical system (1.14) and a completion of the physical region by a boundary at infinity with respect to the coordinates z, m and V

In the new variable τ_0 the surface of non-extendability of solutions $L = 0$ becomes a non-critical surface of the dynamical system (1.14) everywhere except at the line of critical points I. The eigenvalues $\lambda_{1,2}$ of the system (1.14) at the points of the line I (in the variable τ_0) satisfy the following characteristic equation:

$$
\begin{aligned}
\lambda^2 &- \frac{\gamma v_0^2}{\omega}(3\omega - 10)\lambda \\
&+ \gamma^2 v_0^3 \left\{ 2(3\gamma - 5)v_0^3 + \frac{4}{\omega}\left(-7 + 3\gamma + \frac{4(\omega - 1)}{\gamma} + \omega\gamma - 2\omega\right)v_0^2 \right. \\
&+ \frac{2}{\omega^2}\left[\gamma\omega^2 - \gamma\omega - 4\omega^2 - 3\omega - 2 + 6\gamma + \frac{16(\omega - 1)}{\gamma}\right]v_0 \\
&\left. + 4(\omega - 2)(\omega - 3)(\gamma + 1)\omega^{-3} \right\} = 0
\end{aligned}
\tag{1.37}
$$

$(\lambda_3 = 0$ because the line I is one-dimensional).

Notice that the trajectories of the system (1.14) going through (in the variable τ_0) the surface $L = 0$ must pass through the critical points contained in the line I. In fact they can only pass through those points which by virtue of (1.37) are nodes or saddles. Examples of such trajectories will be given below.

VI. Concluding Remarks. A calculation of the eigenvalues at the critical points of the dynamical system (1.14) (which is done below in Sects. 2–4) shows that all critical points (except for the line $J_1 J_2$) are non-degenerate. This means that when we constructed the manifold S we resolved the singularities of the initial dynamical system (1.14) along the way.

A resolution of critical points of the line $J_1 J_2$ by means of a sequence of standard transformations of the coordinates (called σ-processes along the manifold in algebraic geometry) leads to a peculiar cyclic process. A repeated application of such transformations any finite number of times leads to an appearance (along with the lines of non-degenerate (and unstable) critical points M_i) of a new line of degenerate critical points, whose properties are similar to those of the initial line of critical points $J_1 J_2$. The above non-degenerate critical points M_i have no separatrices approaching them from the physical region. Their separatrices belong to the components of the boundary Γ and do not correspond to any physical solutions. As a consequence of this, the initial line of degenerate critical points $J_1 J_2$ has no physical applications either.

Six isolated critical points A, B, C, C_1, D, H and the critical points contained in the segments EF and $D_1 G$ are unstable (saddles). All their separatrices lie in the components of the boundary Γ and therefore do not correspond to any exact physical solutions. The ten isolated critical points Z_i are the most important ones in the considered system. These critical points have separatrices which pass through the interior of the manifold S and thus correspond to important physical asymptotic forms of solutions. In subsequent sections we calculate the eigenvalues λ_x of the critical points Z_i (using the variables τ_j determined according to (1.18)–(1.33) by the coordinates of the critical point; the letter subscript x denotes the corresponding eigenvector) and show the corresponding asymptotic forms. For convenience we shall group the critical points $Z_i, \mathscr{L}_k$ in accordance with the physical meaning of the asymptotic forms determined by them.

2. Asymptotics of Gas Moving Away from the Center

Asymptotic forms of gas moving away from the center fall into two classes. The first class comprises the asymptotic forms in which the motion of the gas extends to the center of symmetry $r = 0$. The second class comprises the asymptotic forms with the formation of an expanding cavity inside the gas. Asymptotic forms of the first class correspond to separatrices of the critical points

$$Z_1, Z_9, Z_6 \, (\omega < 2), \quad \mathscr{L}_1, \mathscr{L}_2 \, (\gamma = 4/3), Z_7.$$

Here we point out the eigenvalues of the dynamical system on the manifold S at these critical points and the explicit form of the corresponding asymptotic forms:

I. $\quad Z_1 \left(v_0 = \dfrac{4(\omega - 1) - 6\gamma}{3\gamma\omega}, \, m_2 = \rho_0 = 0 \right): \lambda_{v_0} = \dfrac{4(\omega - 1) - 6\gamma}{\gamma\omega},$

$$\lambda_{m_2} = \dfrac{4(4 - 3\gamma)(\omega - 1)}{3\gamma\omega}, \, \lambda_{\rho_0} = \dfrac{2(\omega - 1) + 3\gamma(2 - \omega)}{3\gamma\omega}.$$

The critical point Z_1 belongs to the manifold S if $\gamma > 2(\omega - 1)/3$. This critical point Z_1 has separatrices going through the interior of the manifold S in the following two cases:

1) $2(\omega - 1)/3 < \gamma < 4/3$. The point Z_1 is unstable

$$\left(\lambda_{v_0} < 0, \lambda_{m_2} > 0, \lambda_{\rho_0} > 0\right)$$

and has a two-dimensional separatrix leaving it (for $\lambda \to 0$).

2) $\gamma > 2(\omega - 1)/3(\omega - 2)$ (with $\omega > 2$). The point Z_1 is attracting for $\lambda \to \infty$.

The corresponding asymptotic forms (for $\lambda \to 0$ and $\lambda \to \infty$) have the following form:

$$\mathcal{M} = C_1 A r^{3-\omega} \lambda^a, \quad \rho = \frac{3\gamma b}{4\pi} C_1 A \lambda^a r^{-\omega},$$

$$p = C_2 \frac{(AG)^{2/\omega}}{G} t^{-4(\omega-1)/\omega}, \quad v = \frac{4(\omega - 1)}{3\gamma\omega} \frac{r}{t}, \tag{2.1}$$

$$a = 2(\omega - 1)b, \quad b = \frac{3 - \omega}{3\gamma - 2(\omega - 1)}.$$

At the critical point Z_9 $(V = 2/3, z = 0, m = 2/9)$ we have

$$\lambda_{V,m} = \frac{3\omega}{2(3 - \omega)}, \quad \frac{-\omega}{3 - \omega}, \quad \lambda_z = -\frac{\omega(4 - 3\gamma)}{3 - \omega}.$$

For $\gamma < 4/3$ there is a two-dimensional separatrix approaching Z_9 $(\lambda \to \infty)$. For $\gamma > 4/3$ there is a two-dimensional separatrix leaving Z_9 $(\lambda \to 0)$. The corresponding asymptotic forms (for $\lambda \to 0$ and $\lambda \to \infty$) have the following form:

$$\mathcal{M} = \frac{2}{9} \frac{r^3}{Gt^2}, \quad \rho = \frac{1}{6\pi Gt^2}, \quad v = \frac{2}{3} \frac{r}{t},$$

$$p = C_1 \frac{1}{G} (AG)^a r^{2-\omega a} \cdot t^{2a-4}, \quad a = \frac{4 - 3\gamma}{3 - \omega}. \tag{2.2}$$

The asymptotic forms (2.1) describe the motion of gas with an asymptotically constant (with respect to r) pressure. The asymptotic forms (2.2) describe the motion of gas with an asymptotically constant (with respect to r) density. For $\gamma = 2(\omega - 1)/\omega$ there is a one-dimensional separatrix leaving Z_1 and approaching Z_9 along which we have

$$V = 2/3, \quad m = 2/9, \quad 0 \leqslant z \leqslant \infty. \tag{2.3}$$

For this separatrix the asymptotic forms (2.1) and (2.2) are exact solutions which appeared earlier in the book [7].

II. At the critical point Z_6 $(\omega < 2)$ $(z = m = 0, V = 1)$ we have

$$\lambda_m = \frac{\omega}{2 - \omega}, \quad \lambda_z = \frac{3\omega(\gamma - 1)}{2 - \omega}, \quad \lambda_V = \frac{\omega}{2 - \omega}.$$

The critical point Z_6 $(\omega < 2)$ is repelling. Separatrices leaving this point have the following asymptotic form for $\lambda \to 0$:

$$\mathcal{M} = C_1 A r^{3-\omega} \lambda^{\omega(3-\omega)/(2-\omega)}$$

$$\rho = \frac{3-\omega}{2\pi(2-\omega)} C_1 A r^{-\omega} \lambda^{\omega(3-\omega)/(2-\omega)}, \tag{2.4}$$

$$p = C_1 A^2 Gr^{-2(\omega-1)} \lambda^{\omega(3\gamma-2(\omega-1))/(2-\omega)}, \quad v = \frac{r}{t}.$$

In this stable (for $t \to \infty$) asymptotic region, the matter moves away from the center with constant velocity. The Mach number of the flow of gas tends to infinity, $M \to \infty$.

III. The line of critical points

$$\mathcal{L}_1 \, (\gamma = 4/3) \, (z = 4/27 - 2m/3, \, V = 2/3, \, 0 \leqslant m \leqslant 2/9)$$

lies in the interior of the manifold S. Therefore the critical points of the line $\mathcal{L}_1$ correspond to exact solutions mentioned for the first time in the book [7]. For

$$0 \leqslant m \leqslant B = 2/9 - 2(3-\omega)^2/3\omega^2$$

(the segment of $\mathcal{L}_1$ in the subsonic region $L = 0$) these solutions are stable for $\lambda \to \infty$. For $B < m < 2/9$ these solutions are unstable. The critical points of the line

$$\mathcal{L}_2 \, (\gamma = 4/3): \, \rho_0 = 0, \, v_0 = -((3-\omega)/\omega + m_2/3), \, 0 \leqslant m_2 \leqslant \infty$$

have the following eigenvalues:

$$\lambda_{\rho 0} = \frac{1}{2}\left(\frac{(3-\omega)}{\omega} + m_2\right), \quad \lambda_1 = 0, \quad \lambda_2 = \frac{m_2}{3} - \frac{3(3-\omega)}{\omega}.$$

For $m_2 > 9(3-\omega)/\omega$ these critical points are repelling. They correspond to a three-dimensional separatrix (leaving as $\lambda \to 0$). For $m_2 < 9(3-\omega)/\omega$ they are unstable and the separatrix leaving them is two-dimensional. For $\lambda \to 0$ the corresponding asymptotic forms are:

$$\mathcal{M} = A_1 r^{6(3-\omega)\alpha_2}, \quad \rho = A_1 \frac{3-\omega}{2\pi\omega|v_0|} r^\alpha,$$

$$p = A_1 \frac{G}{m_2} \frac{3-\omega}{2\pi\gamma\omega} r^{-4\omega m\alpha_2}, \quad v = \frac{\alpha_1}{3} \frac{r}{t},$$

$$\alpha = -3 \frac{3-\omega+\omega m_2}{3(3-\omega)+\omega m_2}, \quad \alpha_1 = \frac{3(\omega-1)-\omega m_2}{\omega}, \tag{2.5}$$

$$\alpha_2 = (3(3-\omega)+\omega m_2)^{-1}, \quad A_1 = (AG)^{-\alpha/\omega} \frac{m_2}{G|v_0|} t^{-2(3-\omega)\alpha_1\alpha_2}.$$

In the asymptotic region (2.5) the density and the pressure at the center $(r \to 0)$ are infinite. The critical point

$$Z_3\left(m_2 = 3(\omega - 1)/\omega,\ v_0 = -2/\omega,\ \rho_0 = 0\right)$$

divides the line $\mathscr{L}_2$ into two parts. For $m_2 > 3(\omega - 1)/\omega$ in the asymptotic form (2.5) the gas falls $(v < 0)$ to the center in infinite time. For $m_2 < 3(\omega - 1)/\omega$ the gas moves away from the center and the lines of flow emerge from the center at $t = 0$.

IV. The critical point Z_7 belongs to the manifold S for all $\omega \leqslant 2$ and for $\omega > 2$ if

$$\frac{1 + \omega}{3} < \gamma < \frac{2(\omega - 1)}{3(\omega - 2)}.$$

The point Z_7 belongs to the subsonic region $L = 0$ if

$$3\gamma(\gamma - 1)\omega^2 - 2(3\gamma - (1 + \omega))(2(\omega - 1) + 3\gamma(2 - \omega)) > 0.$$

For $\omega > 2$ this condition is satisfied for all γ. If the critical point Z_7 lies in the subsonic region, then it is attracting in the plane $m = 0$. Since

$$\lambda_m = \omega\frac{4 - 3\gamma}{3\gamma - (1 + \omega)}.$$

for $\gamma < 4/3$ a one-dimensional separatrix leaves Z_7 $(\lambda \to 0)$. For $\gamma > 4/3$ a three-dimensional separatrix approaches Z_7 $(\lambda \to \infty$ and the point Z_7 is attracting$)$. If (for $\omega < 2$) the critical point Z_7 is contained in the supersonic region $L < 0$, then it is a saddle in the plane $m = 0$. Therefore for $\gamma < 4/3$ there is a two-dimensional separatrix leaving Z_7 $(\lambda \to 0)$. For $\gamma > 4/3$ there is a two-dimensional separatrix approaching Z_7 $(\lambda \to \infty)$. The corresponding asymptotics forms for $\lambda \to 0$ and $\lambda \to \infty$ are

$$\mathscr{M} = C_1 A r^{3-\omega}\lambda^a, \quad \rho = \frac{(3\gamma - 1)\alpha}{4\pi\omega}C_1 A r^{-\omega}\lambda^a,$$

$$p = \frac{3(\gamma - 1)}{2\pi(3\gamma - 1)[2(\omega - 1) - 3\gamma(\omega - 2)]}C_1 A^2 G r^{-2(\omega-1)}\lambda^{\omega+a}, \tag{2.6}$$

$$v = \frac{2}{3\gamma - 1}\frac{r}{t}, \quad a = \omega\frac{3 - \omega}{3\gamma - (1 + \omega)}.$$

V. Solutions with an expanding cavity inside the gas for $\omega = 5/2$ were analyzed in detail in the book [7]. Here we point out the conditions for the existence of such solutions for arbitrary γ, ω and show their asymptotic form at the internal boundary. Solutions with an expanding cavity correspond to separatrices of the critical points Z_4, Z_5, Z_6 $(\omega > 2)$ and the line of critical points $\mathscr{L}_3$ $(\gamma = \gamma_1)$:

$$Z_4\left(v_0 = m_2 = \rho_0 = 0\right): \lambda_{v_0} = -\frac{4(\omega - 1) - 6\gamma}{\gamma\omega},$$

$$\lambda_{m_2} = 4\frac{3-\omega}{\omega}, \quad \lambda_{p_0} = \frac{2(\omega-1)-\gamma\omega}{\gamma\omega};$$

$$Z_5\,(v_4 = m_2 = u = 0): \lambda_{v_4} = 2\frac{3-\omega}{\omega}, \quad \lambda_{m_2} = 4\frac{3-\omega}{\omega},$$

$$\lambda_u = \frac{\gamma\omega - 2(\omega-1)}{\gamma\omega};$$

$$\mathscr{L}_3\left(\gamma = \gamma_1 = \frac{2(\omega-1)}{\omega}\right)(v_4 = m_2 = 0, 0 \leqslant u \leqslant \infty):$$

$$\lambda_{v_4} = 2\frac{3-\omega}{\omega}, \quad \lambda_{m_2} = 4\frac{3-\omega}{\omega}, \quad \lambda_u = 0.$$

The critical point Z_4 is repelling for $2(\omega-1)/3 < \gamma < \gamma_1$. The critical point Z_5 is repelling for $\gamma > \gamma_1$. The line of critical points $\mathscr{L}_3$ is repelling for $\gamma = \gamma_1$. For such values of the parameters the separatrices leaving the critical points Z_4, Z_5 and $\mathscr{L}_3$ have the following stable asymptotic form for $\lambda \to \lambda_1$:

$$\mathscr{M} = \frac{r^3}{Ct^2}C_1\left(\frac{\lambda}{\lambda_1} - 1\right)^a, \quad \rho = \frac{C_1}{Ct^2}\frac{3-\omega}{2\pi\omega b}\left(\frac{\lambda}{\lambda_1} - 1\right)^{a-1},$$

$$p = \frac{r^2}{Ct^4}C_2, \quad v = \frac{r}{t}\left(\frac{2}{\omega} - b\left(\frac{\lambda}{\lambda_1} - 1\right)\right), \tag{2.7}$$

$$a = \gamma\frac{3-\omega}{3\gamma - 2(\omega-1)}; \quad b = 2\frac{3\gamma - 2(\omega-1)}{\gamma\omega}.$$

The critical point

$$Z_6\,(\omega > 2)\left(v_4 = -\frac{\omega(3-\omega)}{\gamma(\omega-2)}, m_2 = u = 0\right)$$

is unstable:

$$\lambda_{v_4} = -\lambda_{m_2} = \frac{2(3-\omega)}{\omega}, \quad \lambda_u = \frac{3\gamma - (1+\omega)}{\omega}.$$

For $\gamma > (1+\omega)/3$ there is a two-dimensional separatrix leaving Z_6 which corresponds to the following asymptotic form as $\lambda \to \lambda_1$:

$$\mathscr{M} = \frac{r^3}{Gt^2}C_1\left(\frac{\lambda}{\lambda_1} - 1\right)^c, \quad \rho = \frac{C_1}{Gt^2}\frac{3-\omega}{2\pi\omega d}\left(\frac{\lambda}{\lambda_1} - 1\right)^{c-1},$$

$$p = \frac{r^2}{Gt^4}\frac{\omega-2}{2\pi\omega^2}C_1\left(\frac{\lambda}{\lambda_1} - 1\right)^c, \quad v = \frac{r}{t}\left(\frac{2}{\omega} - d\left(\frac{\lambda}{\lambda_1} - 1\right)\right), \tag{2.8}$$

$$c = \gamma\frac{3-\omega}{3\gamma - (1+\omega)}, \quad d = 2\frac{3\gamma - (1+\omega)}{\gamma\omega}.$$

In the asymptotic forms (2.7) and (2.8) the mass of gas inside the expanding sphere $\lambda = \lambda_1$ is equal to zero, i.e. this sphere contains nothing but vacuum. In the asymptotic form (2.7) the pressure of gas at the internal boundary is not zero $p \neq 0$, i.e. the gas is being pushed out by some "spherical piston" [7]. For example an infinitely heated gas $(T = \infty$ and $\rho = 0)$ serves as a model of such a piston. For $\gamma > \gamma_1$ at the internal boundary we have $\rho \to \beta$, so $T \sim p/\rho \to 0$ for $\lambda \to \lambda_1$. Therefore for a real non-self-similar solution an intensive heat exchange takes place at the internal boundary with an infinitely hot gas inside the cavity, which leads to a filling of the cavity by a gas with non-zero density.

Another type of instability of the internal boundary is connected to the well-known mechanism of appearance of the Taylor instability [128, 129]. Indeed for $\omega < 2$ the acceleration of the internal boundary is positive $(a_1 > 0)$, i.e. in the system connected with the internal boundary the acceleration is directed towards vacuum which leads to the Taylor instability of the gas. For $\omega > 2$ this instability does not occur. For $\gamma < \gamma_1$ (in this case $\omega > 2$ and $\gamma < 4/3$) we have $\rho \to 0$ and $T \to \infty$ for $\lambda \to \lambda_1$. Therefore the internal boundary is stable relative to the considered perturbations.

In the asymptotic region (2.8) at the internal boundary we have $p \to 0$, $\rho \to 0$ and $T \to 0$ for $\lambda \to \lambda_1$, i.e. this asymptotic form describes the expansion of gas with a free internal boundary, which is stable by virtue of $\omega > 2$ $(a_1 < 0)$. From the conditions of existence of the asymptotic forms (2.7) and (2.8) follows an important corollary. For $\gamma < 2(\omega - 1)/3$ (in which case $\omega > 5/2$) there are no self-similar solutions with the formation of a cavity inside the gas. In particular in this region of values of the parameters such solutions do not exist also for the problem of stellar explosions.

3. Analysis of the Dynamical System on the Components of the Boundary Γ_2 and Γ_8

Based on the resolution of singularities of the system of self-similar equations (Sect. 1) it turns out to be possible to analyze in detail the asymptotics of self-similar accretion of self-gravitating gas to the center and to solve the problem of stellar explosions. The behavior of corresponding self-similar solutions substantially depends on the behavior of the dynamical system on the components of the boundary Γ_2 and Γ_8. It is this behavior that we will consider in this section.

I. Analysis of the Dynamical System on the Component of the Boundary Γ_2. In the component of the boundary Γ_2 $(\rho_0 = 0)$ the dynamical system has six non-degenerate critical points: Z_1, Z_2, Z_3, Z_4, A and B. The critical points A and B were considered in Sect. 2. As mentioned in Sect. 1 these critical points A and B are saddles and have no separatrices passing into the interior of the physical region. In the component of the boundary Γ_2 $(\rho_0 = 0)$ the critical point A is still a saddle and the critical point B is attracting. The critical point

$$Z_2 \left(v_1 = 0, m_3 = (\gamma - 1)^{-1}, \rho_2 = 0\right)$$

has the following eigenvalues:

$$\lambda_{v_1} = \frac{5 - 3\gamma}{2(\gamma - 1)}, \quad \lambda_{m_3} = 1, \quad \lambda_{\rho_2} = -\frac{4 - 3\gamma}{\gamma - 1}. \tag{3.1}$$

For $4/3 < \gamma < 5/3$ this critical point has separatrices going inside the physical region and is repelling. Separatrices leaving the critical point Z_2 for $4/3 < \gamma < 5/3$ determine a stable asymptotic form for the accretion of gas to the center which is discussed in Sec. 4.

The dynamical system on the component of the boundary $\Gamma_2 \left(\rho_0 = 0\right)$ is most conveniently considered in the coordinates v_0 and m_2, where it has the following form:

$$\frac{dv_0}{d\tau_3} = v_0 \left(\frac{6\gamma - 4(\omega - 1)}{\gamma \omega} + 3v_0 + m_2\right) = P,$$

$$\frac{dm_2}{d\tau_3} = m_2 \left(4\frac{3 - \omega}{\omega} + 4v_0 + \gamma m_2\right) = Q. \tag{3.2}$$

The critical point

$$Z_3 \left(v_0 = -2/\omega, m_2 = 4(\omega - 1)/\gamma\omega, \rho_0 = 0\right)$$

has the following eigenvalues:

$$\lambda_{v_0, m_2} = \lambda_\pm = \{2\omega - 5 \pm [(2\omega - 5)^2 - 8(\omega - 1)(4 - 3\gamma)/\gamma]^{1/2}\}\omega^{-1},$$

$$\lambda_{\rho_0} = 1. \tag{3.3}$$

The critical point Z_3 is repelling for $\gamma < 4/3$, $5/2 < \omega < 3$ and unstable for $\gamma < 4/3$, $1 < \omega < 5/2$ and $\gamma > 4/3$, $1 < \omega < 3$. In these cases the separatrices leaving the critical point Z_3 have dimensions 3, 1 and 2 respectively. The following fact plays a substantial role in the analysis of the model of stellar explosions. The one-dimensional separatrix leaving the critical point Z_3 which corresponds to the eigenvalues $\lambda_{\rho_0} = 1$ can be integrated explicitly and in fact is the following trajectory X of the system (1.14):

$$V = 0, \quad z = \frac{\gamma}{2(\omega - 1)} m. \tag{3.4}$$

The trajectory X corresponds to an exact solution of the equations of gas dynamics describing an equilibrium state of the gas:

$$\rho = \frac{c_1 A}{r^\omega}, \quad p = \frac{2\pi c_1^2 A^2 G}{(\omega - 1)(3 - \omega)} r^{2(1-\omega)}, \quad \mathcal{M} = \frac{4\pi c_1 A}{3 - \omega} r^{3-\omega}, \quad v = 0. \tag{3.5}$$

The remaining separatrices leaving the critical point Z_3 describe approximations of the gas for $\lambda \to 0$ $(t \to \infty)$ to the equilibrium state (3.5).

For $\gamma < \gamma_2 = 4[3 + (2\omega - 5)^2/8(\omega - 1)]^{-1} < 4/3$, the eigenvalues $\lambda_\pm$ (3.3) are

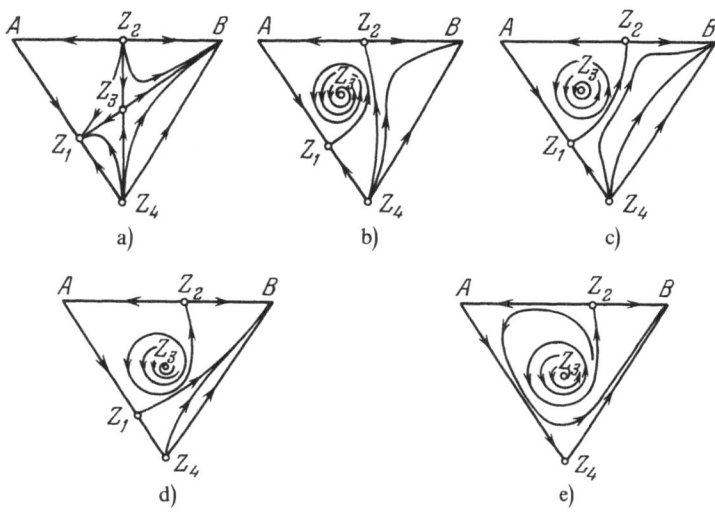

Fig. 27. Phase portraits for the dynamical system (3.2) in the component of the boundary Γ_2:
a) $\gamma > 4/3$, $1 < \omega < 3$; b) $\gamma < 4/3$, $\omega < 5/2$; c) $\gamma < 4/3$, $\omega = 5/2$; d) $\gamma < 4/3$, $5/2 < \omega < 1 + 3\gamma/2$;
e) $\gamma < 4/3$, $1 + 3\gamma/2 < \omega < 3$

complex, i.e. the critical point Z_3 in the component of the boundary Γ_2 ($\rho_0 = 0$) is a focus and all trajectories in its neighborhood (except for the emerging separatrix X) rotate around this point. For $1 < \omega < 5/2$, $\gamma < \gamma_2$ the focus Z_3 is attracting (in the plane $\rho_0 = 0$) and for $5/2 < \omega < 3$, $\gamma < \gamma_2$ it is repelling. In the corresponding solutions the velocity of gas v oscillates about zero (because at the critical point Z_3 we have $V = 0$) and the gas undergoes radial oscillations.

According to (3.3), for $\gamma > 4/3$ the critical point Z_3 in the component of the boundary Γ_2 is a saddle. In this case the phase portrait of the dynamical system (3.2) in Γ_2 is uniquely determined by the properties of the critical points and is pictured in Fig. 27(a). For $\gamma < 4/3$ the real parts of the eigenvalues $\lambda_{\pm}$ (3.3) have the same sign. Therefore, in order to determine the phase portrait of the dynamical system (3.2), it is necessary to analyze the possibility of existence of limit cycles for this system.

The system (3.2) belongs to a general class of two-dimensional dynamical systems of the type $\dot{x} = xL_1$, $\dot{y} = yL_2$, where L_1 and L_2 are linear functions of x and y, which was studied in [130]. Following the methods of [130] we find that for the function

$$B = v_0^{k-1} m_2^{h-1}, \quad k = \frac{\gamma}{3\gamma - 4}, \quad h = -\frac{3(\gamma - 1)}{3\gamma - 4},$$

the following identity holds:

$$\frac{\partial BP}{\partial v_0} + \frac{\partial BQ}{\partial m_2} = \frac{2(2\omega - 5)}{\omega} B.$$

According to the Dulac-Bendixon critierion this implies that the system (3.2) has no closed trajectories (and separatrix cycles as well) in the component of the boundary Γ_2 for $\omega \neq 5/2$. This fact along with the information about the critical points Z_1, Z_2, Z_3, Z_4, A and B obtained above allows us to reconstruct the phase portrait of the dynamical system (3.2) completely. The phase portraits (which depend on the values of the parameters γ and ω) are shown in Fig. 27.

The case $\omega = 5/2$ is special. In this case the dynamical system (3.2) has a first integral:

$$F_1 = \left| \frac{4}{5\gamma} + \frac{v_0}{\gamma - 1} + m_2 \right|^{4-3\gamma} |v_0|^{-\gamma} m_2^{3(\gamma-1)}. \tag{3.6}$$

For $\gamma < 4/3$, $\omega = 5/2$ the critical point Z_3 in the component of the boundary Γ_2 is a center, i.e. Re $\lambda_\pm = 0$. The integral F_1 has a minimum at the point Z_3. Therefore all trajectories of the dynamical system (3.2) which coincide with the level lines of the integral F_1 in the region $4/(5\gamma) + v_0/(\gamma - 1) + m_2 < 0$ are closed curves (see Fig. 27(b)).

Thus for $5/2 \leqslant \omega < 3$, $\gamma < 4/3$ the critical point Z_3 is unstable as $\lambda \to 0$. Trajectories of the dynamical system (1.14)–(1.34) moving in the neighborhood of the critical point Z_3 determine self-similar solutions for $5/2 \leqslant \omega < 3$, $\gamma < \gamma_2$, where as $\lambda \to 0$ an infinite number of radial oscillations of the gas takes place. Let us derive asymptotic formulas (as $\lambda \to 0$) for such self-similar solutions. To do this it is sufficient to integrate the linear part of the system (1.34) in the neighborhood of the point Z_3. When the system (1.34) is linearized in the neighborhood of the point

$$Z_3 \left(v_0 = -\frac{2}{\omega}, m_2 = \frac{4(\omega - 1)}{\gamma \omega}, \rho_0 = 0 \right)$$

after changing to the variable $\tau = \ln \lambda$ and the coordinates

$$x_0 = v_0 + \frac{2}{\omega} = V, \quad y_0 = m_2 - \frac{4(\omega - 1)}{\gamma \omega}$$

it takes the following form:

$$\dot{x}_0 = -3x_0 - y_0,$$

$$\dot{y}_0 = \frac{2(\omega - 1)}{\gamma}(4x_0 + \gamma y_0),$$

$$\dot{\rho}_0 = \rho_0 \frac{\omega}{2} \left(1 + \left(\frac{\omega}{2} - 1 \right) x_0 + \frac{\gamma - 1}{2} y_0 \right).$$

This system is easily integrated:

$$\frac{\omega}{2} x_0 = x = C_1 \lambda^{\omega - 5/2} \sin(\beta(\ln \lambda) + \theta),$$

$$\frac{\gamma\omega}{4(\omega-1)}y_0 = y = -\frac{\gamma}{2(\omega-1)}C_1\lambda^{\omega-5/2}[(\omega+1/2)\sin(\beta(\ln\lambda)+\theta)$$

$$+ \beta\cos(\beta\ln\lambda+\theta)], \tag{3.7}$$

$$\rho_0^2 = C_2\lambda^\omega(1+u),$$

$$u = 2C_1\lambda^{\omega-5/2}\left[\left((\omega-5/2)\alpha - \frac{\gamma-1}{2}\right)\sin(\beta\ln\lambda+\theta) - \alpha\beta\cos(\beta\ln\lambda+\theta)\right].$$

where we use the following notation:

$$\alpha = \gamma\frac{\omega-3\gamma+1}{4(\omega-1)(4-3\gamma)}, \quad \beta = \left(\frac{2(\omega-1)(4-3\gamma)}{\gamma} - (\omega-5/2)^2\right)^{1/2},$$

β is the imaginary part of the eigenvalues $\lambda_\pm$ (3.3) (for $\gamma < \gamma_2$) and C_1, C_2 and θ are arbitrary constants. After changing back to the initial coordinates V, m, R and P:

$$V = \frac{2}{\omega}x, \quad m = \frac{m_2}{\rho_0^2|v_0|} = \frac{2(\omega-1)(1+y)}{\gamma\rho_0^2(1-x)},$$

$$R = \frac{(1-3/\omega)m}{2\pi(V-2/\omega)}, \quad P = \frac{1}{\gamma}zR = \frac{R}{\gamma\rho_0^2}$$

and a substitution into (1.8), the obtained solutions (3.7) lead to the following asymptotic formulas for self-similar solutions with $5/2 \leqslant \omega < 3$, $\gamma < \gamma_2$, $\lambda \to 0$:

$$v = \frac{r}{t}\frac{2}{\omega}x, \quad \rho = \frac{C_0A}{r^\omega}(1+2x+y-u),$$

$$p = \frac{2\pi C_0^2 A^2 G}{(\omega-1)(3-\omega)}r^{2(1-\omega)}(1+2x+y-2u), \tag{3.8}$$

$$m = \frac{4\pi}{3-\omega}C_0Ar^{3-\omega}(1+x+y-u).$$

where $C_0 = C_2(\omega-1)(3-\omega)/2\pi\gamma$ and the functions u, x and y are defined in (3.7). The formulas (3.7)–(3.8) describe the asymptotic behavior of self-similar oscillations of self-gravitating gas in the neighborhood of the equilibrium state (3.5) which is a limit case of (3.8) as $x^2 + y^2 + u^2 \to 0$. The region where the formulas (3.7)–(3.8) are applicable is determined by the following conditions:

$$\omega = 5/2, \quad \gamma < 4/3 \quad C_1 \ll 1, \quad \lambda \to 0,$$

$$\omega > 5/2, \quad \gamma < \gamma_2 \quad C_1\lambda^{\omega-5/2} \ll 1, \quad \lambda \to 0.$$

For $\omega > 5/2$, $\gamma < \gamma_2$ the self-similar oscillations of self-gravitating gas are damped out, while in the entire space $r > 0$ the energy does not dissipate because the motion of gas is adiabatic. Let us show that for these self-similar solutions the energy does not flow to the center of symmetry $r = 0$ (where the solution (3.8) has a singularity) either. The energy density (per unit mass) of an ideal self-gravitating gas is of the following form:

$$\varepsilon = \frac{v^2}{2} + \frac{p}{(\gamma - 1)\rho} - \frac{G\mathcal{M}}{r}.$$

The entire amount of energy transferred by the gas across a sphere of radius r_0 and of the work performed against the forces of pressure during the infinite interval of time $t_0 < t < \infty$) is given by the following formula:

$$\Phi(r_0) = \int_{t_0}^{\infty} 4\pi r_0^2(\rho\varepsilon + p)v\,dt = 4\pi r_0^2 \int_{t_0}^{\infty} \left(\rho\frac{v^2}{2} + \frac{\gamma}{\gamma - 1}p - \frac{\rho G\mathcal{M}}{r_0}\right)v\,dt.$$

After a substitution of the asymptotic formulas (3.7)–(3.8) the above quantity has the following form:

$$\Phi(r_0) = \frac{8\pi^2\gamma C_0^2 C_1 A^2 G[\gamma - 2(\omega - 1)(\gamma - 1)]}{\omega(\omega - 1)^2(3 - \omega)(\gamma - 1)(4 - 3\gamma)}[r_0(AGt_0^2)^{1/\omega}]^{(5/2-\omega)}.$$

$$\cdot\,[-(\omega - 5/2)\sin(\beta\ln\lambda_0 + \theta) + \beta\cos(\beta\ln\lambda_0 + \theta)].$$

where $(\lambda_0 = r_0(AGt_0^2)^{-1/\omega})$.

For $r_0 \to 0$ the function $\Phi(r_0)$ oscillates and changes sign infinitely many times. This means that in the damping process of the self-similar oscillations of self-gravitating gas the flow of energy to the center of symmetry $r = 0$ is absent.

II. Analysis of the Dynamical System on the Component of the Boundary Γ_8.

In the component of the boundary Γ_8 ($z = 0$) the dynamical system describes the self-similar motion of self-gravitating dust $(p = 0)$. In this case the energy density $\varepsilon = (v^2/2) - (G\mathcal{M}/r)$ remains constant along the lines of flow, i.e. $d\varepsilon/dt = 0$. For self-similar solutions we have $\varepsilon = (r^2/t^2)(V^2/2 - m)$. Let $\varepsilon_1 = V^2/2 - m$. The law of conservation of ε turns into the following equation:

$$\frac{d\varepsilon_1}{d\tau} = -2\varepsilon_1\frac{V - 1}{V - 2/\omega}. \tag{3.9}$$

In the component of the boundary Γ_8 the system of self-similar equations (1.14) has the following form:

$$V' = -\frac{m + V^2 - V}{V - 2/\omega}, \quad m' = m\frac{2 - 3V}{V - 2/\omega}. \tag{3.10}$$

This system has a monotone function $f = \varepsilon_1 m^\alpha$, $\alpha = (2 - \omega)/(\omega - 3)$:

$$\frac{df}{d\tau} = \frac{\omega}{\omega - 3}f, \quad f = f_0\exp\left(\frac{\omega}{\omega - 3}\tau\right).$$

From (3.9) we easily obtain one exact trajectory of the system (3.10) explicitly:

$$\varepsilon_1 = 0, \quad \text{or} \quad m = V^2/2.$$

For $\omega > 2$ the dynamical system has five isolated non-degenerate critical points in the component of the boundary Γ_8: Z_8, Z_9, Z_{10}, C_1, H (and another critical point Z_6 for $\omega < 2$) and a segment of saddle-like critical points $D_1 G$. As mentioned in Sect. 1 the critical points contained in the segment $D_1 G$ and the critical points C_1 and H are saddles and all their separatrices lie in the various components of the boundary Γ. The critical points Z_6 ($\omega < 2$) and Z_9 were considered in Sect. 2. Note that the critical points Z_8, Z_9 and Z_{10} belong to the parabola $m = V^2/2$ which thus consists of separatrices of these critical points.

The critical point Z_{10} ($\eta = u_2 = 0$, $m_4 = 2^{-1/3}$) has the following eigenvalues:

$$\lambda_{m_4} = 1, \quad \lambda_\eta = \frac{3}{2}, \quad \lambda_{u_2} = \frac{5 - 3\gamma}{4}. \tag{3.11}$$

For $1 < \gamma < 5/3$ the critical point Z_{10} is repelling. Separatrices emerging from this critical point correspond to the stable (for $\lambda \to 0$) asymptotic form of accretion of the gas to the center which is discussed in Sect. 4. For $\gamma > 5/3$ the critical point Z_{10} is a saddle and the two-dimensional separatrix leaving Z_{10} lies in the component of the boundary Γ_8, whereas the one-dimensional separatrix approaching Z_{10} lies in the component of the boundary Γ_1.

The critical point Z_8 ($V = z = m = 0$) is attracting. Its eigenvalues are $\lambda_m = \lambda_z = -\omega$, $\lambda_V = -\omega/2$. The trajectory X (3.4) emerging from the critical point Z_3 as $\lambda \to 0$ approaches the critical point Z_8 as $\lambda \to \infty$. Other separatrices approaching this critical point as $\lambda \to \infty$ have the following asymptotic form:

$$\mathcal{M} = \frac{4\pi A_1}{3 - \omega} r^{3-\omega}, \quad \rho = A_1 r^{-\omega}, \quad v = (AG)^{1/2} C_3 r^{1-\omega/2}.$$

$$p = \frac{4\pi}{(3 - \omega)\gamma} \frac{C_1}{C_2} A_1^2 G r^{2(1-\omega)}, \quad A_1 = \frac{3 - \omega}{4\pi} AC_2. \tag{3.12}$$

For the separatrices tangent to the plane $V = 2(\omega - 1)z/\gamma - m$ the velocity of gas is $v = AG(2(\omega - 1)C_1/\gamma - C_2)r^{1-\omega}$. Obviously the above asymptotic form (3.12) describes some perturbation of the equilibrium state (3.5).

The above analysis of critical points allows us to completely reconstruct the phase portrait of the dynamical system in the component of the boundary Γ_8. It is shown in Fig. 28.

4. Self-Similar Accretion of Self-Gravitating Gas to the Center

Self-similar solutions describing (for $t \to \infty$) the accretion (fall) of self-gravitating gas to the center correspond to separatrices of the critical points Z_2, Z_{10}, $\mathscr{L}$ ($\gamma = 5/3$) and $\mathscr{L}_2$ ($\gamma = 4/3$). We shall find the asymptotic forms of accretion corresponding to the separatrices of these critical points.

Separatrices leaving the repelling (for $4/3 < \gamma < 5/3$) critical point Z_2 have the

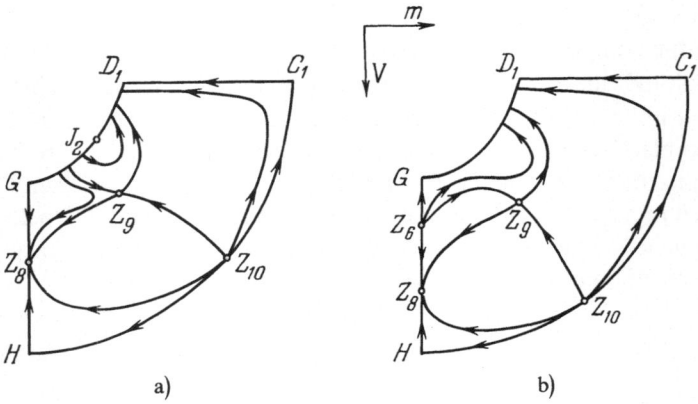

Fig. 28. Phase portraits of the dynamical system (3.10) in the component of the boundary Γ_8: a) $\gamma > 1$, $2 < \omega < 3$; b) $\gamma > 1$, $1 < \omega < 2$

following stable asymptotic form as $\lambda \to 0$ $(t \to \infty)$:

$$\mathcal{M} = A^{3/\omega} G^{(3-\omega)/\omega} C_1 t^{2(3-\omega)/\omega},$$

$$\rho = (AG)^{1/(\gamma-1)\omega} \frac{3-\omega}{2\pi\omega G} C_1 C_2 r^{-1/(\gamma-1)} t^{2(1/(\gamma-1)\omega-1)},$$

$$p = (AG)^{(3\gamma-2)/(\gamma-1)\omega} \frac{(3-\omega)(\gamma-1)}{2\pi\gamma\omega G} C_1^2 C_2 r^{-\gamma/(\gamma-1)} t^{-2(2-(3\gamma-2)/(\gamma-1)\omega)}, \tag{4.1}$$

$$v = -\frac{1}{C_2}(AG)^{(3\gamma-4)/(\gamma-1)\omega} r^{(3-2\gamma)/(\gamma-1)} t^{-1-2(4-3\gamma)/(\gamma-1)\omega}.$$

Separatrices leaving the repelling (for $1 < \gamma < 5/3$) critical point Z_{10} correspond to the stable (for $\lambda \to 0$ $(t \to \infty)$) asymptotic form

$$\mathcal{M} = (AG)^{3/\omega} \frac{C_1^2}{2G} t^{2(3-\omega)/\omega},$$

$$\rho = (AG)^{3/2\omega} \frac{(3-\omega)C_1}{4\pi\omega G} r^{-3/2} t^{(3-2\omega)/\omega},$$

$$p = (AG)^{(4+3\gamma)/2\omega} \frac{C_2}{G} r^{-3\gamma/2} t^{(4+3\gamma)/\omega-4}, \tag{4.2}$$

$$v = -(AG)^{3/2\omega} C_1 r^{-1/2} t^{(3-\omega)/\omega}.$$

Critical points in the line

$$\mathcal{L} \ (\gamma = 5/3)(\rho_0 = 0, \ m_1^3 = -v_1(3 + v_1^2)/2, \ -\infty \leqslant v_1 \leqslant 0)$$

are repelling and have the following eigenvalues:

$$\lambda_{p0} = -3v_1/2, \quad \lambda_2 = -v_1, \quad \lambda_3 = 0.$$

Separatrices leaving these critical points have the following stable asymptotic form as $\lambda \to 0 \, (t \to \infty)$:

$$\mathcal{M} = (AG)^{3/\omega} \frac{3 + v_1^2}{2G} C_1^2 t^{2(3-\omega)/\omega},$$

$$\rho = (AG)^{3/2\omega} a r^{-3/2} t^{(3-2\omega)/\omega}, \quad a = \frac{(3 - \omega)(3 + v_1^2) C_1}{4\pi\omega G |v_1|}, \tag{4.3}$$

$$p = (AG)^{9/2\omega} \frac{a}{\gamma} C_1^2 r^{-5/2} t^{(9-4\omega)/\omega}, \quad v = v_1 (AG)^{3/2\omega} C_1 r^{-1/2} t^{(3-\omega)/\omega}.$$

In the asymptotic forms (4.1)–(4.3) the gas falls to the center in finite time. At the center $r = 0$ a point mass forms which grows with time. Therefore these asymptotics describe the accretion of gas to a black hole in the classical theory. When this happens, at the center of symmetry the pressure, density, temperature and velocity of the gas are infinite. In the asymptotic form (4.1) the Mach number of the flow of gas tends to 0. In the asymptotic form (4.2) it tends to infinity and in the asymptotic form (4.3) we have $M \to |v_1|$ for $\lambda \to 0$. Critical points on the line $\mathcal{L}_2$ ($\gamma = 4/3$) correspond to the stable asymptotic form of accretion (2.5) (for $m_2 > 3(\omega - 1)/\omega$). In this asymptotic form the gas falls to the center in infinite time, the mass at the center $\mathcal{M}(0)$ is zero and a black hole does not form.

The full analysis of the critical points of the dynamical system in the manifold S conducted in Sects 2–4 shows that the list of all possible asymptotics of self-similar accretion is exhausted by the asymptotic forms (4.1)–(4.3) and (2.5). Thus for $1 < \gamma < 4/3$ the self-similar accretion of self-gravitating gas toward the center has the asymptotic form (4.2) for $t \to \infty$. For $\gamma = 4/3$ two asymptotic forms are possible: (2.5) and (4.2). For $4/3 < \gamma < 5/3$ the asymptotic forms (4.1) and (4.2) can occur. For $\gamma = 5/3$ we have the asymptotic form (4.3). For $\gamma > 5/3$ the self-similar accretion of self-gravitating gas does not happen.

With the help of the phase portraits of the dynamical system in the components of the boundary Γ_2 and Γ_8 constructed in Sect. 3 we can obtain a full (for all $0 < \lambda < \infty$) qualitative description of some self-similar solutions for which shock waves do not exist. Indeed according to the phase portrait shown in Fig. 27 for $\omega > 2$ and $4/3 < \gamma_3 = 2(\omega - 1)/3(\omega - 2) < \gamma < 5/3$ (in this case $\omega > 8/3$) there is a whole region of the manifold S which is filled by stable trajectories emerging from the repelling critical point Z_2 as $\lambda \to 0$ and approaching the attracting critical point Z_1 as $\lambda \to \infty$. In the corresponding self-similar solutions the gas moves away from the center with the asymptotic form (2.1) for $t \ll 1$. Then as t increases the motion of gas reverses direction and for further increases of t the gas drops to the center (to the black hole) in finite time with the asymptotic form of accretion (4.1).

For $4/3 < \gamma < \gamma_3$ the critical point Z_1 is a saddle (see Sect. 2) and has one-dimensional separatrix leaving it. This separatrix lies in the component of the boundary Γ_7 and approaches (for $\omega > 2$) the attracting critical point Z_7 (see Fig. 26). Therefore for $4/3 < \gamma < \gamma_3$, $\omega > 2$ there is a region in S filled by stable trajectories leaving (for $\lambda \to 0$) the repelling critical point Z_2 and approaching (for $\lambda \to \infty$) the attracting critical point Z_7 after moving along the separatrices $Z_2 Z_1$ and $Z_1 Z_7$. For $t \ll 1$ these solutions have the asymptotic form of expansion (2.6). In all other respects they are similar to the solutions described above.

For the same values of the parameters γ and ω we find that there is a whole region in the manifold S filled by stable trajectories leaving the repelling critical point Z_2 for $\lambda \to \lambda_1$ and after moving along the separatrices $Z_5 Z_4$ and $Z_4 Z_1$ for $\gamma_3 < \gamma < 5/3$ (and along the separatrices $Z_5 Z_4$, $Z_4 Z_1$ and $Z_1 Z_7$ for $4/3 < \gamma < \gamma_3$) approaching the attracting critical point Z_1 (respectively Z_7). In the corresponding solutions the gas moves away from the center monotonically and as r goes to infinity we have the asymptotic form of expansion (2.1) (respectively (2.6)) while inside the gas an expanding (with the asymptotic form (2.7)) cavity forms.

The two classes of solutions described above are separated by an unstable two-dimensional separatrix, which leaves the saddle Z_3 for $\lambda \to 0$ and approaches the attracting critical point Z_1 (Z_7) for $\lambda \to \infty$. The corresponding self-similar solutions describe the unstable process of formation of the equilibrium state (3.5) from the initially dispersing (with the asymptotic form (2.1) (respectively (2.6))) gas.

Completely analogously, with the help of the phase portraits in Fig. 28, we see that for $1 < \gamma < 5/3$ a whole region of the manifold S is filled by stable trajectories leaving the repelling critical point Z_{10} for $\lambda \to 0$ and approaching the attracting critical point Z_8 for $\lambda \to \infty$. The corresponding self-similar solutions describe intense accretion (with the Mach number going to infinity) of self-gravitating gas to the center (asymptotic form (4.2)) from the initial state (3.12), which is some perturbation (due to the presence of motion of the gas at $t = 0$) of the equilibrium state (3.5). In these solutions the velocity of gas at $t = 0$ can have either sign. For $\omega > 2$ in the asymptotic form (3.12) we have $v \to 0$ for $r \to \infty$. Therefore in the classical theory the above solutions describe the collapse of a star with the formation of a black hole at the center from an asymptotically (for $r \to \infty$) equilibrium state. Note that for $\gamma = 5/3$ and $\gamma = 4/3$ there also exist self-similar solutions analogous to the ones described above and having the asymptotic forms of accretion (4.3) and (2.5).

5. New Solutions in the Model of Stellar Explosions

I. Statement of the Problem. Let us recall the physical formulation of the model of stellar explosions according to [7, 116]. In this model it is assumed that at the initial moment of time the star is formed by a mass of ideal self-gravitating gas in a state of equilibrium:

$$\rho = \frac{c_1 A}{r^\omega}, \quad p = \frac{2\pi c_1^2 A^2 G}{(\omega - 1)(3 - \omega)} r^{2(1-\omega)}, \quad \mathcal{M} = \frac{4\pi c_1 A}{3 - \omega} r^{3-\omega}, \quad v = 0, \quad (5.1)$$

where $1 < \omega < 3$ and c_1 is a dimensionless constant. Then from the center of symmetry $r = 0$ as a result of an emission of energy (for example an explosion) or a breakdown of equilibrium a shock wave starts propagating. This is manifested by a discontinuity of the physical parameters of the gas (density, pressure, velocity of gas and entropy). It is further assumed that the motion of gas beyond the shock wave is self-similar, i.e. the parameters of the gas have the following form:

$$\rho = \frac{1}{Gt^2} R(\lambda), \quad p = \frac{r^2}{Gt^4} P(\lambda),$$

$$\mathscr{M} = \frac{r^3}{Gt^2} m(\lambda), \quad v = \frac{r}{t} V(\lambda),$$

(5.2)

where $\lambda = r/(AGt^2)^{1/\omega}$. Note that the equilibrium state (5.1) is also self-similar. This solution corresponds to the following dimensionless functions:

$$R(\lambda) = c_1 \lambda^{-\omega}, \quad P(\lambda) = \frac{2\pi v_1^2}{(\omega - 1)(3 - \omega)} \lambda^{-2\omega},$$

$$m(\lambda) = \frac{4\pi c_1}{3 - \omega} \lambda^{-\omega}, \quad V(\lambda) = 0.$$

(5.3)

After a transformation into the variables $z = \gamma P/R, m$ and V the equilibrium state (5.1) corresponds to the trajectory X of the system of self-similar equations (1.14):

$$V = 0, \quad z = \frac{\gamma}{2(\omega - 1)} m.$$

(5.4)

In a self-similar solution the law of motion of the shock wave is $\lambda = \lambda_*$. The speed of propagation of the shock wave is $v_* = (2/\omega) \lambda_* (AG)^{1/\omega} t^{(2-\omega)/\omega}$. Therefore for $\omega > 2$ the shock wave slows down as it moves away from the center and speeds up for $\omega < 2$. At the front of the shock wave the conditions of conservation of mass, momentum and the flow of energy are satisfied. After changing to dimensionless variables these conditions take the following form (subscripts 1 and 2 denote the two sides of the discontinuity):

$$R_1(V_1 - 2/\omega) = R_2(V_2 - 2/\omega), \quad m_1 = m_2,$$

$$V_1 - \frac{2}{\omega} + \frac{z_1}{\gamma(V_1 - 2/\omega)} = V_2 - \frac{2}{\omega} + \frac{z_2}{\gamma(V_2 - 2/\omega)},$$

$$\left(V_1 - \frac{2}{\omega}\right)^2 + \frac{2z_1}{\gamma - 1} = \left(V_2 - \frac{2}{\omega}\right)^2 + \frac{2z_2}{\gamma - 1}.$$

(5.5)

Under the transformation (5.5) (quantities with subscript 2 change into quantities with subscript 1) the trajectory X (5.4) changes into the line $Y(q)$:

$$z = \frac{4(2\gamma - (\gamma - 1)q)(\gamma - 1 + 2q)}{\omega^2(\gamma + 1)^2}, \quad m = \frac{8(\omega - 1)}{\gamma\omega^2} q, \quad V = \frac{4(1 - q)}{(\gamma + 1)\omega}.$$

(5.6)

Self-similar solutions, which are conjugate to the equilibrium state of the gas (5.1) across the shock wave, correspond to the trajectories of the system (1.14) which approach the points in the line $Y(q)$ $(0 < q < 1)$ for some $\lambda = \lambda_*$. For $\lambda_* < \lambda < \infty$ the solution is determined by a segment of the trajectory X (5.4). (For $\lambda < \lambda_*$ the trajectory goes through the subsonic region $L = z - (v - 2/\omega)^2 > 0$ in the manifold S, whereas the segment of the trajectory X for $\lambda > \lambda_*$ lies in the supersonic region $L < 0$). Since at the shock wave we have $m_1 = m_2$, by comparing the expressions for m in (5.3) and (5.6) we find that

$$q = \frac{\pi \gamma \omega^2 c_1}{2(\omega - 1)(3 - \omega)} \lambda_*^{-\omega}. \tag{5.7}$$

Thus the solutions of the problem of stellar explosions are determined by only one parameter $q = M^{-2}$, where $M > 1$ is the Mach number of motion of the shock wave over the stationary gas.

For the equilibrium state (5.1) the total energy (gravitational plus thermal) of a spherical layer $r_1 < r < r_2$ has the following form:

$$E = \int_{r_1}^{r_2} \left(\frac{p}{\gamma - 1} - \frac{\rho G \mathcal{M}}{r} \right) 4\pi r^2 \, dr,$$

$$\omega \neq 5/2: \; E = 8\pi^2 \frac{1 - 2(\omega - 1)(\gamma - 1)}{(\gamma - 1)(\omega - 1)(3 - \omega)} c_1^2 G A^2 \frac{1}{5 - 2\omega} r^{5 - 2\omega}|_{r_1}^{r_2}, \tag{5.8}$$

$$\omega = 5/2: \; E = c_1^2 G A^2 \frac{32\pi^2(4 - 3\gamma)}{3(\gamma - 1)} \ln \frac{r_2}{r_1}.$$

For $\omega < 5/2$ the total energy E of a ball of radius r is finite. For $\omega \geqslant 5/2$ it is infinite. For $\gamma > \gamma_4$ we have $E < 0$ and for $\gamma \leqslant \gamma_4$ we have $E \geqslant 0$, where $\gamma_4 = (2\omega - 1)/2(\omega - 1)$.

Solutions to the problem of stellar explosions in the class of motions of the gas under consideration are some intermediate asymptotic forms which are applicable outside a small neighborhood of the center $r = 0$ (for $\omega > 5/2$) as a consequence of divergence of the energy at the lower limit (see (5.8)). Solutions with $\gamma \leqslant \gamma_4$ $(E \geqslant 0)$ describe the breakdown of unstable equilibrium of a star. The law of liberation of energy in a self-similar solution has the following form:

$$E = \alpha G^{5/\omega - 1} A^{5/\omega} t^{2(5 - 2\omega)/\omega}. \tag{5.9}$$

The emitted energy E does not depend on time only when $\omega = 5/2$ or $\alpha = 0$. The constant α is calculated from the solution itself by way of comparison of the total energy (5.8) of a ball of gas of radius r in the equilibrium state to its energy at the moment when the shock wave hits the surface of the ball. Thus α is a function of the Mach number of motion of the shock wave. In some cases the constant α is infinite. The corresponding solutions are asymptotic forms of a very strong explosion $(E \gg 1)$.

II. Analysis of the Critical Point $Y_1 = Y(1)$. In order to study the solutions of the problem of stellar explosions for $M \approx 1$ it is necessary to analyze the behavior of trajectories of the dynamical system (1.14) in the neighborhood of the point $Y_1 = Y(1)$:

$$z = \frac{4}{\omega^2}, \quad m = \frac{8(\omega - 1)}{\gamma\omega^2}, \quad V = 0. \tag{5.10}$$

This point Y_1 belongs to the surface of non-extendability of solutions $L = 0$ and is a fixed point of the transformation (5.5). At this point the following three lines intersect: the line $Y(q)$ (5.6), the trajectory X (5.4) and the line of critical points $I_1 I_4$. The point Y_1 divides the line $Y(q)$ and the trajectory into two parts: the subsonic $(L > 0)$ and the supersonic $(L < 0)$ regions in the manifold S. Below we analyze the position of the segment of $Y(q), 0 < q < 1$ contained in the subsonic region.

The eigenvalues (1.37) at the critical point Y_1 are

$$\lambda_1 = 20\gamma(\omega - 2)\omega^{-3}, \quad \lambda_2 = -8\gamma\omega^{-2}, \quad \lambda_3 = 0. \tag{5.11}$$

The eigenvalues of critical points in the line $I_1 I_4$ near Y_1 are of course obtained from (5.11) by a continuous change. The system (1.14) has two invariant two-dimensional surfaces L_1 and L_2, which pass through the line of critical points $I_1 I_4$ and are filled by the separatrices of these critical points corresponding to the eigenvalues λ_1 and λ_2. The trajectory X in the subsonic region $L > 0$ is a separatrix approaching the critical point Y_1 and lies in the surface L_2.

Let us show that the segment of the line $Y(q), 0 < q < 1$ in the neighborhood of the critical point Y_1 for $\omega > 10/7$ lies (in the subsonic region) between the surfaces L_1 and L_2. Let l_1, l_2 and l_3 denote the eigenvectors corresponding to the eigenvalues λ_1, λ_2 and λ_3 (5.11) (vector l_3 is tangent to the line of critical points $I_1 I_4$). Let y be a vector tangent to the line $Y(q)$ at the point Y_1. The coordinates $(\delta z, \delta m, \delta V)$ of these vectors have the following form:

$$l_1: \left(\frac{2}{\omega} \cdot \frac{5(\gamma - 1)(\omega - 2) - 4\omega}{7\omega - 10}, \quad -\frac{8(\omega - 1)(\gamma + 1)}{\gamma(7\omega - 10)}, \quad 1 \right),$$

$$l_2: \left(\frac{\gamma}{2(\omega - 1)}, \quad 1, \quad 0 \right),$$

$$l_3: \left(-\frac{4}{\omega}, \quad -\frac{4(\omega - 1) + 6\gamma - \gamma\omega}{\gamma\omega}, \quad 1 \right),$$

$$y: \left(-\frac{3 - \gamma}{\omega}, \quad -\frac{2(\omega - 1)(\gamma + 1)}{\gamma\omega}, \quad 1 \right). \tag{5.12}$$

At the point Y_1 the surface L_1 is tangent to the two-dimensional plane P_1, which passes through the vectors l_1 and l_3. The surface L_2 is tangent to the plane P_2,

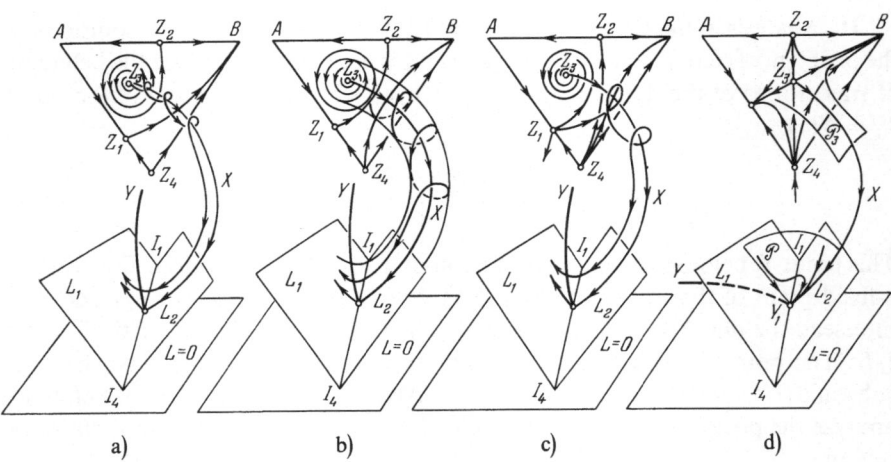

Fig. 29. The problem of stellar explosions with the Mach number of motion of the shock wave $M \approx 1$: a) $\gamma < \gamma_2$, $\omega > 5/2$; b) $\gamma < 4/3$, $\omega = 5/2$; c) $\gamma < \gamma_2$, $\omega < 5/2$; d) $4/3 < \gamma < 5/3$, $1 < \omega < 10/7$

which passes throught the vectors l_2 and l_3. It is not difficult to calculate the following determinants:

$$D_1 = \det(l_1, l_3, y) = \frac{5(\omega - 2)(\gamma + 1)}{\omega^2(7\omega - 10)}(3\omega - 10),$$

$$D_2 = \det(l_1, l_3, l_2) = -\gamma\frac{5(\omega - 2)}{2\omega(\omega - 1)}, \qquad (5.13)$$

$$D_3 = \det(y, l_3, l_2) = D_2.$$

Since $D_2 = D_3$, the vector y lies to the same side of the plane P_2 as the vector l_1. Since for $\omega > 10/7$ we have sign $D_1 =$ sign D_2, the vector y lies to the same side of the plane P_1 as the vector l_1. Therefore for $\omega > 10/7$ the vector y lies between the planes P_1 and P_2, i.e. the line $Y(q)$ in the neighborhood of the point Y_1 lies between the surfaces L_1 and L_2 (Fig. 29 a, b, c).

All trajectories of the dynamical system(1.14), which enter a neighborhood of the trajectory X and do not lie in the surface L_2 when approaching the line $I_1 I_4$ for $\omega > 10/7$, diverge from the surface L_2 and end up in a small neighborhood of the surface L_1 (this follows because $\lambda_1 > \lambda_2$ for $\omega > 10/7$). Since for $\omega > 10/7$ the line Y is located between the surfaces L_1 and L_2, some segment $Y_1 Y(q_0)$ ($q_0 < 1$, $M \approx 1$) in the line Y is completely crossed by trajectories, which earlier moved in the neighborhood of the trajectory X. This and the fact that the trajectory X leaves the critical point Z_3 contained in the component of the boundary Γ_2 allow us to analyze in detail the solutions of the problem of stellar explosions for the Mach numbers of motion of the shock wave $M \approx 1$. To do this we extensively use the phase portraits (Fig. 27) of the dynamical system in Γ_2 (see Sect. 3).

III. Self-Similar Solutions with Infinite Damping of Radial Oscillations of the Gas. For $\gamma < 4/3$, $\omega > 5/2$ the critical point Z_3 is repelling (see (3.3)). Therefore an entire neighborhood of the trajectory X is filled by the separatrices of this critical point. Consequently all solutions of the problem of stellar explosions for $M \approx 1$ tend to the equilibrium state (5.1) as $\lambda \to 0$ $(t \to \infty)$ (see Fig. 29 a). For $\gamma < \gamma_2 = 4(3 + (2\omega - 5)^2/8(\omega - 1))^{-1}$ during the approach to the equilibrium state the velocity of gas has an infinite number of oscillations near zero (see (3.6)). In this case the critical point Z_3 contained in the component of the boundary $\Gamma_2 (\rho_0 = 0)$ is a repelling focus (see Fig. 29 c). Thus in the solutions of the problem of stellar explosions for $M \approx 1$ and for $\gamma < \gamma_2$, $\omega > 5/2$ after the passing of the shock wave the gas starts pulsing and all gas particles undergo an infinite number of damped oscillations. The asymptotic formulas for these oscillations are given in Sect. 3 (see (3.7)–(3.8)). For $\omega \to 5/2$ the period (in the variable τ_0) of each oscillation is much smaller than the time of relaxation. The solutions obtained are self-similar perturbations, which do not destroy the star's equilibrium (on the whole for $\gamma < 4/3$ this equilibrium is unstable (see [73])). According to (5.8) in these solutions the total energy E of a gaseous ball of radius r is equal to $\pm \infty$ (depending on the relationship between γ and γ_4). After the passing of the shock wave the energy of the same ball is still equal to $\pm \infty$. Therefore the question about the amount of energy emitted (according to (5.9)) at the center of symmetry cannot be resolved in this case by comparing the energy of a gaseous ball before and after the explosion. However with the help of the asymptotic formulas (3.7)–(3.8) it can be shown that in the self-similar solutions under consideration there is no directed flow of energy to the center of symmetry (see part I of Sect. 3).

IV. Analysis of Solutions of the Problem of Stellar Explosions for $\omega = 5/2$. This case is of special interest, because it corresponds to an instantaneous release of energy at the center of symmetry (an explosion) (see (5.9)). For $\omega = 5/2$, besides the monotone function (1.36) connected with the adiabatic integral (1.12), the dynamical system (1.14) has another monotone function. This monotone function is the energy integral [7]:

$$H = -m\left[\frac{z}{\gamma - 1} + \frac{v_0^2}{2} - m + \frac{4}{5}\left(v_0 + \frac{z}{v_0}\right) + \frac{1}{2}\left(\frac{4}{5}\right)^2\right] = Ce^{-5\tau}. \quad (5.14)$$

For $\omega = 5/2$ all critical points Z_i, except Z_3, Z_4 and Z_5, belong to the surface $H = 0$, which is invariant relative to the system (1.14) and is a two-dimensional separatrix emerging from the critical point Z_6. The critical points Z_4 and Z_5 belong to the region $H > 0$ (at the critical point Z_3 we have $H < 0$ for $\gamma < 4/3$ and $H > 0$ for $\gamma > 4/3$). For $\gamma < 4/3$ the line $Y(q)$ (5.6) lies in the region $H < 0$ and for $\gamma > 4/3$ in the region $H > 0$. From this we obtain the following corollary.

Corollary. *The problem of a stellar explosion for $\gamma < 4/3$, $\omega = 5/2$ and any Mach number $(1 < M < \infty)$ of motion of the shock wave has no solutions with an expanding cavity inside the gas.*

For $\gamma = 4/3$, $\omega = 5/2$ the line $Y(q)$ lies in the surface $H = 0$, which is a separatrix of the critical point Z_6. This implies that the behavior of solutions of the problem of a stellar explosion for $\gamma = 4/3$ (integrated in [7]) is exceptional. In this case all solutions with an expanding cavity inside the gas $(M > 6)$ leave the critical point Z_6 and have the asymptotic form (2.8), which cannot occur for solutions with $\gamma \neq 4/3$, $\omega = 5/2$. Besides for $\gamma = 4/3$, $\omega = 5/2$ the solutions for Mach numbers $6 > M > 1$ are separatrices of the line of critical points $\mathscr{L}_2$ which is absent for $\gamma \neq 4/3$.

From the two monotone functions (1.36) and (5.14) we can form an integral of the system (1.14) (not used earlier) which in the coordinates (1.21) has the following form:

$$F = |H|^{4-3\gamma}\Phi_{\mu 2}^{-1},$$

$$F = \left| \frac{4}{5\gamma} + \frac{v_0}{\gamma - 1} + m_2 + 2^{-1}v_0\left(v_0 + \frac{4}{5}\right)^2 \rho_0^2 \right|^{4-3\gamma} |v_0|^{-\gamma}m_2^{3(\gamma-1)}. \tag{5.15}$$

On the trajectory X the integral F has constant value

$$F = F_0 = \gamma^{-1}(4/5)^{1-\gamma}3^{3(\gamma-1)}|(4-3\gamma)/(\gamma-1)|^{4-3\gamma},$$

and the line X consists of extrema of the function F, i.e. the first differential of F is zero $(dF = 0)$. The second differential d^2F on the line X has the following form:

$$d^2F = -F_0 3^{-1}\gamma^2(\gamma - 1)(5/4)^2(4 - 3\gamma)^{-1}$$
$$\times [(3dv_0 + dm_2)^2 + 3\gamma^{-1}(4 - 3\gamma)(1 - (4/5)^2\rho_0^2)dv_0^2]. \tag{5.16}$$

Let us consider the two cases $\gamma < 4/3$ and $\gamma > 4/3$ separately. According to (5.16) for $\gamma < 4/3$ the segment of the line X from the point $Z_3 (\rho_0 = 0)$ to the point $Y_1 (\rho_0 = 5/4)$ consists of maxima of the function F. Therefore the level surfaces of the integral F in the neighborhood of the segment $Z_3 Y_1$ of the line X are two-dimensional cylinders and intersect the component of the boundary $\Gamma_2 (\rho_0 = 0)$ at cycles (closed integral curves of the dynamical system in Γ_2). In Γ_2 the integral F coincides with the integral F_1 (3.6). As mentioned in section 3, in the region $m_2 + v_0/(\gamma - 1) + 4/(5\gamma) < 0$ in the component of the boundary Γ_2 all trajectories of the dynamical system (3.2) $(\gamma < 4/3, \omega = 5/2)$ (level curves of the integral F) are cycles. As λ decreases $(\lambda \to 0)$ all cycles (according to (1.34)) are attracting $(\rho_0 \to 0)$. From this it is not difficult to deduce that all solutions of the problem of a stellar explosion for Mach numbers $M \approx 1$ and $\lambda \to 0$ wind onto the cycles $F = \text{const}$, $\rho_0 = 0$ (see Fig. 29 b).

Solutions of the problem of a stellar explosion for Mach numbers $M \gg 1$ correspond to the trajectories of the system (1.14) passing through the line $Y(q)$ in the neighborhood of the point

$$Y_0 (q = 0): V = 4/(\gamma + 1)\omega, z = 8\gamma(\gamma - 1)(\gamma + 1)^{-2}\omega^{-2}, m = 0.$$

This point Y_0 lies in the integral curve $H = 0$ of the system (1.14) in the component of the boundary Γ_7:

$$z = -\gamma(\gamma - 1)\frac{(V - 4/5)V^2}{2(\gamma V - 4/5)}, \quad m = 0. \tag{5.17}$$

For $9/7 < \gamma < 4/3$ the trajectory (5.17) leaving the point Y_0 approaches the critical point Z_6 as λ decreases. Therefore all trajectories leaving the line $Y(q)$ $(q \approx 0, M \gg 1)$ (as λ decreases) for some finite $\lambda = \lambda_1$ approach the surface $L = 0$ (because for $\gamma < 4/3$ the line $Y(q)$ lies in the region $H < 0$ (see Fig. 26)). These trajectories have no physical meaning, because they cannot be continued for $\lambda < \lambda_1$ (continuous extension is clearly impossible and an extension with a discontinuity requires the introduction of a shock wave of evacuation which is also impossible in matter with normal properties (see [115])). Therefore in the class of self-similar motions of the gas considered for $9/7 < \gamma < 4/3, \omega = 5/2$ there are no solutions of the problem of a stellar explosion for the Mach number of the shock wave $M \gg 1$.

For $\gamma = 9/7$ the point Y_0 coincides with the critical point Z_7 which also always lies on the line (5.17). For $1 < \gamma < 9/7$ the trajectory (5.17) leaving the point Y_0 approaches the critical point Z_1 for $\lambda \to 0$. Therefore all trajectories leaving the line $Y(q)$ $(q \approx 0, M \gg 1)$ as λ decreases also turn up in the neighborhood of the critical point Z_1. Since they are contained in the region $H < 0$ they wind onto limit cycles $F = \text{const}, \rho_0 = 0$. In these solutions $(M \gg 1)$ the function $V(\lambda)$ for $\lambda \to 0$ varies from a finite maximum $V_* < 4/(5\gamma)$ to an arbitrarily large negative minimum.

In solutions corresponding to trajectories winding onto limit cycles the gas after the passing of the shock waves undergoes undamped oscillations. Indeed along the lines of flow we have

$$\frac{d\ln r}{d\ln t} = V = v_0 + \frac{4}{5}, \quad \frac{d\ln \lambda}{d\ln t} = v_0, \quad \frac{d\tau_3}{d\ln t} = -1; \tag{5.18}$$

so $d\ln r/d\tau_3 = -(v_0 + 4/5)$. In the neighborhood of a cycle the change in $\ln r$ over a period of one oscillation is

$$\ln r|_0^T \approx -\oint (v_0 + 4/5)\,d\tau_3 = A.$$

Let $B = \oint (m_2 - 12/5\gamma)\,d\tau_3$. From the system (3.2) for $\omega = 5/2$ it follows that $-3A + B = 0$ and $-4A + \gamma B = 0$, so for $\gamma \neq 4/3$ we have $A = B = 0$ for all cycles. Thus in the process of oscillations of the gas there is no drift.

The period of oscillations in the neighborhood of the critical point Z_3 in the variable τ_3 is approximately equal to $T_0 = 5\pi(\gamma/12(4 - 3\gamma))^{1/2}$. According to (5.18) in time t the oscillations of the gas slow down. In fact the periods of successive oscillations increase in a geometric progression with the coefficient $\exp T_0$. The amplitude A_0 of oscillations of $\ln r$ for solutions with $M \approx 1$ has order $A_0 \sim T_0(M - 1)^{3/2}$. In solutions with $M \gg 1$ $(1 < \gamma < 9/7)$ we have $A_0 \sim 2(3(\gamma - 1)(4 - 3\gamma))^{-1} \ln M$ for $M \to \infty$.

Note that the oscillations described above occur in the class of self-similar

solutions and qualitatively differ from the strictly periodic oscillations of homogeneous gaseous balls used as models of pulsations of variable stars [132]. In
particular in the solutions obtained, for a fixed t the velocity of gas for $r \to 0$ has
infinitely many zeroes and the density of gas ρ tends to infinity. However, the total
mass of the gas in the neighborhood of the center $r = 0$ is finite and varies like
$Cr^{1/2}$ (see the asymptotic formulas for self-similar oscillations of the gas (3.7)–
(3.8)).

The above solutions just as in the case $\gamma < 4/3$, $\omega > 5/2$ analyzed earlier are
self-similar perturbations for which the star pulses in the neighborhood of the
equilibrium distribution (5.1) even though the total energy of the gaseous ball
$E = +\infty$ (see (5.8)). The obtained solutions in any finite interval of t and in $0 <
r_1 < r < r_2$ are stable and can be used (not even necessarily in connection with
the problem of a stellar explosion) as a model of slowing pulsations in the depths
of the star (with constant amplitude).

According to (5.16), for $\gamma > 4/3$ the segment $Z_3 Y_1$ of the line X consists of
saddle points of the function F. The neighborhood of the segment $Z_3 Y_1$ is divided
into four parts by the level surface $F = F_0$: two regions D_1 and D_2, where $F \leqslant F_0$
and two regions D_3 and D_4, where $F \geqslant F_0$. The surface $F = F_0$ is single sheeted,
because from (5.15) it follows that on this surface ρ_0 is a single-valued function of
v_0 and m_2. The intersection of the surface $F = F_0$ with the component of the
boundary Γ_2 consists of four separatrices of the critical point Z_3 pictured in Fig.
26. For $\rho_0 = 0$ the region D_1 slices out a triangle $\triangle$ bounded by the separatrices
$Z_4 Z_3$, $Z_3 Z_1$ and $Z_1 Z_4$. Let us show that the projection of any trajectory in the
region D_1 to the v_0, m_2 plane does not intersect the separatrices $Z_3 Z_1$ and $Z_4 Z_3$.
Indeed since at the points of these separatrices we have $F = F_0$ and $H > 0$, in any
point over them (where $\rho_0 > 0$) we obtain $F > F_0$ from (5.15). Now in the triangle
$\triangle$ the function V is positive, so it is positive in the entire region D_1 (analogously it
can be shown that $V < 0$ in the region D_2).

On the line $Y(q)$ the integral F monotonically increases from 0 to F_0:

$$F(Y(q)) = F_0 \frac{q(\gamma + 1)^{\gamma+1}}{(2\gamma - (\gamma - 1)q)(\gamma - 1 + 2q)^\gamma} \tag{5.19}$$

and the function V is positive. Therefore the line $Y(q)$, $0 < q < 1$ lies entirely in the
region D_1 for $H > 0$. Let us describe the geometry of this region. The intersection
of the region D_1 with the boundary Γ is bounded by the separatrices $Z_3 Z_1$,
$Z_1 Z_7 Z_6$, $Z_6 Z_5$, $Z_5 Z_4$ and $Z_4 Z_3$ (see Fig. 26). The region D_1 is a closed invariant
(relative to the system (1.14)) region of the manifold S, which belongs entirely to
the subsonic part $L > 0$. The last assertion follows from the fact that on the
surface $L = 0$: $\rho_0 = |v_0|$ over the triangle $\triangle$ in the region $H > 0$ the function F
(5.15) has a unique extremum: a minimum at the point Y_1 $(E(Y_1) = F_0)$.

In the region D_1 there is a unique repelling critical point Z_5 and a unique
attracting critical point Z_7. All other critical points in this region (Z_1, Z_3, Z_4 and
Z_6) are unstable and their separatrices lie in the boundary of the region D_1.
Therefore as a consequence of the existence of the monotone function Φ_μ (1.36) all

trajectories in the region D_1 and in particular all trajectories passing through the line $Y(q)$ emerge from the critical point Z_5. Thus we have just proved the following theorem.

Theorem. *All solutions of the problem of a star's explosion for $\gamma > 4/3$, $\omega = 5/2$ and all Mach numbers $(1 < M < \infty)$ of motion of the shock wave have an expanding cavity inside the gas with the asymptotic form* (2.7).

In all these solutions the dispersion of gas from the center is monotonic, because $V > 0$ in the region D_1. Solutions with an expanding cavity inside the gas were first found (numerically for $\gamma = 5/3$ and analytically for $\gamma = 4/3$) in [7, 116]. Note that as shown above in the problem of a stellar explosion for $\gamma < 4/3$, $\omega = 5/2$ there are no solutions with the formation of a cavity inside the gas.

In the above solutions for $\gamma > 4/3$ the total energy E emitted during the explosion is infinite. In other words these solutions should be considered as asymptotic forms describing the explosion of a star for $E \rightarrow \infty$.

The analysis conducted above shows that the behavior of solutions of the problem of a stellar explosion in the general case $\gamma \neq 4/3 \, (\omega = 5/2)$ is qualitatively different from the case $\gamma = 4/3$, $\omega = 5/2$ (integrated in [7]), where there are no oscillations of the gas at all and for Mach numbers $1 < M \leqslant 6$ all solutions can be continued to the center of symmetry.

V. Self-Similar Solutions with an Arbitrary Finite Number of Radial Oscillations of the Gas. For $\gamma < 4/3$, $\omega < 5/2$ the critical point Z_3 in the component of the boundary $\Gamma_2 \, (\rho_0 = 0)$ is attracting (for $\gamma < \gamma_2$ it is an attracting focus (see (3.3))). A small neighborhood of the critical point Z_3 is filled by separatrices of the critical points Z_4 (or Z_5 depending on the relationship between γ and γ_1 (see Sect. 2)) and Z_1. For $\gamma < \gamma_2$ the two-dimensional separatrix Z leaving the critical point Z_1 intersects the component of the boundary Γ_2 along a spiral, which infinitely winds onto the critical point Z_3 (see Fig. 29 c). Therefore the entire two-dimensional separatrix Z winds infinitely many times onto the separatrix X leaving the critical point Z_3. Consequently in the neighborhood of the critical point Y_1 the separatrix Z intersects the line Y infinitely many times at the points $Y_i(M_i) \, (M_i \rightarrow 1$ for $i \rightarrow \infty)$ (see Fig. 29 c). Thus we have just proved that for $\gamma < \gamma_2$, $10/7 < \omega < 5/2$ there exists an infinite sequence of Mach numbers of motion of the shock wave $M_i \rightarrow 1$ for which the solutions of the problem of a star's explosion can be continued to the center of symmetry and for $\lambda = 0$ have the asymptotic form (2.1). For these solutions the energy of a mass of gas beyond the shock wave is equal to the energy of the same mass of gas in the state of equilibrium, because at the center $r = 0$ the asymptotic form (2.1) has no singularity and the flow of gas is everywhere adiabatic. Therefore the solutions obtained correspond to an explosive type of breakdown of equilibrium of the star without the loss of energy. The first solution of this type (dynamical explosion of the equilibrium) was given in an explicit form for $\gamma = 7/6$, $\omega = 12/5$ in [7]. For this solution we have $M^2 = 15/2$ and the oscillations of gas are absent.

All solutions for Mach numbers M: $M_i > M_{i+1}$ are separatrices of the critical point Z_4 (or Z_5) and consequently have an expanding cavity inside the gas with the asymptotic form (2.7). The corresponding trajectories of the dynamical system (1.14) perform some finite number $N(M)$ of turns around the trajectory X, where $V = 0$. Therefore, for these solutions the dimensionless velocity of the gas becomes zero $2N(M)$ times. Along the lines of flow of the gas (according to (5.18)) we have $d \ln r/d\tau_3 = -V$, so in the considered solutions all gas particles undergo $N(M)$ oscillations. $N(M) \to \infty$ for $M \to 1$, so for $\gamma < \gamma_2$, $\omega < 5/2$ there exist solutions of the problem of a stellar explosion with an arbitrary finite number of oscillations of the gas after the passing of the shock wave. The oscillations of the gas give way to a monotone expansion of the gas away from the center, which for a discrete set of values of Mach numbers of the shock wave takes place with the asymptotic form (2.1) and for all other values of Mach numbers with the asymptotic form (2.7).

VI. Self-Similar Solutions with a Weak Discontinuity. In the preceding parts we studied self-similar solutions describing the various types of loss of stability of a star after the passing of a shock wave. Now let us consider a new type of breakdown of equilibrium of a star, in which there are no shock waves, but a weak discontinuity propagates over the stationary gas in the state (5.1), beyond which the solution is self-similar. By definition (see [7]) at the points of weak discontinuity all physical parameters of the gas (ρ, p, v, $\mathcal{M}$ and the entropy S) remain continuous, but some of their derivatives have a discontinuity. Trajectories of the dynamical system (1.14) corresponding to the considered solutions approach the critical point Y_1 for $\lambda = \lambda_*$ (λ_* is determined from (5.7) for $q = 1$) and for $\lambda > \lambda_*$ are continued by a segment of the trajectory X (5.4). The law of motion of the surface of weak discontinuity is $\lambda = \lambda_*$.

For $\omega > 2$ according to (5.11) the critical point Y_1 is a saddle. Therefore it is approached by a unique trajectory X, i.e. for $\omega > 2$ the desired solutions with a weak discontinuity do not exist.

For $1 < \omega < 2$ the critical point Y_1 and all critical points contained in the $I_1 I_4$ in the neighborhood of Y_1 are attracting. Therefore there exists some two-dimensional surface $\mathscr{P}$, filled by trajectories of the dynamical system (1.14) approaching the critical point Y_1. According to the earlier (part II) analysis, for $10/7 < \omega < 2$ trajectories in the neighborhood of the point Y_1 approach the line of critical points $I_1 I_4$ tangentially to the surface L_1 (Fig. 30 a) and for $1 < \omega < 10/7$ tangentially to the surface L_2, where in particular the trajectory X is contained (Fig. 30 b). As a consequence of this, for $1 < \omega < 10/7$ in all solutions with a weak discontinuity, except for the one contained in the surface L_1, the first derivatives of the physical quantities remain continuous and the discontinuity occurs in their second derivatives.

Let us consider self-similar solutions with a weak discontinuity for $1 < \omega < 10/7$, $4/3 < \gamma < 5/3$. For $4/3 < \gamma < 5/3$ the neighborhood of the trajectory X is divided by the two-dimensional separatrix $\mathscr{P}_3$ leaving the critical point Z_3 (see

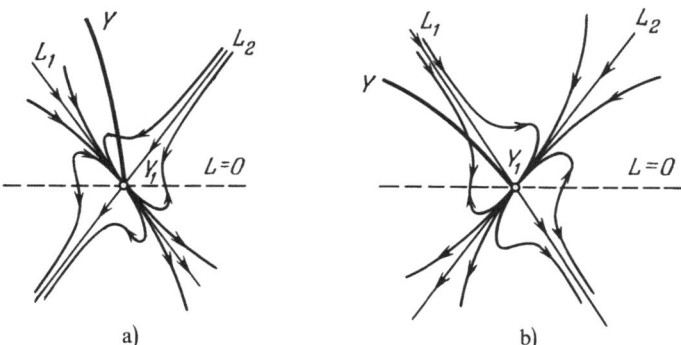

Fig. 30. The behavior of the trajectories of the dynamical system (1.14) in the neighborhood of the critical point Y_1: a) $\gamma > 1$, $10/7 < \omega < 2$; b) $\gamma > 1$, $1 < \omega < 10/7$

Fig. 29 d) into two regions U_1 and U_2. Trajectories of the system (1.14) in the region U_1 emerge from the critical point Z_2 and the corresponding solutions describe the accretion of the gas to a black hole (see Sect. 4). Trajectories in the region U_2 emerge from the critical point Z_5 and describe the dispersal of the gas with the formation of an expanding cavity inside the gas (see Sect. 2). In the general case, the two two-dimensional surfaces $\mathscr{P}$ and $\mathscr{P}_3$ invariant relative to the system (1.14) intersect along the unique trajectory X, while the surface $\mathscr{P}$ intersects both regions U_1 and U_2. For $1 < \omega < 10/7$ all trajectories in the surface $\mathscr{P}$ in the neighborhood of the trajectory X approach the critical point Y_1 (see Fig. 29d). Therefore for $1 < \omega < 10/7$, $4/3 < 5/3$ the intersection of the surface $\mathscr{P}$ with the region U_1 determines a one-parameter family of trajectories approaching (for $\lambda = \lambda_*$) the critical point Y_1 and leaving (for $\lambda = 0$) the critical point Z_2. These trajectories after their continuation by a segment of the trajectory X (for $\lambda > \lambda_*$) determine self-similar solutions with a weak discontinuity. They serve as models of breakdown of the equilibrium state of the star (5.1), after which the gas collapses to the center with the asymptotic form (4.1) for $\lambda \to 0$ resulting in the formation of a black hole (a growing point mass) at the center. Such type of breakdown of equilibrium of a star is accompanied (for $4/3 < \gamma < 3/2$) by the radiation of infinitely large amounts of energy. Indeed in the equilibrium state (5.1) a gaseous ball of radius r for $1 < \omega < 10/7$ has finite positive energy (5.8). The total energy of a spherical layer $r_1 < r < r_2$ in the asymptotic form (4.1) has the following form:

$$E = \int_{r_1}^{r_2} \left(\frac{\rho v^2}{2} + \frac{p}{\gamma - 1} - \frac{\rho G \mathscr{M}}{r} \right) 4\pi r^2 \, dr$$

$$= -\frac{K_1}{2\gamma - 3} r^{(2\gamma-3)/(\gamma-1)} \Big|_{r_1}^{r_2} + \frac{K_2}{2 - \gamma} r^{(2-\gamma)/(\gamma-1)} \Big|_{r_1}^{r_2},$$

(5.20)

where the positive quantities K_1 and K_2 are expressed in terms of the constants of the asymptotic form (4.1) and time t. For $\gamma < 3/2$ the energy (5.19) diverges at the lower limit, i.e. for $4/3 < \gamma < 3/2$ the gaseous ball has infinite negative energy in the asymptotic form (4.1). Consequently for $4/3 < \gamma < 3/2, 1 < \omega < 10/7$ in the above solutions during the breakdown of the equilibrium state (5.1) there is a loss (emission) of an infinitely large positive amount of energy. (For these solutions the constant α in the law of emission of energy (5.9) is equal to $-\infty$).

Another interesting class of self-similar solutions with a weak discontinuity is $1 < \omega < 2, \gamma < \gamma_2 < 4/3$. In this case the two-dimensional surface $\mathscr{P}$ intersects the two-dimensional separatrix Z which winds around the trajectory X infinitely many times along an infinite number of trajectories, which therefore leave the critical point Z_1 for $\lambda = 0$ and approach the critical point Y_1 for $\lambda = \lambda_*$. These trajectories after their continuation for $\lambda > \lambda_*$ by a segment of the trajectory X correspond to solutions with a weak discontinuity, in which the breakdown of equilibrium of the star is not accompanied by an emission of energy and which can be smoothly continued to the center of symmetry without a singularity. All conclusions of part V about the presence of a finite (arbitrarily large) number of radial oscillations of the gas are also applicable to these solutions.

For all ω such that $1 < \omega < 2$ and $\gamma > 1$ there are one-parameter families of solutions with a weak discontinuity and an expanding cavity inside the gas (with the asymptotic form (2.7)). For $\gamma > 4/3$ these solutions correspond to trajectories contained in the intersection of the surface $\mathscr{P}$ with the region U_2 (see above). For $\gamma < 4/3$ the existence of the solutions considered follows from the fact that almost all trajectories in the neighborhood of the trajectory X are separatrices of the critical point Z_5.

VII. Physical Interpretation. In the equilibrium state (5.1) a self-gravitating gas fills the whole space $0 < r < \infty$ and the density of gas ρ tends to ∞ as $r \to 0$. Therefore the model of a stellar explosion considered in this section can be used to model the non-stationary motion of the gas in real stars only in the region $0 < r_0 < r < R_0 < R$ (where R is the radius of the star) and in the time interval $t < T_0$ (where T_0 is the time when the shock wave hits the surface of the star). The power law (5.1) of distribution of density and pressure is of course only an approximation of reality. A necessary condition of admissibility of such an approximation is the stability of the distribution (5.1) relative to the convective perturbations, which according to the Schwarzschild criterion [73] $(dS/dr > 0$, where S is entropy density) is the same as

$$\gamma > 2 - \frac{2}{\omega}. \tag{5.21}$$

For $\gamma > 4/3$ this condition is satisfied in the entire interval $1 < \omega < 3$.

Self-similar solutions with $\omega = 5/2, \gamma > 4/3$ (see part IV) are models of the most powerful explosions of supernovae, where most of the mass of the star disperses into the surrounding space and the remaining mass (a neutron star) is negligible in comparison with the total initial mass of the star. The condition of

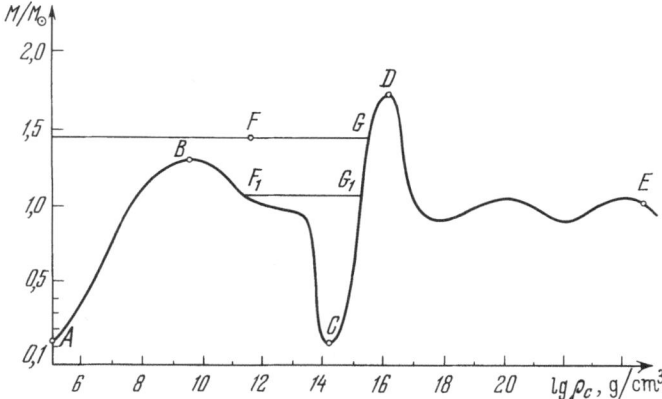

Fig. 31. Dependence of the mass M of a cold star on the density ρ_c at the center. Data from Saakyan/Vartanyan [114]

admissibility of this model is $E \gg E_1, E_2$, where E is the total energy emitted (in the neighborhood of the center of the star) during the explosion, E_1 is the energy transmitted by way of emission of neutrinos and their subsequent absorption in the outer layers of the star (such a form of transmission of energy is called deposition (see [73])) and E_2 is the energy emitted at the front of the shock wave as a result of detonation of the gas. Modern estimates of the role of the neutrino mechanism of transmission of energy in the process of a stellar explosion can be found in [73, 134]. Self-similar solutions with neutrino emission were analyzed in [135]. Detonation waves in the self-similar motion of a self-gravitating gas can only be introduced for $\omega = 2$ (see [118]). To qualitatively analyze self-similar solutions with detonation waves with Mach numbers $M \approx 1$, it is sufficient to resolve the critical point Y_1, which for $\omega = 2$ is degenerate $(\lambda_1 = 0$ (see (5.11))), and use the results obtained above about the behavior of the trajectories of the dynamical system (1.14) in the neighborhood of the component of the boundary Γ_2.

To establish the physical applicability of self-similar solutions with $\gamma < 4/3$ let us recall the dependence of the mass M of the star (in its final stage of evolution) on the density ρ_c at the center (Fig. 31) (this dependence was derived in [114] as a result of numerical integration of the equations of equilibrium with the use of the real equation of state of matter). A necessary condition for the stability of the equilibrium configuration is $dM/d\rho_c > 0$ (see [73]). This condition corresponds to two segments AB and CD of the curve in Fig. 31, which determine respectively white dwarfs and neutron stars. The points B and D correspond to the Chandrasekar limit and the Oppenheimer-Volkoff limit. The segment BC corresponds to unstable configurations in which the equation of state of matter of the star is approximately adiabatic with the coefficient $\gamma < 4/3$ (see [73]). Equilibrium configurations with $\log \rho_c < 14$ are accurately described by the Newtonian theory of gravitation. To study configurations with higher density $(\log \rho_c \geqslant 15)$ it is necessary to take into account the effect of the general theory of relativity. All equilibrium configurations with $\log \rho_c > 16$ are unstable. By virtue of the equations of the general theory of relativity the mass of an equilibrium configuration M is a periodic function of ρ_c for $\rho_c \to \infty$ [113, 112].

Consider a star, which corresponds to the point F or F_1 on the graph of $M(\rho_c)$. These configurations are unstable and adiabatic with $\gamma < 4/3$ (see [73]). In the process of evolution the star in the $M(\rho_c)$ graph moves along a horizontal line (because the mass M is constant) until it reaches the neighborhood of a stable equilibrium G or G_1. In the neighborhood of a stable equilibrium the star will undergo (in first approximation) damped oscillations (see [73]).

The process of evolution of a star described above is modelled by self-similar solutions with

$$\omega > 5/2, \quad \gamma < \gamma_2 = 4[3 + (2\omega - 5)^2/8(\omega - 1)]^{-1} < 4/3$$

(see part III). In these solutions for $t \to \infty$ the gas undergoes damped oscillations in the neighborhood of the equilibrium state (5.1). The total energy of a spherical layer $(r_1 < r < r_2)$ in the equilibrium state (5.1) is negative if $\gamma > \gamma_4 = (2\omega - 1)/2(\omega - 1)$ and positive if $\gamma < \gamma_4$ (see (5.8)). On the whole, arbitrary equilibrium configurations with $\gamma < 4/3$ are considered to be unstable [73, 76] and the equilibrium distributions (5.1) with $\omega \geqslant 5/2$, $\gamma < 4/3$ can be called metastable relative to a certain class of perturbations (including self-similar perturbations).

Note that there are no analogous self-similar solutions in the general theory of relativity. Self-similar solutions in relativity can have only a finite number of radial oscillations of the gas, after which the entire mass of gas disperses from the center (see Sect. 4 of chapter IV). This is completely consistent with the fact that in the particularly relativistic region $\ln \rho_c > 16$ (see segment DE in Fig. 31) there are no stable equilibrium configurations.

In the equilibrium state (5.1) with $1 < \omega \leqslant 5/2$, $\gamma < 4/3$ the total energy of a spherical layer $(r_1 < r < r_2)$ is positive. Configurations with positive energy cannot occur in diffusive matter, however in principle they can exist and even be metastable (see [73]). Self-similar solutions in the model of a stellar explosion with $1 < \omega < 5/2$, $\gamma < \gamma_2 < 4/3$ (see part V) describe a special explosive type of breakdown of an unstable configuration with positive energy. In fact there exist arbitrarily small perturbations corresponding to Mach numbers of motion of the shock wave $M_i \to 1$ for which there is no emission of energy, however, (after a finite number of radial oscillations) there is a total dispersion of the entire mass of gas (the first solution of this type was obtained for $\gamma = 7/6$, $\omega = 12/5$ in [7]) (for the self-similar solutions considered here with $1 < \omega < 10/7$, $\gamma < \gamma_2$ instead of the shock wave there is a weak discontinuity (see part VI)). Self-similar solutions with $1 < \omega < 10/7$, $\gamma < \gamma_2$ having an arbitrarily large but necessarily finite number of radial oscillations of the gas can also be considered as models of oscillatory instability, which occurs in some massive stars [73].

Another type of breakdown of an equilibrium configuration with positive energy is represented by self-similar solutions with $1 < \omega < 10/7$, $4/3 < \gamma < 3/2$ (see part VI). As a result of emission of an infinitely large amount of energy from the center in these solutions the entire mass of the star starts collapsing and the equilibrium distribution (5.1) is glued to the self-similar solution describing the accretion of gas to the center across a weak discontinuity. The self-similar solutions just described represent the process of implosion of the central part of the star accompanied by powerful deposition [73], e.g. by virtue of neutrino emission.

6. Analysis of Models of Explosions in Stellar Envelopes

I. Statement of the Problem. According to modern data (see [73, 152]), during explosions of some novae the emission of energy and motion of gas occur only in the region surrounding the star's surface. This region is called the star's envelope. During such explosions the matter inside the star's nucleus is not involved in the interactions, however by virtue of its gravitational attraction it exerts a substantial influence on the motion of gas in the envelope. The mass of a star's envelope $\mathcal{M}_0$ is much smaller than the mass of the nucleus $\mathcal{M}$. Therefore the self-gravitation of gas in the envelope can be considered negligible compared to the force of gravitational attraction of the nucleus. In the spherically symmetric case this force is equal to the force of gravity due to a point mass $\mathcal{M}$ located at the center of symmetry. Thus in order to study explosions in stellar envelopes it is necessary to analyze the spherically symmetric motion of an ideal gas (neglecting self-gravitation) in the field of an attracting center.

In [124, 125] the following model of explosions in envelopes of stars was proposed for the first time. Initially the gas in the envelope is in the state of equilibrium subject to forces of pressure and the force of gravitational attraction of the point mass $\mathcal{M}$. The parameters of the gas in the state of equilibrium have the following form:

$$\rho = \frac{c_1 a (G\mathcal{M})^{s/2}}{r^\omega}, \quad p = \frac{c_1 a (G\mathcal{M})^{1+s/2}}{(\omega + 1) r^{\omega+1}}, \quad v = 0, \tag{6.1}$$

where a is a constant with dimensions $M L^k T^s$, $\omega = (3/2)(s + 2) + k$ and c_1 is a dimensionless constant. Later as a result of liberation of energy (e.g. an explosion) or a loss of equilibrium, a shock wave emerges from the center of symmetry $r = 0$, beyond which the motion of gas is adiabatic and self-similar. The motion of the gas satisfies the equations of gas dynamics (1.4)–(1.6) for $\mathcal{M} = \text{const}$. These equations contain a constant $G\mathcal{M} = b^3$ with dimensions $L^3 T^{-2}$. Therefore the self-similar variable λ is defined uniquely: $\lambda = r/(bt^{1/2})$. The parameters of self-similar motion of the gas have the following form:

$$\rho = \frac{a}{r^{k+3} t^s} R(\lambda), \quad p = \frac{a}{r^{k+1} t^{s+2}} P(\lambda), \quad v = \frac{r}{t} V(\lambda). \tag{6.2}$$

The study of the model of explosions in envelopes of stars is reduced to the analysis of self-similar solutions of the type (6.2) which by virtue of Hugoniot

conditions (see (5.5)) are conjugate to the equilibrium state of the gas (6.1) across
the shock wave.

In this section we qualitatively analyze the model of explosions in envelopes
of stars, which in many ways is analogous to the analysis in Sect. 5. It is assumed
that the parameter ω satisfies $0 \leqslant \omega < 3$ while in the equilibrium state (6.1) the
density of gas ρ is not higher farther away from the center and the mass of gas in
the neighborhood of the center $r = 0$ is finite. The Schwarzschild criterion for the
stability of equilibrium of the gas relative to the convective perturbations (see part
VII of Sect. 5) leads to the condition $\omega > 1/(\gamma - 1)$, from which for $\omega < 3$ it
follows that $\gamma > 4/3$. Therefore solutions in the model of explosions in stellar
envelopes can have a relation to reality only for $\gamma > 4/3$, $1/(\gamma - 1) < \omega < 3$.
However the analysis of self-similar solutions with $\gamma < 4/3$, $\omega < 3$ is of independent interest in connection with the problem of the existence of self-similar oscillations of an ideal gas in the field of an attracting center analogous to self-similar
oscillations of a self-gravitating gas studied in Sects. 3 and 5 of Chap. V and Sect. 4
of Chap. IV.

II. General Properties of the Dynamical System. The system of equations of
gas dynamics (1.4)–(1.6) for self-similar solutions of the type (6.2) in the new
variables $z = \gamma P/R$, R, V and $\tau = \ln \lambda$ has the following form (prime means
differentiation with respect to τ):

$$\frac{R'}{R} = \frac{1}{V - \delta}(-V' + s + kV), \tag{6.3}$$

$$V'(V - \delta) + \frac{z'}{\gamma} + \frac{z}{\gamma}\frac{R'}{R} = -\frac{2}{\gamma}z - V^2 + V + \frac{k + 3}{\gamma}z - \lambda^{-3}, \tag{6.4}$$

$$\frac{z'}{z}(V - \delta) - (\gamma - 1)\frac{R'}{R}(V - \delta) = 2 - s(\gamma - 1) - V(2 + (\gamma - 1)(k + 3)). \tag{6.5}$$

The system of equations (6.3)–(6.5) contains an explicit dependence on the
variable τ in the form of the λ^{-3} term. To obtain an autonomous dynamical
system we introduce a new variable $m = \lambda^{-3}$. After eliminating R'/R (by virtue of
(6.3)) from the equations (6.4) and (6.5) and solving the obtained equations for the
derivatives we arrive at a three-dimensional dynamical system in the variables z,
V and m:

$$V' = \frac{1}{z - (V - \delta)^2}\left[z\left(\frac{2(\omega + 1)}{3\gamma} - 3V\right) + (m + V^2 - V)(V - \delta)\right],$$

$$z' = \frac{z}{(V - \delta)(z - (V - \delta)^2)}\left[-(\gamma - 1)\left\{z\left(\frac{2(\omega + 1)}{3\gamma} - 3V\right)\right.\right. \tag{6.6}$$

$$\left.\left. + (m + V^2 - V)(V - \delta)\right\} + (2 + (1 - 3\gamma)V)(z - (V - \delta)^2)\right],$$

$$m' = -3m,$$

where

$$\omega = \frac{3}{2}(s + 2) + k, \quad \delta = \frac{2}{3}.$$

The algebraic structure of the equations (6.6) is analogous to the structure of the equations (1.14). Therefore the qualitative analysis of the dynamical system (6.6) is most naturally conducted in the context of the constructions developed in Sects. 1 through 5. We will consider the most important points in this analysis.

The equilibrium state of the gas (6.1) corresponds to the following trajectory X of the dynamical system (6.6):

$$z = \frac{\gamma}{\omega + 1} m, \quad V = 0. \tag{6.7}$$

Self-similar solutions which are conjugate to the equilibrium state of the gas (6.1) across the shock wave correspond to the trajectories of the system (6.6) passing through the line Y, which under the Hugoniot transformation (5.5) goes to the line X. The points on the line Y are parametrized by the variable $q = 1/M^2$ (where M is the Mach number of motion of the shock wave) and have the following coordinates:

$$z = \frac{4(2\gamma - (\gamma - 1)q)(\gamma - 1 + 2q)}{9(\gamma + 1)^2}, \quad m = \frac{4(\omega + 1)}{9\gamma} q,$$

$$V = \frac{4(1 - q)}{3(\gamma + 1)}. \tag{6.8}$$

The law of motion of the shock wave is $\lambda = \lambda_*$, where the constant λ_* is determined from the expression (6.8) for the variable $m = \lambda^{-3}$, i.e. $\lambda_*^3 = 9\gamma/(4q(\omega + 1))$. The trajectory X and the line Y intersect at the point

$$Y_1(q = 1, z = 4/9, m = 4(\omega + 1)/9\gamma, V = 0)$$

contained in the surface of non-extendability of solutions $L = z - (V - \delta)^2 = 0$. The point Y_1 belongs to the line of critical points I determined by the following conditions:

$$z = (V - \delta)^2, \quad m = (V - \delta)\left(3V - \frac{2(\omega + 1)}{3\gamma}\right) + V - V^2.$$

The resolution of degenerate critical points of the dynamical system (6.6) and the construction of the compact manifold S are done by means of the same transformations of coordinates (1.17)–(1.33). We will write the dynamical system (6.6) in the coordinates (1.21) $(v_0 = V - \delta, m_2 = -v_0 m z^{-1}, \rho_0 = z^{-1/2})$. After a change of variable $d\tau_3/d\tau = -1/v_0$ in the coordinates (1.21) we obtain the following system:

$$\dot{v}_0 = -\frac{v_0}{1 - \rho_0^2 v_0^2}\left[-3v_0 - m_2 + \frac{2(\omega + 1)}{3\gamma} - 3\delta + v_0(v_0 + \delta)\right.$$

$$\left. \times (v_0 + \delta - 1)\rho_0^2\right],$$

$$\dot{m}_2 = \frac{m_2}{1 - \rho_0^2 v_0^2}\left[3v_0(1 - \rho_0^2 v_0^2) - \gamma\left\{-3v_0 - m_2 + \frac{2(\omega + 1)}{3\gamma}\right.\right.$$

$$\left.\left. - 3\delta + v_0(v_0 + \delta)(v_0 + \delta - 1)\rho_0^2\right\} + (2 + (v_0 + \delta)(1 - 3\gamma))(1 - \rho_0^2 v_0^2)\right],$$

$$\dot{\rho}_0 = \frac{\rho_0}{2(1 - \rho_0^2 v_0^2)}\left[-(\gamma - 1)\left\{-3v_0 - m_2 + \frac{2(\omega + 1)}{3\gamma} - 3\delta\right.\right.$$

$$\left.\left. + v_0(v_0 + \delta)(v_0 + \delta - 1)\rho_0^2\right\} + (2 + (v_0 + \delta)(1 - 3\gamma))(1 - \rho_0^2 v_0^2)\right].$$

On the component $\Gamma_2\,(\rho_0 = 0)$ of the boundary the system (6.9) has three critical points:

$$Z_1\left(v_0 = \frac{2(\omega + 1)}{9\gamma} - \delta = \alpha,\, m_2 = 0,\, \rho_0 = 0\right),$$

$$Z_3\left(v_0 = -\delta,\, m_2 = \frac{2(\omega + 1)}{3\gamma},\, \rho_0 = 0\right),\quad Z_4\,(v_0 = m_2 = \rho_0 = 0).$$

The critical point Z_1 belongs to the physical region $v_0 < 0$ for $\alpha < 0$ or $\gamma > (\omega + 1)/3$. The eigenvalues of the system (6.9) at the critical points Z_1, Z_3 and Z_4 are

$$Z_1: \lambda_{v_0} = 3\alpha < 0,\, \lambda_{m_2} = \frac{2(\omega + 1)(4 - 3\gamma)}{9\gamma},\, \lambda_{\rho_0} = \frac{1 + \omega + 3\gamma(2 - \omega)}{9\gamma}; \quad (6.10)$$

$$Z_3: \lambda_{1,2} = \frac{\omega - 2}{3} \pm \left[\frac{\omega - 2}{3} - \frac{4(\omega + 1)}{3\gamma}\left(\frac{4}{3} - \gamma\right)\right]^{1/2},\, \lambda_{\rho_0} = 1; \quad (6.11)$$

$$Z_4: \lambda_{v_0} = -3\alpha,\, \lambda_{m_2} = \frac{2(3 - \omega)}{3},\, \lambda_{\rho_0} = \frac{\omega + 1 - \gamma\omega}{3\gamma}. \quad (6.12)$$

where the subscripts of the eigenvalues represent the corresponding eigenvectors. According to (6.10) for $(\omega + 1)/3 < \gamma < 4/3$ the critical point Z_1 is unstable and has a two-dimensional separatrix leaving it. The corresponding self-similar solutions continue to the center of symmetry and for $\lambda \to 0$ have the following asymptotic form:

$$v = \frac{r}{t}V_1,\quad R = C_1\lambda^{(s + kV_1)/(V_1 - 2/3)},$$

$$P = C_2\lambda^{(2 + s + (1 - 3\gamma + k)V_1)/(V_1 - 2/3)},\quad V_1 = \frac{2(\omega + 1)}{9\gamma}. \quad (6.13)$$

According to (6.11) for $\gamma > 4/3$ the critical point Z_3 is a saddle and has a two-dimensional separatrix leaving it. In the coordinates (1.21) the trajectory X is determined by the following conditions:

$$v_0 = -\frac{2}{3}, \quad m_2 = \frac{2(\omega + 1)}{3\gamma}, \quad \rho_0 > 0.$$

As $\lambda \to 0$ this trajectory leaves the critical point Z_3 and therefore it is a separatrix corresponding to the eigenvalue λ_{ρ_0}.

For

$$\gamma < \gamma_1 = \frac{4/3}{1 + (\omega - 2)^2/(12(\omega + 1))}$$

the eigenvalues $\lambda_{1,2}$ (6.11) are complex conjugate. In this case the critical point Z_3 in the component of the boundary Γ_2 $(\rho_0 = 0)$ is a focus. For $\omega < 2$ it is an attracting focus and for $\omega > 2$ it is repelling. For $\omega = 2$ it is a center. For $\gamma < \gamma_1$ trajectories moving in the neighborhood of the critical point Z_3 rotate around the separatrix X (6.7), where $V = 0$. Therefore in the corresponding self-similar solutions the velocity of the gas V oscillates around zero, i.e. the gas undergoes radial oscillations. For $\gamma < \gamma_1$, $\omega > 2$ self-similar solutions have an arbitrarily large but necessarily finite number of radial oscillations of the gas. For $\omega = 2$ the oscillations of the gas are asymptotically periodic in time $\ln t$. The asymptotic formulas for self-similar oscillations of an ideal gas in the field of an attracting center are derived just like in Sect. 3. The problem of explosions in stellar envelopes for $\gamma < \gamma_1, 0 \leqslant \omega < 3$ and Mach numbers of motion of the shock wave $M \approx 1$ has solutions with radial oscillations of the gas completely analogous to the ones studied in Sect. 5.

The critical point Z_4 is repelling for $(\omega + 1)/3 < \gamma < (\omega + 1)/\omega, \omega < 3$. Separatrices leaving this critical point describe self-similar solutions with an expanding cavity inside the gas having the following asymptotic form at the internal boundary $(\lambda \to \lambda_1)$:

$$V = \frac{2}{3} - \frac{2(\gamma - (\omega + 1)/3)}{\gamma}\left(\frac{\lambda}{\lambda_1} - 1\right),$$

$$R = C_1\left(\frac{\lambda}{\lambda_1} - 1\right)^{(-\gamma\omega + 1 + \omega)/(3\gamma - 1 - \omega)}, \quad P = C_2. \tag{6.14}$$

The same asymptotic form for $\lambda \to \lambda_1$ occurs in the solutions corresponding to separatrices leaving the repelling (for $\gamma > (\omega + 1)/\omega$) critical point Z_5 $(v_4 = v_0/z = 0, z = m_2 = 0)$. In the asymptotic form (6.14) at the internal boundary $\lambda = \lambda_1$ the pressure is not zero $(p \neq 0)$, i.e. the gas is being pushed out from the inside by a spherical piston, which in reality can be represented by a very hot gas filling the region $\lambda < \lambda_1$ and having small density.

For $\gamma > 4/3$ there is one more class of solutions with an expanding cavity inside the gas. These solutions correspond to separatrices emerging from the line of unstable critical points $\mathscr{L}$ contained in the component of the boundary Γ_4 (see Fig. 26) and determined by the following conditions:

$$z = 0, \quad m_2 + \frac{2}{9}v_4 + \frac{2(3 - \omega)}{3\gamma} = 0. \tag{6.15}$$

The asymptotic form of such solutions for $\lambda \to \lambda_1$ is unstable and has the following form:

$$V = \frac{2}{3} + \frac{2(4 - 3\gamma)}{3\gamma}\left(\frac{\lambda}{\lambda_1} - 1\right), \quad R = C_1\left(\frac{\lambda}{\lambda_1} - 1\right)^{(\gamma\omega - 4)/(4 - 3\gamma)},$$

$$P = C_2\left(\frac{\lambda}{\lambda_1} - 1\right)^{\gamma(\omega - 4)/(4 - 3\gamma)}. \tag{6.16}$$

For $(\omega + 1)/3 < \gamma < 4/3$ the asymptotic form (6.14) is the only possible one for the self-similar solutions with an expanding cavity inside the gas. For $\gamma < (\omega + 1)/3$ such solutions do not exist at all. For $\gamma > 4/3$, $\omega < 3$ self-similar solutions with an expanding cavity inside the gas have either the stable asymptotic form (6.14) or the unstable asymptotic form (6.16) at the internal boundary.

The acceleration a_0 of gas particles in the system of coordinates connected with the moving boundary $\lambda = \lambda_1$ is determined by the following formulas:

$$a_0 = a_1 - a_2 = \frac{2}{9}\frac{\lambda_1^3 G\mathcal{M}}{r^2} - \frac{G\mathcal{M}}{r^2},$$

where $a_1 = -d^2r/dt^2$, $r = \lambda_1 bt^{2/3}$, and the acceleration a_2 is caused by the gravitational attraction of the point mass $\mathcal{M}$. For $\lambda_1^3 > 9/2$ the acceleration a_0 is directed towards the gas. In this case the boundary between the gas and vacuum is stable (see [128]). If some solution of the problem of explosions in stellar envelopes has an expanding cavity inside the gas, then the necessary condition for the stability of the internal boundary has the following form:

$$\frac{9}{2} < \lambda_1^3 < \lambda_*^3 = \frac{9\gamma}{4q(\omega + 1)},$$

where λ_* is the value of λ at the front of the shock wave. From this we obtain that the internal boundary of the gas in such solutions can be stable only if

$$M > M_0 > \left(\frac{2(\omega + 1)}{\gamma}\right)^{1/2},$$

where $M = q^{-1/2}$ is the Mach number of motion of the shock wave.

III. Analysis of the Special Case $\omega = 2$.

In the self-similar adiabatic motion of an ideal gas energy can be injected at the center of symmetry (if the solution has a singularity at $r = 0$) or at the boundary of the expanding cavity $\lambda = \lambda_1$. The law of liberation of energy is

$$E = \alpha a(G\mathcal{M})^{(2-k)/3}t^{2(2-\omega)/3}. \tag{6.17}$$

The constant α depends on the solution itself. In order to calculate α it is necessary to compare the energy of a ball of gas of radius r in the state of equilibrium (6.1) and at the moment of passing of the shock wave through its surface. The energy of a spherical layer $r_1 < r < r_2$ in the equilibrium state (6.1) is determined by the following formulas:

$$E_0 = \int_{r_1}^{r_2} \left(\frac{p}{\gamma - 1} - \frac{\rho G \mathcal{M}}{r} \right) 4\pi r^2 \, dr,$$

$$\omega \neq 2: \quad E_0 = \frac{1 - (\gamma - 1)(\omega + 1)}{2 - \omega} 4\pi c_1 a (G \mathcal{M})^{1+s/2} r^{2-\omega} |_{r_1}^{r_2}, \tag{6.18}$$

$$\omega = 2: \quad E_0 = \frac{4 - 3\gamma}{\gamma - 1} 4\pi c_1 a (G \mathcal{M})^{1+s/2} \ln \frac{r_2}{r_1}.$$

From this we see that $E_0 > 0$ for $\gamma < \gamma_2 = (\omega + 2)/(\omega + 1)$ and $E_0 < 0$ for $\gamma > \gamma_2$. From the formulas (6.17) and (6.18) it follows that the value $\omega = 2$ is exceptional [124, 125], because in this case the amount of freed energy E does not depend on time.

Self-similar solutions of the type (6.2) have a monotone function (the adiabatic integral [125]):

$$\Phi = z(V - \delta)^{\gamma - 1 + \beta} R^\beta = C_1 \lambda^{1 - 3\gamma + k\beta}, \quad \beta = \frac{4 - 3\gamma}{3 - \omega}. \tag{6.19}$$

For $\omega = 2$ there exists one more monotone function (the energy integral [7, 125]):

$$H = R \left[\frac{zV}{\gamma} + (V - \delta) \left(\frac{V^2}{2} + \frac{z}{\gamma(\gamma - 1)} - \lambda^{-3} \right) \right] = C_2 \lambda^{k-2}. \tag{6.20}$$

For $\omega = 2$ it is possible to eliminate the variable R from the two monotone functions Φ and H and thus obtain a first integral $F = |H|^{4-3\gamma} \Phi^{-1}$ of the dynamical system (6.6), where the variable λ is replaced by $m = \lambda^{-3}$. In the coordinates (1.21) the integral F has the following form:

$$F = \left| \frac{2}{3\gamma} + \frac{v_0}{\gamma - 1} + m_2 + \frac{v_0}{2} \left(v_0 + \frac{2}{3} \right)^2 \rho_0^2 \right|^{4-3\gamma} |v_0|^{-\gamma} m_2^{3(\gamma - 1)}. \tag{6.21}$$

On the trajectory X the integral F has constant value:

$$F(X) = F_0 = \frac{1}{\gamma} \left(\frac{3^4}{2} \right)^{\gamma - 1} \left(\frac{|4 - 3\gamma|}{\gamma - 1} \right)^{4-3\gamma}.$$

At the points of the line X the first differential DF is equal to zero, i.e. the line X consists of extrema of the function F. The second differential $d^2 F$ has the following form:

$$d^2 F = -F_0 \left(\frac{3}{8} \right) \gamma^2 (\gamma - 1)(4 - 3\gamma)^{-1} \left[(3 dv_0 + dm_2)^2 \right.$$

$$\left. + 3\gamma^{-1}(4 - 3\gamma) \left(1 - \left(\frac{2}{3} \right)^2 \rho_0^{\,2} \right) dv_0^2 \right]. \tag{6.22}$$

According to (6.22), for $\gamma < 4/3$ the segment of the line X from the point Z_3 to the point $Y_1\,(\rho_0 = 3/2)$ consists of maxima of the function F. Therefore for $\gamma < 4/3$ the level surfaces $F = \text{const}$ in the neighborhood of the segment $Z_3 Y_1$ are two-dimensional cylinders and intersect the component $\Gamma_2\,(\rho_0 = 0)$ of the boundary along cycles (closed trajectories of the system (6.9)) in the region

$$\rho_0 = 0, \quad \frac{2}{3\gamma} + \frac{v_0}{\gamma - 1} + m_2 < 0.$$

Trajectories winding onto these cycles determine self-similar solutions having an infinite number of undamped radial oscillations of the gas in the field of an attracting center, which are asymptotically periodic in time $\ln t$.

For $\gamma > 4/3$, the segment $Z_3 Y_1$ consists of inflection points of the function F. The level surface $F = F_0$ divides the neighborhood of the line X into four regions. In the first two regions D_1 and D_2 we have $F \leqslant F_0$. In the region D_1 we have $V > 0$ and in the region D_2 we have $V < 0$. In the last two regions D_3 and D_4 we have $F \geqslant F_0$.

The values of the function F on the line $Y(q)$ (6.8) are determined by the same formula (5.19) as in part IV of Sect. 5. On the segment $Y(q)\,(0 < q < 1, \infty > M > 1)$ we have $V > 0$ and $F < F_0$. Consequently this segment lies in the region D_1 for $\gamma > 4/3$. In complete analogy with the reasoning in part IV of Sect. 5 it can be shown that all trajectories in the region D_1 (as λ decreases) emerge from the repelling critical point Z_5 or (for $\gamma < (\omega + 1)/\omega = 3/2$) from the repelling critical point Z_4. In particular all trajectories passing through the line segment $Y(q)$, $0 < q < 1$ have this property. Thus we have just proved that for $\omega = 2, \gamma > 4/3$ and all Mach numbers of motion of the shock wave $M > 1$ all solutions in the model of explosions in stellar envelopes have an expanding cavity inside the gas with the asymptotic form (6.14) at the internal boundary.

The energy E_0 of a ball of gas with radius r for $\omega = 2, \gamma > 4/3$ is equal to $-\infty$ in the equilibrium state (6.1) according to (6.18). After the passing of the shock wave the energy of the gas ball (with a cavity inside) is finite. Consequently in the considered self-similar solutions for $\omega = 2, \gamma > 4/3, M > 1$ an infinite amount of energy is emitted at the time $t = 0$ at the center of symmetry $r = 0$. In other words these solutions are models of a powerful explosion of a star accompanied by a shedding of the light envelope of the star with the remaining mass much larger than the mass of the discarded envelope.

7. Self-Similar Solutions with Converging Shock Waves

I. Consider the problem of the motion of a converging shock wave in a gas initially at rest. Without taking gravity into account this problem was first studied in the well-known works [136, 137]. In the presence of self-gravitation of the gas the solutions of this problem are self-similar solutions with a converging

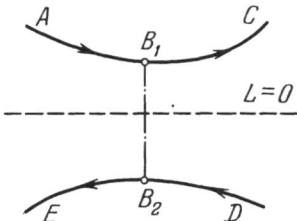

Fig. 32. Construction of dual solutions

shock wave, for which the gas is in the state of equilibrium (5.1) in the region before the shock wave and beyond the shock wave the gas is set in motion with zero velocity at infinity.

In some cases self-similar solutions with converging shock waves can be connected to solutions with diverging shock waves. It is natural to call such solutions dual to each other. Indeed suppose that B_1 and B_2 are two points in the manifold S conjugate to each other under the transformation (5.5) such that the point B_1 is contained in the subsonic region $L > 0$ and the point B_2 is contained in the supersonic region $L < 0$. Suppose that $AB_1 C$ and $DB_2 E$ are trajectories of the system of self-similar equations (1.14) passing through the points B_1 and B_2 (Fig. 32). A self-similar solution with a diverging shock wave corresponding to a discontinuous step from the point B_1 to the point B_2 is determined by the segments of trajectories AB_1 and $B_2 E$, where the segment $B_2 E$ describes the solution in the region before the shock wave. A self-similar solution with a converging shock wave defined for $t < 0, r > 0$ corresponds to the segment of trajectories DB_2 and $B_1 C$ after the transformation $t \to -t, v \to -v$. In this solution the region before the shock wave is described by the segment DB_2. If both of the above solutions exist (i.e. are defined for all $0 < \lambda < \infty$) then they are called dual to each other.

A simple example of dual solutions can be easily constructed using the exact solution called "the dynamical explosion of equilibrium" found in [7] for $\gamma = 7/6$, $\omega = 12/5$. In this solution the law of motion of the shock wave is $\lambda = \lambda_0 = (6\pi/5)^{8/12}$. For $\lambda > \lambda_0$ the gas is in the state of equilibrium (5.1). For $0 < \lambda < \lambda_0$ the solution has the following form:

$$\rho = \frac{1}{6\pi G t^2}, \quad p = \frac{K}{G t^{7/3}}, \quad K = \frac{4(30\pi G A)^{5/6}}{189\pi},$$

$$\mathcal{M} = \frac{2}{9}\frac{r^3}{G t^2}, \quad v = \frac{2}{3}\frac{r}{t}.$$

(7.1)

This solution corresponds to the separatrix (2.3) going from the critical point Z_1 to the critical point Z_9 and intersecting the line $I_1 I_4$ at the critical point I_0 ($V = 2/3$, $m = 2/9$, $z = 1/6^2$) (Fig. 33). For $\gamma = 7/6$, $\omega = 12/5$ the trajectory (2.3) intersects the line Y (5.6) at the point $B_1 = Y(2/15)$ conjugate by means of

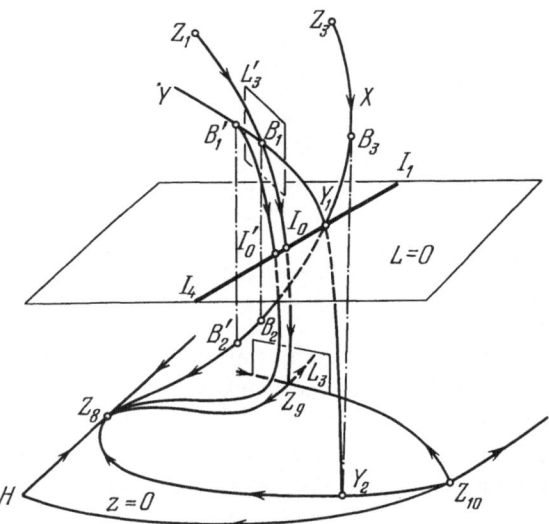

Fig. 33. Construction of some self-similar solutions with converging shock waves

the transformation (5.5) to the point B_2 contained in the trajectory X (5.4). The complete "dynamical explosion of equilibrium" [7] is described by the segments of trajectories

$$Z_1 B_1 \left(0 < \lambda < \lambda_0\right), \quad B_2 Z_8 \left(\lambda_0 < \lambda < \infty\right).$$

In this case the dual solution corresponds to the trajectories

$$Z_3 B_2 \left(0 < \lambda < \lambda_0 \, (5.1)\right), \quad B_1 Z_9 \left(\lambda_0 < \lambda < \infty \, (7.1)\right)$$

and after a change $t \to -t$, $v \to -v$ describes the collapse of a star initially at rest (segment $Z_3 B_2$) as a result of its compression by the shock wave. In this process the density of the gas becomes constant in space (segment $B_1 Z_9$) and growing with time as $t \to -0$ according to (7.1).

Let us use the solution obtained above to analyze the problem of the motion of a converging shock wave in a gas initially at rest. For $\gamma = 7/6$, $\omega = 12/5$ the eigenvalues (1.37) at the critical point I_0 $(V = 2/3,\ m = 2/9,\ z = 1/6^2)$ are $\lambda_1 = -14/6^2$, $\lambda_2 = -35/6^2$, $\lambda_3 = 0$. The trajectory (2.3) is a separatrix approaching the critical point I_0. It corresponds to the eigenvalue λ_1 and lies in the smooth two-dimensional surface L_1 passing (in the neighborhood of the point I_0) through the line $I_1 I_4$ (see Sect. 5). All trajectories near the trajectory (2.3) outside the surface $L = 0$ are continued through the line of critical points $I_1 I_4$, because the critical points in the neighborhood of I_0 are attracting $(\lambda_1 < 0,\ \lambda_2 < 0)$. These trajectories are tangent to the surface L_1, because $\lambda_1 > \lambda_2$. Thus they remain close to the trajectory (2.3) even after the continuation through the critical points in the line $I_1 I_4$. In particular all trajectories passing in the neighborhood of the critical point

Z_9 end up (after going through the line of critical points $I_1 I_4$) in the neighborhood of the point B_1 for a decrease of λ.

The neighborhood of the critical point Z_9 is divided by the two-dimensional separatrix L_3 (see Sect. 2) approaching (for $\lambda \to \infty$) this critical point into two regions V_1 and V_2. Trajectories emerging from the region V_1 approach the critical point Z_8 as λ increases. Trajectories emerging from the region V_2 for an increase of λ approach the surface of non-extendability of solutions $L = 0$ (see Figs. 26 and 33) and therefore have no physical applications. As λ decreases the two-dimensional separatrix L_3 determines some surface L_3' intersecting the line Y (5.6) at the point B_1. All trajectories emerging from the region V_1 stay to one side of the surface L_3 as λ decreases. Trajectories emerging from the region V_2 are on the other side. Consequently the segment of the line Y to one side of the point B_1 is completely crossed by trajectories emerging (as λ decreases) from the region V_1. The segment of Y to the other side of the point B_1 is crossed by trajectories that left the region V_2.

The above reasoning proves the existence of trajectories $B_1' I_0' Z_8$ (see Fig. 33) passing in the neighborhood of the trajectory (2.3) through the line Y and through the region V_1. It is these trajectories that determine the solution of the problem of motion of a converging shock wave over a gas initially at rest. Indeed suppose that B_2' is a point conjugate to B_1' under the transformation (5.5) (B_2' belongs to the trajectory X). The self-similar solution determined by the segments of trajectories $Z_3 B_2'$ and $B_1' I_0' Z_8$ after the change $t \to -t$, $v \to -v$ satisfies all conditions of the considered problem. Note that by virtue of continuity of behavior of the trajectories of the dynamical system (1.14) such solutions exist also for all values of γ, ω close to $\gamma = 7/6$, $\omega = 12/5$.

The solutions obtained (for γ, ω close to 7/6, 12/5 and Mach numbers of motion of the shock wave $M \approx (15/2)^{1/2}$) describe the process of compression of a star by a shock wave. In this process after the passing of the shock wave for $t \to -0$ the state (3.12) sets in. This state is a perturbation of the initial equilibrium state (5.1) and we have $v \to 0$ for $r \to \infty$. The obtained solutions have a weak discontinuity (they lose analyticity) during the continuation through the line of critical points $I_1 I_4$ at the points I_0'. It can be shown that in the neighborhood of $\gamma = 7/6$, $\omega = 12/5$ there exist infinitely many values of γ, ω for which there exists some everywhere (except at the shock wave) analytic solution of the considered problem.

II. Here is another example of exact dual solutions. Consider the point of intersection of the line Y (5.6) with the component of the boundary Γ_8 ($z = 0$):

$$Y_2 \left(z = 0, m = \frac{16(\omega - 1)}{(\gamma - 1)\omega^2}, V = -\frac{4}{(\gamma - 1)\omega} \right).$$

For $\omega = \omega_0 = (2\gamma - 1)/2(\gamma - 1)$ the trajectory of the dynamical system (1.14) passing through the point Y_2 can be integrated explicitly: $z = 0$, $m = V^2/2$ (see

Sect. 3). This trajectory passes through the critical points Z_8 and Z_{10} (see Fig. 33). It corresponds to the following exact solution:

$$\mathcal{M} = \frac{r^3}{2Gt^2}V^2, \quad v = -\frac{r}{t}V, \quad \rho = \frac{(3-\omega)V}{Gt^2 4\pi(2+\omega V)}, \quad p = 0, \qquad (7.2)$$

where the function $V(\lambda) > 0$ is determined by the following equation:

$$V^{-1}(1 + 3V/2)^{1-\omega/3} = \lambda^{\omega/2}. \qquad (7.3)$$

For $t > 0$ the solution (7.2) describes the collapse of a cold gas with the formation of a black hole at the center (see (4.2)).

Let B_3 be a point conjugate to the point Y_2 under the transformation (5.5) (B_3 belongs to the trajectory X). The segments of trajectories $Z_3 B_3$ and $Y_2 Z_8$ (see Fig. 33) determine a solution with a diverging shock wave. In this solution at the shock wave we have $\lambda = \lambda_1$, where λ_1 is determined from (7.3) for $V = 4\omega/(\gamma - 1)$. For $\lambda > \lambda_1$ the solution describes the collapse of a cold gas (7.2) such that for $r \to \infty$ the solution tends to the equilibrium state with the asymptotic form (3.12) and $v \to 0$ for $\gamma < 3/2$. For $0 \leqslant \lambda < \lambda_1$ the gas is in the state of equilibrium (5.1). On the whole this solution describes the formation of a star at rest as a result of heating of the collapsing cold gas by the shock wave.

The dual solution corresponds to the trajectories $Z_{10} Y_2$ and $B_3 Z_8$ and after the change $t \to -t, v \to -v$ describes (for $0 < \lambda < \lambda_1$) (segment $Z_{10} Y_2$) the flow of cold gas away from the center. Note that according to the asymptotic form (4.2), at the center there is a decreasing point mass as $t \to -0$, which is the analogue of a white hole in the classical theory. For $\lambda > \lambda_1$ (segment $B_3 Z_8$) the solution is the state of equilibrium (5.1). Thus this solution describes the compression of the white hole by a shock wave, during which the gas goes into the state of equilibrium. Solutions with analogous properties also exist for all nearby γ, ω and for $p \neq 0$.

Chapter VI
Self-Similar Rotation of an Ideal Gas

In nature there is a wide variety of phenomena in which a substantial role is played by the rotation of a gas or a fluid about a stationary axis. For example in the category of geophysical phenomena it is necessary to consider the rotation of the atmosphere and the oceans caused by the rotation of the earth. Rotation is the main property of such phenomena as cyclones and tornadoes in the atmosphere and whirlpools in water. Rotation must also be taken into account in some types of motion of a plasma in a magnetic field. It is important to note that in the above phenomena the angular velocity of rotation of the gas in general depends on the distance to the axis of rotation. The simplest class of solutions of the equations of gas dynamics having this property is the class of self-similar solutions for the rotation of a gas, to the study of which this chapter is devoted.

Self-similar solutions for the rotation of an ideal gas belong to the class of partially invariant solutions according to the classification in [138]. Self-similar rotation of a viscous incompressible fluid was studied in [139, 140]. The complete classification of invariant and partially invariant solutions of the Navier-Stokes equations including the self-similar rotation of a viscous incompressible fluid was given in [141, 142].

1. Definition of Self-Similar Rotation of an Ideal Gas

In the analysis of self-similar solutions for the rotation of gas around a stationary axis it is most natural to use the cylindrical system of coordinates. In an arbitrary system of coordinates the equations of gas dynamics have the following invariant form:

$$\frac{\partial v^i}{\partial t} + v^i_{;j} v^j = -\frac{1}{\rho} p_{;j} g^{ij}, \tag{1.1}$$

$$\frac{\partial \rho}{\partial t} + (\rho v^i)_{;i} = 0, \tag{1.2}$$

$$\frac{\partial (p/\rho^\gamma)}{\partial t} + \left(\frac{p}{\rho^\gamma}\right)_{;i} v^i = 0. \tag{1.3}$$

where g_{ij} is the metric in the given system of coordinates and we use the standard notation for the covariant derivatives:

$$v^i_{;j} = \frac{\partial v^i}{\partial x_j} + \Gamma^i_{kj} v^k, \quad p_{;j} = \frac{\partial p}{\partial x_j},$$

$$v^i_{;j;k} = \frac{\partial v^i_{;j}}{\partial x_k} - \Gamma^m_{jk} v^i_{;m} + \Gamma^i_{mk} v^m_{;j}. \tag{1.4}$$

The Christoffel symbols have the standard form:

$$\Gamma^i_{kj} = \frac{1}{2} g^{il} \left(\frac{\partial g_{kl}}{\partial x_j} + \frac{\partial g_{lj}}{\partial x_k} - \frac{\partial g_{kj}}{\partial x_l} \right). \tag{1.5}$$

In the cylindrical system of coordinates r, φ, z_1 the metric of Euclidean space is $ds^2 = dr^2 + r^2 d\varphi^2 + dz_1^2$. Among the Christoffel symbols of this metric only three of them are different from zero:

$$\Gamma^1_{22} = -r, \quad \Gamma^2_{12} = \Gamma^2_{21} = r^{-1}.$$

The velocity vector of the gas in cylindrical coordinates has components (v, ω, u), where v is the radial velocity, ω is the angular velocity and u is the velocity of gas along the z_1 axis. From now on let $w = r\omega$.

With the help of the formulas (1.4) we obtain the explicit form of the dynamical equations of the gas (1.1) in cylindrical coordinates:

$$\frac{\partial v}{\partial t} + v \frac{\partial v}{\partial r} + \frac{w}{r} \frac{\partial v}{\partial \varphi} + u \frac{\partial v}{\partial z_1} - \frac{w^2}{r} = -\frac{1}{\rho} \frac{\partial p}{\partial r},$$

$$\frac{\partial w}{\partial t} + v \frac{\partial w}{\partial r} + \frac{w}{r} \frac{\partial w}{\partial \varphi} + u \frac{\partial w}{\partial z_1} + \frac{vw}{r} = -\frac{1}{r\rho} \frac{\partial p}{\partial \varphi}, \tag{1.6}$$

$$\frac{\partial u}{\partial t} + v \frac{\partial u}{\partial r} + \frac{w}{r} \frac{\partial u}{\partial \varphi} + u \frac{\partial u}{\partial z_1} = -\frac{1}{\rho} \frac{\partial p}{\partial z_1}.$$

The continuity and adiabatic equations (1.2) and (1.3) take the following form:

$$\frac{\partial \rho}{\partial t} + \frac{\partial \rho v}{\partial r} + \frac{1}{r} \frac{\partial (\rho w)}{\partial \varphi} + \frac{\partial \rho u}{\partial z_1} + \frac{\rho v}{r} = 0, \tag{1.7}$$

$$\frac{\partial (p/\rho^\gamma)}{\partial t} + v \frac{\partial (p/\rho^\gamma)}{\partial r} + \frac{w}{r} \frac{\partial (p/\rho^\gamma)}{\partial \varphi} + u \frac{\partial (p/\rho^\gamma)}{\partial z_1} = 0. \tag{1.8}$$

The parameters for self-similar solutions for the rotation of an ideal gas have the following form:

$$v = \frac{r}{t} V(\lambda), \quad w = \frac{r}{t} \Omega(\lambda), \quad u = \frac{z_1}{t} U(\lambda),$$

$$\rho = \frac{a}{r^{k+3} t^s} R(\lambda), \quad p = \frac{a}{r^{k+1} t^{s+2}} P(\lambda). \tag{1.9}$$

where $\lambda = r/bt^\delta$ and the constants a and b have dimensions $[a] = ML^k T^s$ and

$[b] = LT^{-\delta}$. In solutions of the type (1.9) along with the radial motion of gas there is also the rotation of gas around the z_1 axis and the movement of gas along this axis. Thus in the self-similar solutions (1.9) the motion of gas particles is intrinsically three-dimensional.

Self-similar solutions with a one-dimensional motion of gas particles with flat, cylindrical and spherical waves were investigated in detail in many works [7, 115, 137, 143–147]. In the case of cylindrical waves these solutions are of the form (1.9) with $w \equiv u \equiv 0$ and describe one-dimensional radial motion of gas. Several exact self-similar solutions with rotation for a viscous incompressible fluid were found in [139, 140, 148]. These solutions have well known applications in the problem of diffusion of a vortex [139, 143] and in modelling tornadoes [140]. Judging from the available publications and monographs [7, 115, 137, 143–147], the self-similar rotation of an ideal gas has never been investigated. In this chapter we proceed with a detailed analysis of the self-similar rotation of an ideal gas.

Note that in the class of self-similar solutions with flat waves an intrinsically three-dimensional motion of the ideal gas does occur. Suppose that in the cartesian coordinates x^1, x^2, x^3 the wave propagates along the $x^1 = r$ axis. In the generalized self-similar solution the pressure and density of gas are determined by the formulas (1.9) and the components v^i of the velocity of the gas have the following form:

$$v^1 = \frac{r}{t} V(\lambda), \quad v^i = \frac{x^j}{t} U_j^i(\lambda) + \frac{r}{t} W^i(\lambda), \tag{1.10}$$

where $i, j = 2, 3$. In solutions of types (1.9)–(1.10) the components of velocity of the gas for fixed $t = \text{const}$ and $r = \text{const}$ are linear functions of the other spatial coordinates. This fact makes the self-similar solutions of types (1.9)–(1.10) seem close to the motion of gas with homogeneous deformation (see Chapt. VII).

2. Algebraic Integrals for the Self-Similar Rotation of an Ideal Gas

I. The System of Self-Similar Equations. The equations for the adiabatic motion of an ideal gas (1.6)–(1.8) for the self-similar solutions (1.9) in the variables

$$V(\tau), \quad \Omega(\tau), \quad z(\tau) = \gamma P/R, \quad R(\tau), \quad U(\tau), \quad \tau = \ln \lambda$$

take the following form:

$$\frac{dz}{d\tau} = z' = \frac{z}{(V - \delta)(z - (V - \delta)^2)} [z(2 - \varkappa(\gamma - 1) - 2V)$$
$$+ (2(\gamma V - 1) + (\gamma - 1)U)(V - \delta)^2$$
$$- (\gamma - 1)(-\Omega^2 + V^2 - V)(V - \delta)],$$

$$V' = \{z(\varkappa - 2V - U) + (-\Omega^2 + V^2 - V)(V - \delta)\}(z - (V - \delta)^2)^{-1},$$

$$\Omega' = \frac{\Omega(1 - 2V)}{V - \delta}, \quad U' = \frac{U(1 - U)}{V - \delta}, \quad \varkappa = \frac{s + 2 + \delta(k + 1)}{\gamma}. \tag{2.1}$$

The equation for the variation of density can be separated:

$$\frac{R'}{R} = \frac{-V' + s + (k + 1)V - U}{V - \delta}. \tag{2.2}$$

Note that the first equation (2.1) can also be written in the following form:

$$\frac{z'}{z} = \frac{-(\gamma - 1)V' + 2(1 - \gamma V) - (\gamma - 1)U}{V - \delta}. \tag{2.3}$$

The system (2.1) has three invariant three-dimensional submanifolds: $\Omega = 0$, $U = 0$ and $U = 1$. On the two-dimensional invariant manifold $\Omega = 0$, $U = 0$ the system (2.1) describes the self-similar radial motion of a gas with cylindrical waves, which was analyzed in detail in the book [7].

The algebraic structure of the dynamical system (2.1) is quite analogous to the algebraic structure of the system (1.14) from Chapt. V describing the self-similar spherically-symmetric motion of a self-gravitating gas with Ω^2 being the formal analogue of the variable m. Just like the system (1.14) in chapter V the system (2.1) has a surface of non-extendability of solutions $L = z - (V - \delta)^2 = 0$. The existence of this surface is formally the reason for the formation of shock waves in some self-similar solutions for the rotation of a gas.

The system (2.1) becomes singular for $V = \delta$. Outside the surface $V = \delta$ (in the region of finite values of z, V, Ω, U), there are six critical points of the system (2.1) (for arbitrary γ). We shall denote these points by Z_i^ε in analogy with the critical points of the system (1.14) in Chapt. V:

$$\bar{Z}_6^\varepsilon: \ V = 1, \Omega = 0, U = \varepsilon = 0; 1; z = 0;$$

$$Z_7^0: \ V = \frac{1}{\gamma}, \Omega = 0, U = 0, z = \frac{(\gamma - 1)(1 - \gamma\delta)}{\gamma^2(\varkappa\gamma - 2)};$$

$$Z_7^1: \ V = \frac{3 - \gamma}{2\gamma}, \Omega = 0, U = 1, z = \frac{3(\gamma - 1)(3 - \gamma)(3 - \gamma - 2\gamma\delta)}{8\gamma^2(\varkappa\gamma - 3)}; \tag{2.4}$$

$$Z_8^\varepsilon: \ V = 0, \Omega = 0, U = \varepsilon = 0; 1; z = 0.$$

All these critical points are non-degenerate. Degenerate critical points of the system (2.1) can be completely resolved by means of a sequence of transformations of the coordinates used in Sect. 1 of Chapt. V (see Sect. 4 below).

The system of equations of gas dynamics for the generalized self-similar motion with flat waves of the form (1.10) is reduced to the following system (see (2.1) for the notation):

$$z' = \frac{z}{(V - \delta)(z - (V - \delta)^2)}[z(2 - \varkappa(\gamma - 1) - 2V) + ((1 + \gamma)V$$

$$- 2 + (\gamma - 1)U_i^i)(V - \delta)^2 - (\gamma - 1)(V^2 - V)(V - \delta)],$$

$$V' = \{z(\varkappa - V - U_i^i) + (V - \delta)(V^2 - V)\}(z - (V - \delta)^2)^{-1},$$

$$U_i^j(V - \delta) = U_i^j - U_k^j U_i^k, \quad W^i(V - \delta) = W^i - U_k^i W^k,$$

$$R'(V - \delta) = R(-V' + s + (k + 2)V - U_i^i).$$

where there is implicit summation over the repeating indices $i, k = 2, 3$. It is not difficult to obtain simple exact solutions with flat waves corresponding to the critical points of the above system, in which the matrix U_j^i is diagonal and its eigenvalues are equal to 0 or 1.

In this section we will list a number of algebraic integrals relating (in finite form) the quantities

$$V, R, P, \Omega, \lambda = \exp \tau,$$

for the system of self-similar equations (2.1)–(2.2).

II. The Angular Momentum Integral. In order to obtain integrals of the self-similar equations (2.1)–(2.2) for $U \equiv 0$ related to various preserved quantities F (the value of F is calculated per unit volume of the gas) we will use the following method from the book [7]. Consider the moving cylindrical surfaces $r'(t)$ and $r''(t)$, on which the self-similar variable λ has constant values λ' and λ''. For any function F the following identity [7] holds:

$$\frac{d}{dt} \int\limits_{r'}^{r''} Fr \, dr = \frac{\tilde{d}}{dt} \int\limits_{r'}^{r''} Fr \, dr + \left[Fr\left(\frac{dr}{dt} - v\right) \right]_{r'}^{r''}. \tag{2.5}$$

Here the symbol $\tilde{d}/dt$ denotes differentiation with respect to time done outside the integral over a moving volume comprised of the same gas particles. The integral in (2.5) for the function F independent of z_1 and φ coincides with the full integral over a column of gas of unit height bounded by cylinders of radii r' and r''.

By virtue of the law of conservation of angular momentum we have

$$\frac{\tilde{d}}{dt} \int\limits_{r'}^{r''} \rho r^2 w \, dr = 0.$$

After a substitution of the formulas (1.10) we obtain

$$\frac{d}{dt} \int\limits_{r'}^{r''} \rho r^2 w \, dr = ab^{1-k}t^{(1-k)\delta-s-2}((1 - k)\delta - (1 + s)) \int\limits_{\lambda'}^{\lambda''} R\Omega\lambda^{-k} \, d\lambda,$$

$$\rho r^2 w \left(\frac{dr}{dt} - v\right)\Big|_{r'}^{r''} = ab^{1-k}t^{(1-k)\delta-s-2}R\Omega\lambda^{1-k}(\delta - V)\Big|_{\lambda'}^{\lambda''}.$$

Therefore for $F = \rho r w$ the identity (2.5) implies that

$$((1 - k)\delta - (1 + s)) \int_{\lambda'}^{\lambda''} R\Omega\lambda^{-k}\,d\lambda = R\Omega\lambda^{1-k}(\delta - V)\Big|_{\lambda'}^{\lambda''}. \tag{2.6}$$

According to (2.6) for $(1 - k)\delta - (1 + s) = 0$ $(\varkappa = (2\delta + 1)/\gamma)$ the values of the function $R\Omega\lambda^{1-k}(V - \delta)$ on an arbitrary trajectory of the system (2.1)–(2.2) coincide for all λ', λ''. Consequently for $\varkappa = (2\delta + 1)/\gamma$ there is an integral of the system (2.1)–(2.2):

$$\Phi_1 = R\Omega(V - \delta) = \text{const } \lambda^{k-1}. \tag{2.7}$$

Note that the condition for the existence of the integral (2.7) $(1 - k)\delta - (1 + s) = 0$ implies that from the constants a and b of the self-similar solution (see (1.9)) we can form a constant with the dimensions of angular momentum per unit length $[ab^{1-k}] = MLT^{-1}$.

III. The Energy Integral. The total energy of a column of gas with unit height bounded by cylinders of radii r' and r'' is equal to

$$\mathscr{E} = 2\pi \int_{r'}^{r''} \left(\varepsilon + \frac{v^2 + w^2}{2}\right)\rho r\,dr.$$

During adiabatic motion the density of internal energy of gas is $\varepsilon = p/(\gamma - 1)\rho$. The change in the energy of gas particles located in a given moment of time between the cylindrical surfaces $r' = \text{const}$ and $r'' = \text{const}$ is equal to the work of the forces of pressure on these surfaces. Therefore we have

$$\frac{\tilde{d}\mathscr{E}}{dt} = -2\pi(p''v''r'' - p'v'r'), \tag{2.8}$$

where v denotes the radial component of the velocity of gas particles.

For $F = p/(\gamma - 1) + \rho(v^2 + w^2)/2$ the identity (2.5) (after substituting formulas (1.10) and (2.8) and cancelling common fractions wherever possible) implies that

$$((1 - k)\delta - (s + 2)) \int_{\lambda'}^{\lambda''} \left(\frac{P}{\gamma - 1} + R\frac{V^2 + \Omega^2}{2}\right)\lambda^{-k}\,d\lambda$$

$$= -\lambda^{1-k}R\left[\frac{P}{R}V + (V - \delta)\left(\frac{V^2 + \Omega^2}{2} + \frac{P}{(\gamma - 1)R}\right)\right]\Big|_{\lambda'}^{\lambda''}. \tag{2.9}$$

Hence for $(1 - k)\delta - (s + 2) = 0$ (or $\varkappa = 2\delta/\gamma$) we obtain the following integral of the system (2.1)–(2.2):

$$H = R\left(\frac{P}{R}V + (V - \delta)\left(\frac{V^2 + \Omega^2}{2} + \frac{P}{(\gamma - 1)R}\right)\right) = \text{const } \lambda^{k-1}. \quad (2.10)$$

Just as in [7] the condition for the existence of the integral (2.10) $(1 - k)\delta - (s + 2) = 0$ implies that from the constants a and b of the self-similar solution we can form a constant with the dimensions of energy per unit length $[ab^{1-k}] = MLT^{-2}$.

IV. Adiabatic and Mass Integrals. These integrals exist for all values of the parameters of the problem and their derivation is completely analogous to the derivation in the book [7]. Let $\mathcal{M}(r, t)$ be the mass of gas (per unit height) inside a cylindrical surface of radius r. In a self-similar solution we have

$$\mathcal{M} = \frac{a}{r^{k+1}t^s} m(\lambda).$$

The mass integral has the following from [7]:

$$(s + \delta(k + 1))m - 2\pi R(V - \delta) = \text{const } \lambda^{1+k}. \quad (2.11)$$

The adiabatic integral has the following form [7]:

$$\Phi_2 = PR^{-\gamma+x}|V - \delta|^x = \text{const } \lambda^y,$$

$$x = \frac{2 - (\gamma - 1)s + \delta(k + 1 - \gamma(k + 3))}{s + \delta(k + 1)}, \quad y = -2\frac{\gamma s + k + 1}{s + \delta(k + 1)}. \quad (2.12)$$

Using the equation (2.3) it is not difficult to verify that the system (2.1) $(U = 0)$ has a monotone function:

$$\Phi_3 = z|V - \delta|^{\gamma-1}\Omega^{2(1-\gamma\delta)/(2\delta-1)} = \text{const} \cdot \lambda^{2(\gamma-2)/(2\delta-1)}. \quad (2.13)$$

The system (2.1)–(2.2) has a first integral:

$$\Phi_4 = R^\alpha z|V - \delta|^{(\gamma-1)+\alpha}\Omega^{-2-\alpha s}, \quad \alpha = \frac{2(\gamma - 2)}{2s + k + 1}. \quad (2.14)$$

For $\gamma \neq 2$ the integral Φ_4 enables us to express the variable R in terms of z, V and Ω. For $\gamma = 2$ the monotone function Φ_3 is a first integral and coincides with Φ_4. For $\gamma = 2$ this integral permits us to reduce the order of the system (2.1) $(U \equiv 0)$ by one, i.e. transform this system onto a plane.

For $U \equiv 0$ the system (2.1)–(2.3) has a monotone function:

$$\Phi_5 = R\Omega^\beta(V - \delta) = \text{const } \lambda^{k+1-2\beta}, \quad \beta = \frac{\varkappa\gamma - 2}{2\delta - 1}. \quad (2.15)$$

A particular case of this monotone function for $\varkappa = (2\delta + 1)/\gamma$, $\beta = 1$ is the angular momentum integral Φ_1 (2.7).

V. Integrals of the Full system of Self-Similar Equations. The system of four equations (2.1) has a monotone function:

$$\Phi_6 = \Omega \left| \frac{U}{1-U} \right|^{2\delta-1} = \text{const} \cdot \lambda^{-2}. \tag{2.16}$$

Using the equation (2.3) it is not difficult to verify that the system (2.1) also has a first integral:

$$\Phi_7 = z |V - \delta|^{\gamma-1} \Omega^{-\gamma} U^{\gamma-2} |1 - U|^{3-2\gamma}. \tag{2.17}$$

The system of equations (2.1)–(2.2) has a first integral:

$$\Phi_8 = R(V - \delta)\Omega^{(k+1)/2} U^x |1 - U|^y,$$
$$x = -s - (k+1)/2, \quad y = s + (k-1)/2. \tag{2.18}$$

The integral Φ_8 enables us to express the density R in terms of the variables V, Ω and U. The integral Φ_7 enables us to reduce the order of the system (2.1) by one. For $\gamma = 2$ this integral is defined on the invariant manifold $U = 0$, where it coincides with the integral (2.13). For $\gamma = 3/2$ the integral Φ_7 is defined on the invariant manifold $U = 1$ and therefore lets us transform the system (2.1) for $U = 1$, $\gamma = 3/2$ onto a plane.

3. Exact Self-Similar Power Solutions

Let us write the self-similar solutions (1.10) in terms of the powers of r and t with indeterminate exponents and numerical coefficients. After substituting these expressions into the equations (1.6)–(1.8) and solving the obtained algebraic equations we obtain all the self-similar power solutions. In some of these exact solutions the gas does not rotate. There are six solutions corresponding to the critical points Z_6^0, Z_6^1, Z_7^0, Z_7^1, Z_8^0, Z_8^1 (2.4) and four solutions, which exist for special values of the parameter $\varkappa$ and correspond to the following simple trajectories of the system (2.1):

$$\varkappa = 0: \ V = 0, \Omega = 0, U = 0, z = C_1 \lambda^{-2/\delta};$$
$$\varkappa = 1: \ V = 0, \Omega = 0, U = 1, z = C_1 \lambda^{(\gamma-3)/\delta};$$
$$\varkappa = 2: \ V = 1, \Omega = 0, U = 0, z = C_1 \lambda^{2(1-\gamma)/(1-\delta)};$$
$$\varkappa = 3: \ V = 1, \Omega = 0, U = 1, z = C_1 \lambda^{3(1-\gamma)/(1-\delta)}. \tag{3.1}$$

For $\gamma = 2$ and $\gamma = 3/2$ there exist power solutions, where the angular velocity of gas does not depend on the coordinate r. These solutions correspond to the following two lines of critical points (parametrized by Ω_0):

$$\gamma = 2: \ V = \frac{1}{2}, \Omega = \Omega_0, U = 0, z = \frac{1-2\delta}{2(\varkappa-1)}\left(\Omega_0^2 + \frac{1}{4}\right); \tag{3.2}$$

$$\gamma = \frac{3}{2}: \ V = \frac{1}{2}, \ \Omega = \Omega_0, \ U = 1, \ z = \frac{1 - 2\delta}{2(\varkappa - 2)}\left(\Omega_0^2 + \frac{1}{4}\right). \tag{3.3}$$

For $\gamma = 2$ let, $\alpha_1 = 4(\varkappa - 1)/(1 - 2\delta)$, $u_1 = 0$ and for $\gamma = 3/2$ let $\alpha_2 = 3(\varkappa - 2)/(1 - 2\delta)$, $u_2 = 1$. Solutions corresponding to the critical points of (3.2) and (3.3) for $\alpha_i > 0$ are determined by the following formulas:

$$v = \frac{1}{2}\frac{r}{t}, \quad w = \Omega_0\frac{r}{t}, \quad u = u_i\frac{z_1}{t},$$

$$p = \frac{C_1}{\alpha_i}\left(\Omega_0^2 + \frac{1}{4}\right)ab^{-\alpha - \alpha_i}r^{\alpha_i}t^{-\delta\alpha_i - \gamma\varkappa}, \tag{3.4}$$

$$\rho = C_1 ab^{-\alpha - \alpha_i}r^{\alpha_i - 2}t^{-\delta\alpha_i - \gamma\varkappa + 2}, \quad \alpha = (\gamma\varkappa - s - 2)/\delta.$$

In the solution (3.4) for $\gamma = 2$ $(i = 1)$ the gas particles move along the logarithmic spirals $\varphi = C_0 \ln r + C_2$, $z_1 = C_3$ and emerge from the axis of rotation $r = 0$ at $t = 0$. In the solution (3.4) for $\gamma = 3/2$ $(i = 2)$ the motion of the gas particles takes place along logarithmic spirals lying on paraboloids $z_1 = Cr^2$ and for $t = 0$ the particles of gas emerge from the center $r = z_1 = 0$.

Just like the solutions corresponding to the critical points (2.4) and the trajectories (3.1), the solutions (3.4) belong to the class of motions of the gas with homogeneous deformation (where the components of velocity of the gas are linear functions of the coordinates).

For $\varkappa/\delta < 0$ there exists a stationary power solution, in which the gas only rotates:

$$v = 0, \quad w = C_1 b^{1/\delta}r^{1 - 1/\delta}, \quad u = 0,$$

$$p = -\frac{\delta}{\gamma\varkappa}C_1^2 C_2 ab^{(2 + s)/\delta}r^{-\gamma\varkappa/\delta}, \quad \rho = C_2 ab^{s/\delta}r^{-\gamma\varkappa/\delta + 2(1 - \delta)/\delta}. \tag{3.5}$$

This solution corresponds to the following trajectory of the system (2.1):

$$V = 0, \quad U = 0, \quad z = -\frac{\delta}{\varkappa}\Omega^2, \quad \Omega = C_1\lambda^{-1/\delta}. \tag{3.6}$$

For $\gamma = 2$ and $\gamma = 5/3$ there exist non-stationary power solutions, in which the angular velocity of rotation of the gas (just like in the solution (3.5)) depends on the distance to the axis of rotation. These solutions correspond to the following two trajectories of the system (2.1):

$$\gamma = 2: \ V = 1, \ U = 0, \ z = \frac{1 - \delta}{\varkappa - 2}\Omega^2, \ \Omega = C_1\lambda^{1/(\delta - 1)}; \tag{3.7}$$

$$\gamma = \frac{5}{3}: \ V = 1, \ U = 1, \ z = \frac{1 - \delta}{\varkappa - 3}\Omega^2, \ \Omega = C_1\lambda^{1/(\delta - 1)}. \tag{3.8}$$

For $\gamma = 2$ let $\beta_1 = 2(-2)/(1 - \delta)$, $u_1 = 0$ and for $\gamma = 5/3$ let $\beta_2 = 5(\varkappa - 3)/3(1 - \delta)$, $u_2 = 1$. For $\beta_i > 0$ the trajectories (3.7) and (3.8) correspond to the following solutions:

$$v = \frac{r}{t}, \quad w = C_1 b^{1/(1-\delta)} r^{\delta/(\delta-1)} t^{-(2\delta-1)/(\delta-1)}, \quad u = u_i \frac{z_1}{t},$$

$$\rho = C_2 ab^{-\beta-\beta_i-2/(1-\delta)} r^{\beta_i+2\delta/(1-\delta)} t^{-\beta_i\delta-\gamma\varkappa-2(2\delta-1)/(1-\delta)}, \tag{3.9}$$

$$p = \frac{C_1^2 C_2}{\beta_i} ab^{-\beta-\beta_i} r^{\beta_i} t^{-\beta_i\delta-\gamma\varkappa}, \quad \beta = \frac{\gamma\varkappa - s - 2}{\delta}.$$

In the solution (3.9) for $\gamma = 2$ $(i = 1)$ the gas particles move along spirals $\varphi = C_3 + C_0/r$, $z_1 = C_4$ and for $t = 0$ emerge from the axis of rotation $r = 0$. In the solution (3.9) for $\gamma = 5/3$ $(i = 2)$ the gas particles move along spirals $\varphi = C_3 + C_0/r$ lying on cones $z_1 = C_4 r$. For $t = 0$ the gas particles emerge from the center $r = z_1 = 0$.

4. Analysis of the Dynamical System

From now on the dynamical system (2.1) will be considered in the following subset S' of the four-dimensional space (z, Ω, V, U):

$$z \geqslant 0, \quad \Omega \geqslant 0, \quad -\infty < V < +\infty, \quad 0 \leqslant U \leqslant 1. \tag{4.1}$$

The restriction of the domain of values of the coordinate U is caused by the fact that outside the segment $(0, 1)$ along the trajectories of the system (2.1) U becomes infinite for finite values of the variable λ. The system (2.1) becomes singular for $V = \delta$ and each trajectory of the system (2.1) stays in one of the regions S_1 $(V < \delta)$ or S_2 $(V > \delta)$ for all λ.

The algebraic structure of the dynamical system (2.1) is (as mentioned in Sect. 2) completely analogous to the algebraic structure of the system (1.14) from Chapt. V, where the variable m is the formal analogue of the variable Ω^2. Hence the full resolution of singularities of the dynamical system (2.1) (i.e. the construction of a compact manifold S, on which the system (2.1) has only non-degenerate critical points) is achieved by means of the transformations of coordinates (1.17)–(1.33) from Chapt. V, where the variable m should be replaced by Ω^2 and the parameter $2/\omega$ should be denoted by δ. For all these transformations the coordinate U remains unchanged. For each of the invariant regions S_1 $(V < \delta)$ and S_2 $(V > \delta)$ the compact manifold S is a product of the three-dimensional compact manifold S_0 (constructed in Chapt. V) and the segment I $(0, 1)$ (the domain of values of the coordinate U). Two new components of the boundary $U = 0$ and $U = 1$ are obviously invariant manifolds of the dynamical system (2.1).

Let us analyze the most interesting critical points of the system (2.1) and the power asymptotic forms of the solutions corresponding to them. Critical points will be denoted by Z_i^ε, where the subscript i is the number of the corresponding critical point in the compact manifold S_0 in Chapt. V and the superscript $\varepsilon = 0, 1$

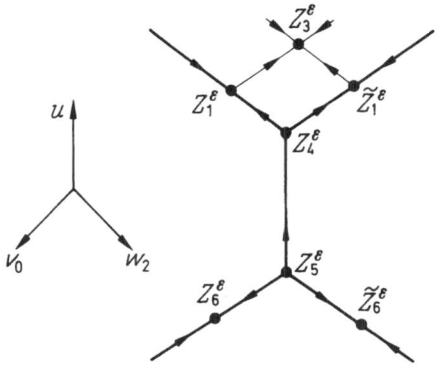

Fig. 34. Resolution of the degenerate critical points 0_1^ε and 0_2^ε

denotes the two copies of the critical point in question contained in $U = 0$ and $U = 1$. In further discussion the system (2.1) will be considered mainly in the region S_1 $(V < \delta)$. Let us make a change of variable $d\tau_3/d\tau = -1/(V - \delta)$ and introduce new coordinates (analogous to the coordinates (1.17) in Chapt. V):

$$u = z^{1/2} > 0, \quad v_0 = V - \delta < 0, \quad w_0 = -(\Omega^2(V - \delta))^{1/3} > 0 \qquad (4.2)$$

(in the region S_2 $(V > \delta)$ the time change is $d\tau_3/d\tau = 1/(V - \delta)$ and the coordinates (4.2) satisfy the conditions $u > 0$, $v_0 > 0$ and $w_0 < 0$). In the coordinates (4.2) the dynamical system (2.1) has an invariant submanifold $v_0 = 0$ (the component of the boundary Γ_6 of the manifold S).

In order to analyze the behavior of trajectories of the dynamical system (2.1) for large values of the coordinates u, v_0 and w_0, we will complete the region S_1 by a boundary at infinity with respect to the coordinates u, v_0 and w_0. Such completion is accomplished by means of the following coordinates (a change of the variable τ is also shown):

$$v_1 = \frac{v_0}{u}, \quad \rho_0 = \frac{1}{u}, \quad w_1 = \frac{w_0}{u}, \quad \frac{d\tau_1}{d\tau} = -\frac{u}{v_0}. \qquad (4.3)$$

The coordinates (4.3) are analogous to the coordinates (1.18) in Chapt. V. In the coordinates (4.2) and (4.3) the system (2.1) has four degenerate critical points:

$$0_1^\varepsilon \ (v_1 = w_1 = \rho_0 = 0, \ U = \varepsilon = 0, 1) \quad \text{and}$$

$$0_2^\varepsilon \ (v_0 = w_0 = u = 0, \ U = \varepsilon = 0, 1).$$

In order to resolve the degenerate critical point 0_1^ε we introduce the following coordinates (see Chapt. V (1.21)):

$$v_0 = \frac{v_1}{\rho_0}, \quad w_2 = \frac{w_1^3}{\rho_0} = -\frac{v_0 \Omega^2}{u^2}, \quad \rho_0 = \frac{1}{u}, \quad \tau_3. \qquad (4.4)$$

In order to resolve the degenerate critical point 0_2^ε we introduce the following coordinates (see Chapt. V (1.27)):

$$v_4 = \frac{v_1}{u} = \frac{v_0}{u^2}, \quad u, \quad w_2 = -\frac{v_0 \Omega^2}{u^2}, \quad \tau_3. \qquad (4.5)$$

Under the conditions (4.2) the coordinates (4.3)–(4.5) cover the part of the manifold shown in Fig. 34, where the critical points of the dynamical system are also pictured (for comparison see Fig. 26).

In the coordinates (4.4) the dynamical system (2.1) has the following form:

$$\dot{v}_0 = \frac{v_0}{1 - \rho_0^2 v_0^2} (X_1 + 2v_0 + 2\delta + U - \varkappa - w_2), \quad \dot{U} = -U(1 - U),$$

$$\dot{w}_2 = \frac{w_2}{1 - \rho_0^2 v_0^2} [\gamma X_1 + X_2 + X_3 - U - 2(1 - 2\delta - 2v_0)(1 - v_0^2 \rho_0^2)], \quad (4.6)$$

$$\dot{\rho}_0 = \frac{\rho_0}{2(1 - \rho_0^2 v_0^2)} [(1 - \gamma)X_1 - X_2 + X_3 + 2v_0 + 2\delta - \varkappa - w_2],$$

where

$$X_1 = -v_0(v_0 + \delta)(v_0 + \delta - 1)\rho_0^2, \quad X_2 = 2 - \gamma(\varkappa + w_2),$$
$$X_3 = v_0^2 \rho_0^2 (2 - (\gamma - 1)U - 2\gamma\delta - 2\gamma v_0).$$

In the coordinates (4.5) the dynamical system (2.1) has the following form:

$$\dot{v}_4 = \frac{v_4}{1 - v_4 u^2} (\gamma Y_1 + Y_2 - Y_3), \quad \dot{U} = -U(1 - U),$$

$$\dot{w}_2 = \frac{w_2}{1 - v_4 u^2} [\gamma Y_1 + Y_2 - Y_3 - 2(1 - 2\delta - 2v_4 u^2)(1 - v_4^2 u^2)], \qquad (4.7)$$

$$\dot{u} = \frac{u}{2(1 - v_4^2 u^2)} [(1 - \gamma)Y_1 - Y_3 + (\gamma - 1)(\varkappa + w_2) + 2\delta - 2 + 2v_2 u^2].$$

where

$$Y_1 = -v_4(v_4 u^2 + \delta)(v_4 u^2 + \delta - 1), \quad Y_2 = 2 + U - \gamma(\varkappa - \omega_2),$$
$$Y_3 = v_4^2 u^2 (2 - (\gamma - 1)U - 2\gamma\delta - 2\gamma v_4 u^2).$$

As a result of the transformations (4.4)–(4.5) on the attached components of the boundary Γ_2 $(\rho_0 = 0)$ and Γ_4 $(u = 0)$ the dynamical systems (4.6)–(4.7) have the following critical points (in all places the coordinate U takes the values $U = \varepsilon = 0, 1$):

$$Z_1^\varepsilon: \rho_0 = 0, \ w_2 = 0, \ v_0 = \frac{\varkappa - \varepsilon}{2} - \delta;$$

$$\tilde{Z}_1^\varepsilon: \rho_0 = 0, \ w_2 = \frac{\varepsilon + 4\delta}{\gamma} - \varkappa, \ v_0 = 0;$$

$$Z_3^\varepsilon: \ \rho_0 = 0, \ w_2 = \frac{\varepsilon}{2-\gamma} - \varkappa, \ v_0 = \frac{\varepsilon(\gamma-1)}{2(2-\gamma)} - \delta;$$

$$Z_4^\varepsilon: \ \rho_0 = w_2 = v_0 = 0; \quad Z_5^\varepsilon: \ u = w_2 = v_4 = 0;$$

$$Z_6^\varepsilon: \ u = 0, \ w_2 = 0, \ v_4 = \frac{\varepsilon + 2 - \varkappa\gamma}{\gamma\delta(\delta-1)};$$

$$\tilde{Z}_6^\varepsilon: \ u = 0, \ w_2 = \frac{\varepsilon + 4\delta}{\gamma} - \varkappa, \ v_4 = 0. \tag{4.8}$$

The tildes denote the critical points which have no analogues in the compact manifold S_0 in Chapt. V (see Figs. 34 and 26). For $\delta \to 1$ at the critical point Z_6^ε we have $v_4 \to -\infty$ (see (4.8)). Therefore in the case $\delta = 1$ the transformation of coordinates (4.5) is insufficient to fully resolve the degenerate critical point O_2. For $\delta = 1$ the necessary additional transformation of coordinates is described below in Sect. 5.

Note that at all the critical points Z_i^0 the eigenvalue λ_U of the system (4.6)–(4.7) corresponding to the coordinate U is equal to -1. At the critical points Z_i^1 we have $\lambda_U = +1$. The eigenvalue λ_U will not be mentioned again.

Now let us analyze in detail two types of critical points, which have applications in the constructions in sections 5 and 6. For $\gamma = 2$ and $2\delta - \varkappa > 0$ on the component of the boundary $\Gamma_2 \, (\rho_0 = 0)$ there is a segment of critical points $\mathscr{L}$, on which we have

$$\rho_0 = 0, \quad U = 0, \quad w_2 - 2v_0 = 2\delta - \varkappa, \quad 0 \leqslant w_2 \leqslant 2\delta - \varkappa. \tag{4.9}$$

The segment $\mathscr{L}$ passes through the critical points Z_1^0, $\tilde{Z}_1^0$ and Z_3^0. Critical points on the segment $\mathscr{L}$ have the following eigenvalues:

$$\lambda_1 = 0, \quad \lambda_U = -1, \quad \lambda_3 = -w_2 + \varkappa - 2\delta < 0, \quad \lambda_{\rho_0} = 1 - u_2 - \varkappa.$$

For $\varkappa > 1$ the segment $\mathscr{L}$ consists entirely of attracting critical points. For $\varkappa < 1$ the segment $\mathscr{L}$ has a subinterval of unstable critical points: $\varkappa/2 < V < 1/2$, from which a two-dimensional separatrix emerges into the physical region of the manifold S. The corresponding solutions have the following asymptotic form for $\lambda \to 0$:

$$v = V\frac{r}{t}, \quad w = \left|\frac{2V-\varkappa}{V-\delta}\right|^{1/2} C_1 b^{-\alpha} r^{1+\alpha} t^{1-\delta\alpha}, \quad u = 0,$$

$$\rho = C_2 ab^{-k-1-\sigma} r^{\sigma-2} t^{-V\sigma}, \quad \sigma = 2\frac{\varkappa-1}{V-\delta} > 0, \quad \alpha = \frac{1-2V}{V-\delta}, \tag{4.10}$$

$$p = \frac{C_1^2 C_2}{\gamma} ab^{-k+3-\beta} r^{2(\varkappa-2V)(V-\delta)} t^{-V\beta}, \quad \beta = 2\frac{\varkappa-2\delta}{V-\delta} > 0.$$

In this asymptotic form $p = 0$ at the axis of rotation. The asymptotic form (4.10) has a physical meaning if for $r = 0$ we also have $w = 0$, i.e. $1 + \alpha > 0$ or $1 - \delta < V < \delta$.

The separatrices of critical points of the line $\mathscr{L}$ determine solutions, which continue to the axis of symmetry $r = 0$. Solutions corresponding to the separatrices of critical points Z_1^ε and Z_3^ε have the same property.

The critical points $\tilde{Z}_1^\varepsilon$, Z_4^ε, Z_5^ε, Z_6^ε and $\tilde{Z}_6^\varepsilon$ belong to the invariant manifold $v_0 = 0$. Separatrices of these critical points determine solutions, in which the region filled with gas has a moving boundary $\lambda = \lambda_0$. The critical point Z_6^ε has the following eigenvalues:

$$\lambda_u = \delta - 1 + (\varepsilon + 2)(\gamma - 1)/2\gamma, \quad \lambda_{u_4} = -(\varepsilon + 2 - \varkappa\gamma),$$
$$\lambda_{w_2} = 2(2\delta - 1). \tag{4.11}$$

Separatrices of the critical point Z_6^ε determine solutions with an inner $(\lambda > \lambda_0)$ or outer $(\lambda < \lambda_0)$ boundary of gas for simultaneously positive or simultaneously negative eigenvalues λ_u, λ_{w_2} (4.11). These solutions have the following asymptotic form for $\lambda \to \lambda_0$:

$$v = \lambda_0 bt^{\delta-1}\left(\delta - \alpha\left(\frac{\lambda}{\lambda_0} - 1\right)\right), \quad u = \varepsilon\frac{z_1}{t},$$

$$w = bt^{\delta-1}C_1\left|\frac{\lambda}{\lambda_0} - 1\right|^{\beta_1}, \quad p = -\frac{\alpha}{\gamma v_4}ab^{-k-1}C_2 t^{-\varkappa\gamma}\left|\frac{\lambda}{\lambda_0} - 1\right|^{-\beta_2},$$

$$\rho = ab^{-k-3}t^{2(1-\delta)-\varkappa\gamma}C_2\left|\frac{\lambda}{\lambda_0} - 1\right|^{-1-\beta_2}, \tag{4.12}$$

$$\alpha = 2\lambda_u = 2(\delta - 1) + (\varepsilon + 2)\frac{\gamma - 1}{\gamma}, \quad \beta_1 = \frac{2\delta - 1}{\alpha} = \frac{\lambda_{w_2}}{4\lambda_u} > 0,$$

$$\beta_2 = \frac{\varkappa\gamma - 2 - \varepsilon}{\alpha} = \frac{\lambda_{v_4}}{2\lambda_u}, \quad v_4 = \frac{\varepsilon + 2 - \varkappa\gamma}{\gamma\delta(\delta - 1)}.$$

In the asymptotic form (4.12), $w \to 0$ as $\lambda \to \lambda_0$. From the necessary condition of boundedness of pressure we obtain $\beta_2 < 0$ for $\lambda \to \lambda_0$. Therefore the asymptotic form (4.12) has a physical meaning only in the case of the unstable critical point Z_6^ε (the eigenvalues λ_{v_4} and λ_u have opposite signs). In this case at the boundary $\lambda = \lambda_0$ we have $p = 0$. For $\lambda \to \lambda_0$ the temperature of gas is $T \sim p/\rho \sim |\lambda/\lambda_0 - 1| \to 0$. For $-1 - \beta_2 > 0$ the density of gas ρ at the boundary tends to 0, i.e. we can assume that this boundary is between the gas and vacuum.

5. Self-Similar Expansion of Rotating Gas

In this section we will analyze three different types of self-similar expansion of a rotating gas existing for the following values of the parameter δ: $\delta > 1$, $\delta = 1$ and $1 > \delta > 1/2$ (in the last case $\gamma = 3/2$). In all three cases the corresponding trajectories of the system (2.1) approach the attracting critical point Z_8^0 $(z = V$

$= \Omega = U = 0$) for $\tau \to \infty$, which ensures the necessary asymptotic form for the solutions as $r \to \infty$. The behavior of the solutions under consideration in the vicinity of the axis of rotation varies and depends on the value of the parameter δ.

I. In order to construct solutions describing the self-similar expansion of a rotating gas for $\delta > 1$, we will analyze the properties of the critical points Z_6^ε $(V = 1, \ \Omega = z = 0, \ U = \varepsilon = 0, \ 1)$ and Z_8^ε $(V = \Omega = z = 0, \ U = \varepsilon = 0, \ 1)$. At the critical point $\bar{Z}_8^\varepsilon$ $(V = \Omega = z = 0, \ U = \varepsilon = 0, \ 1)$. At the critical point $\bar{Z}_6^\varepsilon$ the system (2.1) has the following eigenvalues:

$$\lambda_z = \frac{\gamma - 1}{\delta - 1}(2 + \varepsilon), \quad \lambda_V = \frac{1}{\delta - 1}, \quad \lambda_\Omega = \frac{1}{\delta - 1}, \quad \lambda_U = \frac{2\varepsilon - 1}{\delta - 1}. \tag{5.1}$$

For $\delta > 1$ the critical point $\bar{Z}_6^1$ is repelling and for $\gamma > 4/3$ all separatrices (except one) emerging fom the critical point $\bar{Z}_6^1$ are tangent to the invariant manifold $z = 0$ at this point (because $\lambda_z > \lambda_V = \lambda_\Omega = \lambda_U$ for $\gamma > 4/3$). The corresponding solutions have the following asymptotic form for $\lambda \to 0$:

$$V = \frac{r}{t}, \quad w = C_1 r^{\delta/(\delta-1)} t^{(1-2\delta)/(\delta-1)}, \quad u = \frac{z_1}{t},$$

$$\rho = ab^{(s+k)/(\delta-1)} C_1 r^{(5-2\delta-\varkappa\gamma)/(\delta-1)} t^{(\varkappa\gamma-2-\delta)/(\delta-1)}, \tag{5.2}$$

$$p = ab^{(s+k-3(\gamma-1))/(\delta-1)} C_2 r^{\gamma(3-\varkappa)/(\delta-1)} t^{\gamma(\varkappa-3\delta)/(\delta-1)}.$$

In the asymptotic form (5.2) for $\varkappa < (5 - 2\delta)/\gamma < 3$ we have $\rho \to 0$ and $p \to 0$ when $r \to 0$.

At the critical point Z_8^ε the system (2.1) has the following eigenvalues:

$$\lambda_z = \frac{1}{\delta}(-2 + (\gamma - 1)\varepsilon), \quad \lambda_V = \lambda_\Omega = -\frac{1}{\delta}, \quad \lambda_U = \frac{2\varepsilon - 1}{\delta}. \tag{5.3}$$

The critical point Z_8^0 is attracting for $\delta > 0$. All trajectories (except one) approaching this critical point are tangent to the invariant manifold $z = 0$ (because $|\lambda_z| > |\lambda_V| = |\lambda_\Omega| = |\lambda_U|$). The corresponding solutions have the following asymptotic form for $\lambda \to \infty$:

$$v = b^{1/\delta} C_1 r^{(\delta-1)/\delta}, \quad w = b^{1/\delta} C_2 r^{(\delta-1)/\delta}, \quad u = b^{1/\delta} C_3 z_1 r^{-1/\delta},$$

$$\rho = ab^{s/\delta} C_4 r^{(2-2\delta-\varkappa\gamma)/\delta}, \quad p = ab^{(s+2)/\delta} C_5 r^{-\varkappa\gamma/\delta}. \tag{5.4}$$

In the asymptotic form (5.4) the constant C_1 can have either sign and the other constants C_i are positive.

In the case of a gas expanding into the atmosphere the necessary conditions (for $r \to \infty$) are $p \to p_0$ and $\rho \to \rho_0$. In the asymptotic form (5.4) this is achieved only when $\varkappa = 0, \delta = 1$. In the case of a gas expanding into vacuum for $r \to \infty$ we have $p \to 0$ and $\rho \to 0$, which is achieved for $\varkappa > (0, 2(1 - \delta)/\gamma)$. For solutions with the asymptotic form (5.4) the conditions of finiteness (for $r \gg 1$) of mass and total energy of the gas in a layer of unit height (with respect to the coordinate z_1) are

$$\varkappa > 2/\gamma, \quad \varkappa > 2\delta/\gamma. \tag{5.5}$$

For $\delta > 1$ the system (2.1) has trajectories (separatrices) emerging from the repelling critical point $\bar{Z}_6^1$ and approaching the attracting critical point Z_8^0. Indeed consider the system (2.1) on the invariant manifold $z = 0$:

$$V' = -[(V - 1/2)^2 - \Omega^2 - 1/4](V - \delta)^{-1},$$

$$\Omega' = -2\Omega(V - 1/2)(V - \delta)^{-1}, \tag{5.6}$$

$$U' = -U(U - 1)(V - \delta)^{-1}.$$

Trajectories of the system (5.6) in the V, Ω plane are arcs of circles

$$(V - 1/2)^2 + (\Omega - K)^2 = K^2 + 1/4, \tag{5.7}$$

where $K = [(V - 1/2)^2 + \Omega^2 - 1/4](2\Omega)^{-1}$ is a first integral of the system (5.6). All circles of the family (5.7) pass through the two points $V = 1, \Omega = 0$ and $V = 0$, $\Omega = 0$. Therefore if a circle (5.7) does not intersect the line $V = \delta$ (in this case $K < \delta(\delta - 1)$, $\delta > 1$), then the corresponding trajectory of the system (5.6) emerges from the critical point $\bar{Z}_6^1$ and approaches the critical point Z_8^0 (by virtue of (5.6) in the region $V < \delta$ the coordinate U decreases monotonically from 1 to 0).

For $z \ll 1$ the trajectories of the system (2.1) can be approximated by the trajectories of the system (5.6). Since the critical point $\bar{Z}_6^1$ is repelling and the critical point Z_8^0 is attracting, the above trajectories begin and end at these critical points (analogous solutions also exist for $U \equiv 0$ and correspond to trajectories going from the critical point $\bar{Z}_6^0$ to the critical point Z_8^0). The corresponding solutions for $\varkappa > 0, \delta > 1$ describe the rotating gas flying off into vacuum. The asymptotics of these solutions for $r \to 0$ and $r \to \infty$ are of the form (5.2), (5.4). When the conditions $2\delta\gamma^{-1} < \varkappa < (5 - 2\delta)\gamma^{-1}$ are satisfied, the considered solutions are regular for all $r \geqslant 0$ and the total mass and energy of a column of gas with unit height are finite. The angular velocity of rotation of the gas Ω tends to 0 when $r \to 0, \infty$. It reaches its maximum $\Omega_0 \approx K + (K^2 + 1/4)^{1/2}$ for some $\lambda = \lambda_1$. The region of fastest rotation of the gas $w \approx r\Omega_0/t$ propagates over the particles with the speed $v_1 = \delta\lambda_1 bt^{\delta-1}$. The sign of v for $r \to \infty$ is opposite to the sign of K.

II. For $\delta = 1$ the critical point $\bar{Z}_6^\varepsilon$ coincides with the degenerate critical point O_2^ε ($V = \delta, z = \Omega = 0$). To resolve the degenerate critical point O_2 for $\delta = 1$ we will make (along with the transformation (4.5)) the following transformation of coordinates:

$$v_0 = V - 1, \quad u_2 = \frac{u^2}{v_0^2} = \frac{z}{v_0^2}, \quad w_3 = \frac{w_0^3}{v_0^2} = -\frac{\Omega^2}{v_0}. \tag{5.8}$$

In the coordinates (5.8) and variable τ_3 ($d\tau_3/d\tau = -1/v_0$) the system (2.1) has the following form ($\delta = 1$):

$$\dot{v}_0 = -\frac{v_0}{u_2 - 1}[u_2(\varkappa - U - 2v_0 - 2) + w_3 + v_0 + 1],$$

$$\dot{w}_3 = \frac{w_3}{u_2 - 1}[u_2(\varkappa - U + 2v_0) + w_3 - 1 - 3v_0],$$

$$\dot{u}_2 = -\frac{u_2}{u_2 - 1}[u_2(2v_0 - (\gamma + 1)\varkappa + 4 + 2U) - (\gamma + 1)w_3$$

$$- 3 + \gamma + (\gamma - 1)v_0 + (\gamma - 1)U],$$

$$\dot{U} = -U(U - 1).$$

(5.9)

The system (5.9) has an invariant manifold $v_0 = 0$, which contains the following critical points of this system $(U = \varepsilon = 0, 1)$:

$$X_1^\varepsilon: v_0 = u_2 = w_0 = 0; \quad X_2^\varepsilon: v_0 = u_2 = 0, w_3 = 1;$$

$$X_3^\varepsilon: v_0 = w_3 = 0, u_2 = u_2^0 = \frac{3 - \gamma - \varepsilon(\gamma - 1)}{4 - (\gamma + 1)\varkappa + 2\varepsilon};$$

(5.10)

$$X_4^\varepsilon: v_0 = 0, u_2 = 1, w_3 = 1 + \varepsilon - \varkappa.$$

For $\varkappa < 1 + \varepsilon$ at the critical point X_3^ε we have $u_2^0 < 1$ and at the critical point X_4^ε we have $w_3 > 0$. The eigenvalues of the system (5.10) have the following form:

$$X_1^\varepsilon: \lambda_{v_0} = 1, \lambda_{u_2} = -(3 - \gamma - (\gamma - 1)\varepsilon), \lambda_{w_3} = 1;$$

$$X_2^\varepsilon: \lambda_{v_0} = 2, \lambda_{u_2} = -4 + (\gamma - 1)\varepsilon, \lambda_{w_3} = -1;$$

$$X_3^\varepsilon: \lambda_{v_0} = \alpha = \frac{(2 + \varepsilon)(\gamma - 1)}{\gamma + 1}, \lambda_{u_2} = \frac{3 - \gamma - \varepsilon(\gamma - 1)}{1 - u_2^0},$$

$$\lambda_{w_2} = \beta = \frac{4 + \varepsilon(1 - \gamma)}{\gamma + 1};$$

(5.11)

$$X_4^\varepsilon: \lambda_{v_0} = 0, \lambda_{2,3} = \tfrac{1}{2}(3 + \varepsilon - \gamma\varkappa \pm [(3 + \varepsilon - \gamma\varkappa)^2$$

$$+ 4(1 - \varepsilon - \varkappa)(4 - \varepsilon(\gamma - 1))]^{1/2}).$$

The eigenvalues (5.11) at the critical point X_4 contained in the surface of non-extendability of solutions $z - (V - \delta)^2 = 0$ are calculated after a change of variable $d\tau_4/d\tau_3 = (1 - u_2)^{-1}$. The critical point X_4^ε belongs to some line of critical points I lying in the surface of non-extendability of solutions $u_2 = 1$. For $\varkappa < 2$ and $1 < \gamma < 2$ from (5.11) we see that the critical points X_1^1, X_2^1 and X_4^1 are saddles and the critical point X_3^1 is repelling. The same is true for the critical points X_i^0 when $\varkappa < 1, 1 < \gamma < 3$ (the critical point X_3^0 is repelling in the manifold $U = 0$). The position of critical points and qualitative behavior of the trajectories of the system (5.9) on the three-dimensional invariant manifold $U = 1$ for $\varkappa < 2$, $1 < \gamma < 2$ in the vicinity of the two planes $v_0 = 0$ and $u_2 = 0$ is shown in Fig. 35.

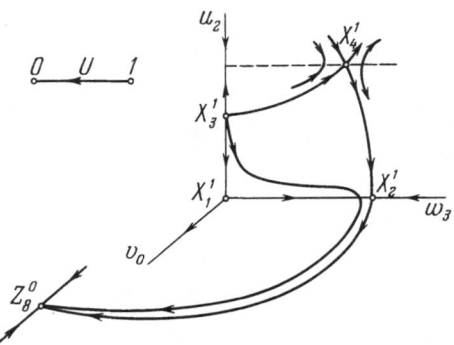

Fig. 35. Construction of solutions describing the self-similar expansion of rotating gas for $\delta = 1$

Separatrices emerging from the critical point X_3^{ξ} determine solutions with an expanding boundary $(\lambda = \lambda_0)$ inside the gas. The asymptotic forms of these solutions for $\lambda \to \lambda_0$ are $(\delta = 1, \varkappa < 2; 1 < \gamma < 2)$

$$v = \lambda_0 b\left(1 - \alpha\left(\frac{\lambda}{\lambda_0} - 1\right)\right), \quad w = \lambda_0 bC_1\left(\frac{\lambda}{\lambda_0} - 1\right)^{(\alpha+\beta)/2\alpha},$$

$$\rho = ab^{-k-3}C_2\left(\frac{\lambda}{\lambda_0} - 1\right)^{\sigma}, \quad u = \frac{z_1}{t}, \tag{5.12}$$

$$p = ab^{-k-1}\lambda_0^2 C_2 \frac{u_2^0\alpha^2}{\gamma}\left(\frac{\lambda}{\lambda_0} - 1\right)^{2+\sigma}, \quad \sigma = \frac{6 - \varkappa\gamma(1 + \gamma)}{3(\gamma - 1)}.$$

The constants u_2, α and β are defined above (see (5.10) and (5.11)). In the asymptotic form (5.12) for $\varkappa < 6\gamma^{-1}(1 + \gamma)^{-1}$ the density of gas, the pressure and the angular velocity w at the inner boundary are equal to zero. Therefore we can assume that in the region $\lambda < \lambda_0$ a cavity forms. The inner boundary of the gas $\lambda = \lambda_0$ expands with constant speed $v_0 = \lambda_0 b$. Since the gas rotates, the inner boundary is stable.

Consider the trajectories of the system (5.9) going along the stable sequence of separatrices $X_3^1 X_1^1$, $X_1^1 X_2^1$, $X_2^1 Z_8^0$ (see Fig. 35). (Note that for $u_2 = 0$ $(z = 0)$ the trajectories of the system (5.9) were analyzed in part I (see (5.7)). In particular the separatrix $X_2^1 Z_8^0$ corresponds to the circle (5.7) for $K = 0$.) For

$$\delta = 1, \quad 2\gamma^{-1} < \kappa < 6\gamma^{-1}(\gamma + 1)^{-1}, \quad \varkappa < 2$$

the corresponding solutions describe a self-similar expansion of gas into vacuum. These solutions are regular for all $r > 0$ and the total mass and energy of a column of gas with unit height are finite. Inside the gas forms a cavity which is expanding with constant speed $v_0 = \lambda_0 b$ $(\lambda < \lambda_0)$ and on whose boundary $(\lambda \to \lambda_0)$ the solution has the asymptotic form (5.12). At this boundary the vertical velocity of gas reaches its maximum. Maximal angular velocity $\Omega \approx 1/2$ is attained for $\lambda \approx \sqrt{2}\lambda_0$. The region of fastest rotation of the gas $\Omega \approx 1/2$ propagates over the particles with constant speed $v_1 \approx \sqrt{2\lambda_0 b}$. The asymptotics of the considered solutions for $r \to \infty$ are of the form (5.4), $\delta = 1$ and the limiting radial velocity of gas v can be arbitrarily small.

III. For a special value of $\gamma = 3/2$ (which is very close to $\gamma = 1.4$ for ordinary air) there exists another class of self-similar solutions describing a rotating gas expanding into space. In order to construct these solutions we shall analyze the properties of the line of critical points (3.3):

$$V = 1/2, \quad \Omega = \Omega_0, \quad U = 1, \quad z_0 = \alpha(\Omega_0^2 + 1/4) \tag{5.13}$$

for $\alpha = (1/2 - \delta)(\varkappa - 2) > 0$. At the critical points of (3.3) the system (2.1) has the following eigenvalues: $\lambda_1 = 0$, $\lambda_4 = \lambda_U = (\delta - 1/2)^{-1}$ and the eigenvalues $\lambda_{2,3}$ satisfy the following characteristic equation:

$$\lambda^2 + \lambda(2z_0 + (4\Omega_0^2 + 3)/8)L_0^{-1} + (3/4 - \Omega_0^2)L_0^{-1} = 0,$$
$$L_0 = z_0 - (1/2 - \delta)^2. \tag{5.14}$$

The eigenvectors corresponding to the eigenvalues λ_1, λ_2 and λ_3 lie in the plane $U = 1$.

The line of critical points (5.13) for $\beta = (1/2 - \delta)(\varkappa - 2) - 1/4 > 0$ is divided by the surface of non-extendability of solutions $L = z - (V - \delta)^2 = 0$ into two parts: $I_1 (L_0 < 0, 0 < \Omega_0 < \beta^{1/2})$ and $I_2 (L_0 > 0, \Omega_0 > \beta^{1/2})$. On the segment I_1 for $1/2 < \delta < 1$, $0 < \varkappa < 2$ we have $\Omega_0^2 < 3/4$. Therefore from (5.14) we see that

$$\lambda_2 \cdot \lambda_3 = (3/4 - \Omega_0^2)L_0^{-1} < 0, \quad \text{i.e. } \lambda_2 > 0, \lambda_3 < 0.$$

Consequently from each point of the segment I_1 emerges a two-dimensional separatrix $\mathscr{L}_0$ corresponding to the eigenvalues $\lambda_2 > 0, \lambda_4 > 0$.

For $\gamma = 3/2$ the system (2.1) on the invariant manifold $U = 1$ has a first integral (se (2.17)):

$$\Phi_7 = z|V - \delta|^{1/2}\Omega^{-3/2} = \text{const.}$$

On a level surface of the integral $\Phi_7 = C$ the system (2.1) for $U = 1$ is transformed onto the z, V plane. Qualitative behavior of the trajectories of the obtained system for $\Phi_7 = C > C_0 - \alpha|1/2 - \delta|^{1/2}(3/4)^{-3/4}$ in the region $L < 0$ is shown in Fig. 36. From each critical point of the segment $I_1 (L < 0)$ of the line (5.13) emerges a unique separatrix l (corresponding to the eigenvalue $\lambda_2 > 0$) approaching (for $\tau \to \infty$) the critical point $Z_8^1 (z = V = \Omega = 0, U = 1)$. According to (5.3), from the critical point Z_8^1 emerges a unique separatrix (along which the coordinate U varies from 1 to 0) approaching the attracting critical point Z_8^0. Therefore there exist separatrices leaving the critical points of the segment I_1 corresponding to the eigenvalues $\lambda_2 > 0, \lambda_4 > 0$ and approaching the critical point Z_8^0.

Self-similar solutions corresponding to the above separatrices for

$$\tfrac{1}{2} < \delta < 1, \quad 2/\gamma < \varkappa < 2, \quad \beta > 0, \quad \Omega_0 < \beta^{1/2}$$

describe the expansion of the rotating gas (for $\gamma = 3/2$) into vacuum. These

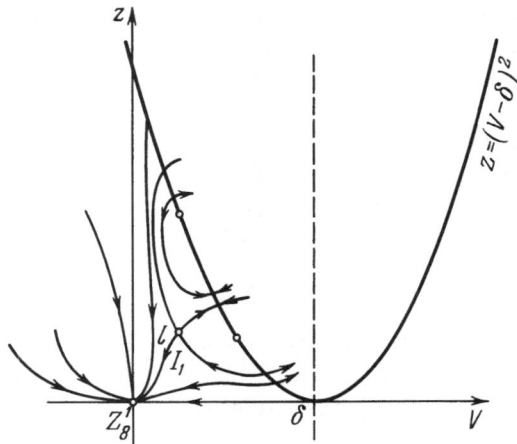

Fig. 36. Qualitative behavior of the trajectories of the dynamical system (2.1) for $\gamma = 3/2$, $U = 1$ on a level surface of the integral $\Phi_7 = C > C_0$

solutions are regular for all $r \geqslant 0$ and for $r \to 0$ have the asymptotic form of the exact solution (3.4). In the neighborhood of the axis of rotation we obtain

$$p \to 0, \quad \rho \to 0, \quad w \to rt^{-1}\Omega_0, \quad u \to z_1 t^{-1}, \quad v \to \tfrac{1}{2}rt^{-1}.$$

For $r \to \infty$ the considered solutions have the asymptotic form (5.4) and the total energy and mass of a layer of gas with unit height are finite. In the considered solutions the angular velocity of rotation of gas Ω and the vertical velocity of gas u are maximal in the vicinity of the axis $r = 0$. For $\delta = 1$, $\varkappa = 0$ there exist analogous solutions describing the self-similar expansion of the gas into the atmosphere (these solutions are applicable in some bounded region $0 \leqslant r < R_0$ for $0 < t_1 < t < t_2$).

6. Several Self-Similar Solutions with $\gamma = 2$

I. Applications of the Energy Integral. The value $\gamma = 2$ is special because the following two important problems are reduced to the adiabatic motion of gas with $\gamma = 2$:
1) motion of an ideal incompressible fluid in the theory of shallow water [110],
2) motion of a gas in magnetic gas dynamics in the following often used idealization [149, 150]: infinite conductivity of the gas and a frozen vertically directed magnetic field orthogonal to the velocity of gas.
Therefore, the self-similar motion of a gas with $\gamma = 2$ can be used to construct models of some phenomena in the above theories.

For $\gamma = 2$ and $U \equiv 0$ the dynamical system (2.1) has a first integral Φ_3 (see (2.13)):

$$z(V - \delta)\Omega^{-2} = C_0. \tag{6.1}$$

For $\varkappa = \delta$ the system (2.1) also has an energy integral (see (2.10)):

$$H = R(zV + (V - \delta)(V^2 + \Omega^2 + z)) = \text{const } \lambda^{k-1}. \tag{6.2}$$

On the zero level surface of the energy integral $H = 0$ from (6.1) and (6.2) we see that

$$z = -\frac{C_0(V - \delta)V^2}{C_0(2V - \delta) + (V - \delta)^2}, \quad \Omega^2 = -\frac{V^2(V - \delta)^2}{C_0(2V - \delta) + (V - \delta)^2}. \tag{6.3}$$

Therefore the integration of the self-similar rotation of gas for $\gamma = 2$, $\varkappa = \delta$ is reduced (after substituting (6.3) into (2.1)) to the following quadrature:

$$\frac{dV}{d\tau} = -V(2V - 1)\frac{C_0(2V - \delta) + (V - \delta)^2}{[C_0(2V - \delta) + (V - \delta)^2](V - \delta) + C_0 V^2}, \tag{6.4}$$

$$\tau = \ln \lambda.$$

After the integration of the equation (6.4) (which can be done explicitly) the functions z and Ω are found from the expressions (6.3) and the function R is found with the help of the monotone function Φ_5 (see (2.15)):

$$R\Omega^{\varkappa}(V - \delta) = \text{const } \lambda^{k+1-2\beta}, \quad \beta = 2\frac{\delta - 1}{2\delta - 1}. \tag{6.5}$$

The solution (6.3)–(6.4) is defined in the region

$$C_0 < 0, \quad V_1 < V < \delta, \quad C_0(2V - \delta) + (V - \delta)^2 \leqslant 0, \tag{6.6}$$

where $V_1 = \delta - C_0 - (C_0^2 - C_0\delta)^{1/2}$ is a root of the equation (6.6). In the above region for

$$1/2 < \delta < 1, \quad C_0 < (1/2 - \delta)^2(\delta - 1)^{-1} < 0 \tag{6.7}$$

there are two trajectories of the equation (6.4): trajectory A going from the point $V = \delta$ to the critical point $V = 1/2$ and trajectory B going from the critical point $V = V_1$ to the critical point $V = 1/2$ ($V_1 < 1/2 < \delta < 1$). Solutions corresponding to these trajectories tend to the exact solution (3.4) as $\lambda \to \infty$ (and $\Omega \to \Omega_0 = \text{const}$). For $\lambda \to \lambda_0$ the solution corresponding to the trajectory A has an expanding cavity inside the gas, on whose boundary we have

$$V \to \delta, \quad P \sim |V - \delta|^{(1-\delta)/(\delta-1/2)}, \quad R \sim |V - \delta|^{(3/4-\delta)/(\delta-1/2)}, \quad \Omega \sim |V - \delta|. \tag{6.8}$$

Thus at the inner boundary ($\lambda \to \lambda_0$) for $1/2 < \delta < 3/4$ we have $p \to 0$ and $\rho \to 0$. The solution corresponding to the trajectory B continues to the center of symmetry ($\lambda \to 0, r \to 0$) and $V \to V_1$.

Note that the surface $H = 0$ in the manifold S (see Sect. 4) passes through the critical points Z_6^0 (4.8), Z_7^0 (2.4), through the line of critical points $\mathscr{L}$ (4.9) and the line of critical points (3.2) (for $1/2 < \delta < 1$). Trajectory A in the manifold S corresponds to the trajectory going from the critical point Z_6^0 to a critical point in the line (3.2). Therefore the corresponding solution for $\lambda \to \lambda_0$ has the asymptotic form (4.12) (which coincides with the asymptotic form (6.8)). Trajectory B in the manifold S corresponds to the trajectory going from a critical point in the line $\mathscr{L}$ (see (4.9)) to a critical point in the line (3.2). Therefore the corresponding solution for $\lambda \to 0$ has the asymptotic form (4.10), which has a physical meaning $(w \to 0)$ when $1 - \delta < V_1$. The other solutions (6.3)–(6.4) defined in the region of parameters different from (6.7) have (for some direction of λ) non-physical asymptotic forms.

II. Self-Similar Rotation in the Theory of Shallow Water. The theory of shallow water is applied mainly when one is considering the flow of an ideal incompressible fluid, where the height of a layer of fluid is $h(x, y) \ll L$, where L is some characteristic horizontal length and the vertical component of velocity $u_3 \ll u_1, u_2$ [110, 151]. Under these assumptions the pressure is $p = \rho_0 g(h - z_1)$, where ρ_0 is the constant density of the fluid, g is the acceleration due to gravity and z_1 is the vertical coordinate. Now we introduce new quantities: effective pressure and density:

$$\tilde{p} = \int_0^h p \, dz_1 = \rho_0 g \frac{h^2}{2}, \quad \tilde{\rho} = \rho_0 h, \tag{6.9}$$

which obviously satisfy the following equation:

$$\tilde{p} = \frac{g}{2\rho_0} \tilde{\rho}^2. \tag{6.10}$$

Effective pressure and density $\tilde{p}$, $\tilde{\rho}$ along with the horizontal components of velocity u_1, u_2 satisfy all the equations of gas dynamics in the two-dimensional case and the equation (6.10) plays the role of the adiabatic equation with $\gamma = 2$. The obtained effective motion of a two-dimensional gas is isentropic because $g/2\rho_0 = \text{const}$.

Thus in order to construct self-similar solutions in the theory of shallow water it is sufficient to find among the self-similar solutions of type (1.9) (for $\gamma = 2, u \equiv 0$) the solutions with constant entropy. According to (1.1) we have

$$\frac{p}{\rho^2} = \frac{1}{a} r^{k+5} t^{s-2} \frac{P}{R^2} = \frac{1}{\gamma^a} \left(\frac{r}{t^{(2-s)/(k+5)}} \right)^{k+5} \frac{z}{R}. \tag{6.11}$$

By virtue of the equations (2.2) and (2.3) we see that $(\gamma = 2, U = 0)$

$$\frac{d \ln(z/R)}{d \ln \lambda} = -(k + 5) \frac{V - (2 - s)/(k + 5)}{V - \delta}. \tag{6.12}$$

For $\delta = (2 - s)/(k + 5)$ (or $\varkappa = 2(1 - \delta)$) from (6.11) and (6.12) it follows that $p/\rho^2 = \text{const}$. Therefore for these values of the parameters the self-similar solutions (1.9) are isentropic.

The self-similar motion of gas (6.3)–(6.4) described above is isentropic for $\varkappa = \delta = 2/3 = 2(1 - \delta)$ and therefore it determines an integrable self-similar motion in the theory of shallow water. For $\delta = 1$ and $\varkappa = 2(1 - \delta) = 0$ in the theory of shallow water there exist self-similar solutions of two types analogous to the solutions described in parts II and III of Sect. 5. Solutions of the first type correspond to the trajectories of the system (5.9) for $U \equiv 0$ going along the sequence of separatrices $X_3^0 X_1^0$, $X_1^0 X_2^0$, $X_2^0 Z_8^0$ (see Fig. 35) and describe self-similar flow away from the center of a thin rotating layer of fluid with constant thickness for $r \to \infty$ and with an expanding (away from the center) inner boundary.

In order to construct a solution of the second type we shall analyze the critical points of the line (3.2) (for arbitrary $\varkappa$, δ):

$$V = 1/2, \quad \Omega = \Omega_0, \quad U = 0, \quad z_0 = \alpha(\Omega_0^3 + 1/4), \tag{6.13}$$

where $\alpha = (1/2 - \delta)/(\varkappa - 1) > 0$. The line (6.13) for $\delta > 1/2$, $\varkappa < 1$ and $\beta = (1/2 - \delta)(\varkappa - 1) - 1/4 > 0$ is divided by the surface of non-extendability of solutions $L = z - (V - \delta)^2 = 0$ into two parts: I_1 $(L < 0, 0 < \Omega_0 < \beta^{1/2})$ and I_2 $(L > 0, \Omega_0 > \beta^{1/2})$. The eigenvalues $\lambda_{2,3}$ of the system (2.1) for $\gamma = 2$, $U \equiv 0$ at the critical points of the line (3.2) satisfy the following characteristic equation $(\lambda_1 = 0$ because the line (3.2) is one-dimensional):

$$\lambda^2 + \lambda(2z_0 + 2\Omega_0^2 + 1/2)L_0^{-1} + L_0^{-1} = 0$$
$$L_0 = z_0 - (1/2 - \delta)^2. \tag{6.14}$$

Thus $\lambda_2 \cdot \lambda_3 = L_0^{-1}$, so the segment I_1 consists of unstable critical points (saddles) and the segment I_2 consists of attracting critical points.

For $\gamma = 2$, $U = 0$ with the help of the integral $\Phi_3 = z(V - \delta)\Omega^{-2}$ the dynamical system (2.1) is reduced to a one-parameter family (depending on the value of the integral Φ_3) of systems of two equations in the z, V plane. The phase portrait of one of these systems for $\beta > 0$ is shown in Fig. 37. There exists a unique trajectory X going from an unstable critical point Z_Ω contained in the segment I_1 of the line (3.2) to the attracting critical point Z_8^0.

The self-similar solution corresponding to the trajectory X is regular for all $r \geqslant 0$ and for $r \to 0$ has the asymptotic form of the exact solution (3.4), whereas for $r \to \infty$ it has the asymptotic form (5.4). For the values of parameters $1/2 < \delta < 1$, $1 - \delta < \varkappa < 1$, $\beta > 0$ all characteristics of the gas p, ρ, v and w tend to zero for $r \to \infty$. In this case the considered solutions describe the spread of a rotating cord of plasma (the angular velocity of rotation is maximal at the axis $r = 0$) and the flow of a thin rotating layer of fluid, whose thickness h goes to 0 for $r \to \infty$. These solutions can be applied in some bounded region $0 \leqslant r < C$ for $0 < C_1 < t < C_2$ (because for $\varkappa < 1$, $\gamma = 2$ the total energy and mass of the gas for $r \to \infty$ diverges (see (5.5))). For $\delta = 1$, $\varkappa = 0$ in the self-similar solution X we have $p \to p_0$ and $\rho \to \rho_0$ for $r \to \infty$. Consequently in the theory of shallow water for this solution we have the thickness of the layer of fluid h approaching a constant for $r \to \infty$. In

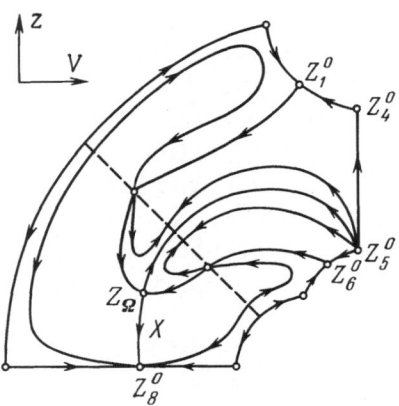

Fig. 37. Phase portrait of the dynamical system (2.1) for $\gamma = 2$, $U = 0$ on a level surface of the integral $\Phi_3 = C$ in the region $V < \delta$ after resolving the degenerate critical points and completing the region $z \geqslant 0$, $V \leqslant \delta$ by a boundary at infinity

this case the solution X can be thought of as a model of damping of a large scale vortex in an ocean.

Let us point out that in the theory of shallow water, there are self-similar solutions with moving discontinuities, which traditionally have been called bores or (in the stationary case) hydraulic jumps [151]. The conditions at the discontinuity in the theory of shallow water are a consequence of the isentropy of motion $(g/2\rho_0 = \text{const}$ (see $(6.10)))$ and the laws of conservation of mass and momentum:

$$z_1/R_1 = z_2/R_2, \quad R_1(V_1 - \delta) = R_2(V_2 - \delta),$$

$$V_1 - \delta + \frac{z_1}{2(V_1 - \delta)} = V_2 - \delta + \frac{z_2}{2(V_2 - \delta)}, \quad \Omega_1 = \Omega_2. \tag{6.15}$$

where the subscripts 1 and 2 designate quantities on the opposite sides of the discontinuity. Note that in the theory of shallow water the energy at the discontinuity is not conserved (in contrast to the Hugoniot conditions (see Chapt. V $(5.5))$ at the shock waves in classical gas dynamics). From the conditions (6.15) we see that

$$V_2 - \delta = \frac{1}{4}\left(\frac{z_1}{V_1 - \delta} - \left(\left(\frac{z_1}{V_1 - \delta}\right)^2 + 8z_1\right)^{1/2}\right),$$

$$z_2 = -\frac{V_1 - \delta}{2}\left(\frac{z_1}{V_1 - \delta} + \left(\left(\frac{z_1}{V_1 - \delta}\right)^2 + 8z_1\right)^{1/2}\right). \tag{6.16}$$

In order to construct a self-similar solution with a moving discontinuity consider for $1/2 < \delta < 1$, $\varkappa = 2(1 - \delta)$ some separatrix Y emerging from the (unstable) critical point Z_6^0 (see Fig. 37). For example for $\gamma = 2/3$ this could be the exact solution (6.3). According to (4.8) and (4.5) at the critical point Z_6^0 we have

$$v_4 = \frac{V - \delta}{z} = \frac{2(\delta - 1/2)}{\delta(\delta - 1)}, \quad V = \delta, \quad z = 0, \quad \Omega = 0.$$

Therefore under the transformation (6.16) the point Z_6^0 goes to a point x with coordinates

$$V = \frac{\delta(5\delta - 3)}{4(\delta - 1/2)}, \quad z = 0, \quad \Omega = 0.$$

Suppose a discontinuity on the separatrix Y is introduced at a point y_1 sufficiently close to the point Z_6^0. Under the transformation (6.16) the point y_1 goes to a point x_1 sufficiently close to the point x. Therefore a trajectory of the system (2.1) passing through the point x_1 moves in the vicinity of the plane $z = 0$ and (as shown in Sect. 5 (see (5.7))) for $\lambda \to \infty$ approaches the critical point Z_8^0 ($z = V = \Omega = 0$). The self-similar solution determined by the segments of trajectories $Z_6^0 y_1$, $x_1 Z_8^0$ describes the flow away from the center of a thin rotating layer of fluid with a discontinuity (the amplitude of the jump is $h_1/h_2 = R_1/R_2 = (V_2 - \delta)/(V_1 - \delta) > 1$), whose motion is determined by the condition $\lambda = \lambda_1 = \text{const}$. For this solution there is an internal boundary ($\lambda = \lambda_0$) expanding from the center at which according to the asymptotic form (4.2), the thickness of the layer of fluid is $h = 0$, the pressure is $p = 0$ and the angular velocity $\Omega = 0$. As $\lambda \to \infty$ ($r \to \infty$) all parameters of the solution (velocity, pressure and the height h) tend to zero according to the asymptotic form (5.4). The above self-similar solutions can probably be used in modelling some phenomena in oceans.

Chapter VII
The Dynamics of a Gaseous Ellipsoid

In this chapter we analyze the adiabatic motion of an ideal gas with homogeneous deformation, where the components of the velocity of the gas particles are linear functions of the coordinates. Such a motion of a continuous medium was studied in a large number of works, the first of which were the classical works of Dirichlet, Dedekind and Riemann in the theory of equilibrium configurations of an ideal incompressible fluid [153–155]. Spherically-symmetric motion of an ideal gas with homogeneous deformation was considered in [156] and with Newtonian gravitation taken into account in [157]. The entire class of motions of an ideal gas with homogeneous deformation was first isolated in [158] and the Hamiltonian formalism of these problems was developed in [159]. The motion with homogeneous deformation is used in the analysis of the expansion into vacuum of a non-rotating gaseous ellipsoid [160, 161] and the compression of an ellipsoid under the action of external pressure. It is used in the analysis of motion of a gravitating dust ellipsoid (with applications to the theory of formation of galaxies and the dynamics of stars in galaxies) [163–165], an ellipsoid of charged fluid [166], and an ellipsoid of incompressible non-gravitating fluid [167]. It is also used in magnetic gas dynamics in the study of oscillations of a plasma cord [168]. The motion of a gravitating gaseous ball was considered as a model for the pulsations of variable stars [132]. The motion of gravitating gaseous ellipsoids was considered in the works [169, 170] (which mainly used numerical methods) as a model for the formation of galaxies and stars from clouds of initially cold gas. In [169] it was observed that the adiabatic motion of a gravitating gaseous ellipsoid with negative energy E (just like the motion of an ellipsoid of incompressible fluid analyzed by Dirichlet in 1860 (see [153]) occurs in an oscillatory mode.

In this chapter it is shown that a general oscillatory mode in the motion of a gravitating gaseous ellipsoid for $E < 0$ and for certain values of the parameters can be approximated by a sequence of simpler motions of a gravitating dust ellipsoid. We also point out a new oscillatory mode of motion of a gaseous ellipsoid for $E > 0$ (the expansion of a rotating gas cloud into vacuum). Finally we analyze various properties of two oscillatory modes in the motion of a gravitating gaseous ellipsoid.

1. Equations of Motion of a Non-Gravitating Gaseous Ellipsoid

I. Motion of Gas with Homogeneous Deformation. The equations for the adiabatic motion of an ideal gas have the following form in cartesian coordinates:

$$\rho\frac{du^i}{dt} = -\frac{\partial p}{\partial x^k}\eta^{ki}, \quad \frac{d\rho}{dt} = -\rho\,\mathrm{div}\,u^i, \quad \frac{d}{dt}\left(\frac{p}{\rho^\gamma}\right) = 0, \tag{1.1}$$

where u^i is the velocity vector of the gas, p is the pressure, ρ is the gas density, $\gamma > 1$ is the adiabatic coefficient and $\eta_{ki} = \delta_k^i$ is the metric tensor of Euclidean space. The motion of gas with homogeneous deformation is represented by solutions of the equations of gas dynamics (1.1), where the Euler coordinates of gas particles x^i are linear functions of the Lagrangian coordinates a^k:

$$x^i = F_k^i(t)a^k, \quad i, k = 1, 2, 3 \tag{1.2}$$

where there is implicit summation over repeating indices. Recall that the Euler coordinates of gas particles are their (variable) coordinates in Euclidean space x^1, x^2, x^3, whereas the Lagrangian coordinates a^1, a^2, a^3 distinguish the gas particles and remain constant (for each particle of gas) during the entire process of motion of a continuous medium. By virtue of definition (1.2) we see that

$$u^i = \frac{dx^i}{dt} = \dot{F}_k^i a^k = \dot{F}_k^i(F^{-1})_j^k x^j,$$

$$\mathrm{div}\,u^i = \frac{\partial u^i}{\partial x^i} = \dot{F}_k^i(F^{-1})_i^k = \frac{d(\ln\det(F_k^i))}{dt}, \tag{1.3}$$

where F^{-1} is the inverse of the matrix $F: F \circ F^{-1} = E$.

The equations of gas dynamics (1.1) for the motions of gas of type (1.2) are reduced to a system of ordinary differential equations, if the pressure and density of the gas have the following form:

$$p = p(t)P(\zeta), \quad \rho = \rho(t)R(\zeta), \quad \zeta = -\tfrac{1}{2}g_{ij}a^i a^j, \tag{1.4}$$

where g_{ij} is some constant symmetric matrix. Let us determine the relations among the functions $\rho(t)$, $p(t)$, $P(\zeta)$ and $R(\zeta)$. After substituting the expressions (1.2)–(1.4) the equations (1.1) turn into the following system of equations:

$$\rho(t)R(\zeta)\dot{F}_k^i(F^{-1})_j^k x^j = p(t)\frac{dP}{d\zeta}\,g_{lk}(F^{-1})_j^k(F^{-1})_n^l\eta^{ni}x^j,$$

$$\frac{d\ln\rho(t)}{dt} = -\frac{d\ln\det(F_i^k)}{dt}, \quad p(t) = C(\rho(t))^\gamma. \tag{1.5}$$

From the system (1.5) we obtain the following necessary conditions:

$$R(\zeta) = C_1\frac{dP}{d\zeta}, \quad \rho(t) = C_2(\det(F_i^k))^{-1}, \tag{1.6}$$

$$p(t) = C_3(\det(F_i^k))^{-\gamma}.$$

After substituting these expressions into the equations of dynamics (1.5) we obtain a system of ordinary differential equations:

$$\ddot{F}_k^i = C(\det F)^{1-\gamma}g_{lk}(F^{-1})_n^l\eta^{ni}, \quad C = C_3/C_1C_2. \tag{1.7}$$

The system (1.7) is a Lagrangian system. Indeed by contracting the equations (1.7) with the tensor $g^{kj}\eta_{im}$ we obtain the following system of equations:

$$\ddot{F}^i_k g^{kj}\eta_{im} = C(\det(F))^{1-\gamma}(F^{-1})^j_m = -\frac{C}{\gamma-1}\frac{\partial(\det(F))^{1-\gamma}}{\partial F^m_j}, \qquad (1.8)$$

which obviously has a Lagrangian form:

$$\frac{d}{dt}\frac{\partial L}{\partial \dot{q}_i} = \frac{\partial L}{\partial q_i}, \quad q_i = F^m_j$$

with the Lagrangian

$$L = \frac{1}{2}\dot{F}^i_k \dot{F}^m_j g^{kj}\eta_{im} - \frac{C}{\gamma-1}(\det(F^i_k))^{1-\gamma}. \qquad (1.9)$$

Thus in the case of an indefinite metric g_{ij} the motion of gas with homogeneous deformation is described by a Lagrangian system (1.8)–(1.9) with an indefinite kinetic energy. In the case of an indefinite metric g_{ij} the surfaces of constant pressure and constant density of the gas are hyperboloids (see (1.4)). Therefore such solutions can be used (for example) to model a tornado type motion of gas in the atmosphere (along with self-similar rotation of gas studied in Chapt. VI).

In further discussion we shall consider two cases: a positive and negative definite metric g_{ij}. Without loss of generality we can assume that

$$g_{ij} = \sigma\delta^j_i, \quad \sigma = \pm 1, \quad C_1 = C_2 = 1, \quad C_3 = C = \gamma - 1.$$

With this normalization the Lagrangian system (1.8) takes the following form:

$$\ddot{F}^i_k = \sigma(\gamma-1)(\det(F))^{1-\gamma}(F^{-1})^k_i = -\sigma\frac{\partial(\det(F))^{1-\gamma}}{\partial F^i_k}. \qquad (1.10)$$

The pressure p and gas density ρ are given by the formulas

$$p = \frac{(\gamma-1)P(\zeta)}{(\det(F^i_k))^\gamma}, \quad \rho = \frac{dP(\zeta)/d\zeta}{(\det(F^i_k))}, \qquad (1.11)$$

where $P(\zeta)$ is an arbitrary function of the parameter $\zeta = -\sigma(a_1^2 + a_2^2 + a_3^2)/2$. From the condition of non-negativity of the gas density ρ follows (see (1.11)) the necessary condition $dP(\zeta)/d\zeta \geqslant 0$. Taking the function $P(\zeta) = \exp(\zeta)$ with $\sigma = +1$, we obtain an isothermal mass of gas filling all space and having a Gaussian density distribution. If however the function $P(\zeta)$ is bounded and has the form shown in Fig. 38, then we obtain a finite mass of gas with a free boundary, where $p = \rho = 0$. For $\sigma = -1$ the pressure grows as the distance to the center $(a_1 = 0)$ increases. In this case the gas is under external pressure varying according to (1.11). Note that the surfaces of constant density and pressure of the gas in Euler coordinates x_i are ellipsoids, whose shape and position in space vary by virtue of the system (1.10).

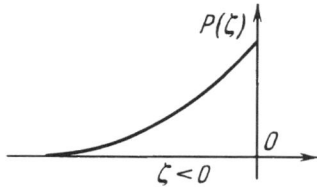

Fig. 38. The form of the function $P(\zeta)$, which ensures that the conditions $p = \rho = 0$ are satisfied at the boundary of the gaseous ellipsoid

II. General Properties of the Dynamics of a Gaseous Ellipsoid. The Lagrangian system (1.10) is invariant relative to a six-dimensional group of transformations

$$F \to O_1 F O_2,$$

where O_1 and O_2 are orthogonal matrices. Hence this system has six first integrals forming two three-dimensional skew-symmetric matrices

$$J = F\dot{F}^t - \dot{F}F^t, \quad K = F^t\dot{F} - \dot{F}^tF \tag{1.12}$$

where F^t is the transpose of F. The integrals J and K were first observed in [159]. These integrals are related to the preservation of total angular momentum of the gas and to the vorticity. If $J = K = 0$, then after a suitable transformation $F \to O_1 F O_2$, the matrix F_j^i is diagonalized for all time.

The total energy of gas filling the ellipsoid is proportional to the energy integral of the Lagrangian system (1.10) (see [159]):

$$E = T + \sigma\big(\det(F)\big)^{1-\gamma}, \quad T = \frac{1}{2}\sum_{i,k}^{3}(\dot{F}_k^i)^2. \tag{1.13}$$

For $\sigma = +1$ the form of the energy integral implies that the solutions of the system (1.10) exist for all $t\,(-\infty < t < +\infty)$ (the trajectories of the system (1.10) do not approach critical points, because from (1.13) it follows that $\det(F) > E^{1/(1-\gamma)}$ and do not go to infinity in finite time, because $(F_k^i)^2 < 2E$). Let us show that as $t \to \pm\infty$, the motion of a gaseous ellipsoid with a free boundary $(\sigma = +1)$ is unbounded. The proof below is a generalization of the reasoning in [161], where for $\gamma = 5/3$ the first integral (1.18) was discovered and with the help of this integral it was proved (for $\gamma = 5/3$) that for $t \to \pm\infty$ there is an infinite dispersal of the gaseous ellipsoid. Let us evaluate the rate of change of the sum of the squares of the semiaxes of the gaseous ellipsoid:

$$D = d_1^2 + d_2^2 + d_3^2 = \mathrm{Tr}(F \circ F^t) = \sum_{i,k}^{3}(F_k^i)^2.$$

By virtue of the equations (1.10) we see that

$$\ddot{D} = 2\sum_{i,k}^{3}(\dot{F}_k^i)^2 + 2\sum_{i,k}^{3}\ddot{F}_k^iF_k^i = 2T + 6(\gamma - 1)\big(\det(F)\big)^{1-\gamma}. \tag{1.14}$$

Replacing successively each of the last two summands in (1.14) by their expressions from the energy integral (1.13) we find that

$$\ddot{D} = 6(\gamma - 1)E + 2(5 - 3\gamma)T,$$
$$\ddot{D} = 4E - 2(5 - 3\gamma)(\det(F))^{1-\gamma}. \tag{1.15}$$

Hence depending on the value of $\gamma > 1$ we obtain the following inequalities:

$$\gamma < 5/3: \ 6(\gamma - 1)E < \ddot{D} < 4E,$$
$$\gamma > 5/3: \ 4E < \ddot{D} < 6(\gamma - 1)E. \tag{1.16}$$

From (1.16) follow the estimates of the rate of growth of the sum of the squares of the semiaxes of the gas ellipsoid:

$$\gamma < 5/3: \ 3(\gamma - 1)Et^2 + A_1 t + B_1 < D < 2Et^2 + A_2 t + B_2,$$
$$\gamma > 5/3: \ 2Et^2 + A_1 t + B_1 < D < 3(\gamma - 1)Et^2 + A_2 t + B_2. \tag{1.17}$$

For $\gamma = 5/3$ both expressions (1.15) coincide and define additional integrals A and B of the system (1.10), which were mentioned for the first time in [161]:

$$D = 2Et^2 + At + B. \tag{1.18}$$

By virtue of (1.17) and (1.18) we see that for all $\gamma > 1$ the sum of the squares of the semiaxes of the gaseous ellipsoid D tends to ∞ as $t \to \pm\infty$.

The absence of bounded (for all t) motion of a gaseous ellipsoid with a free boundary can also be proved with the help of the virial theorem [171]. According to the virial theorem if in a Lagrangian system with the Lagrangian $L = T - U$ (where the kinetic energy T and the potential energy U are homogeneous functions of order 2 and k respectively) there exists a bounded (for all t) trajectory, then the quantities $\bar{T}$ and $\bar{U}$, which are T and U averaged over this trajectory, satisfy the equality $2\bar{T} = k\bar{U}$. For the Lagrangian system (1.10) we have $k = 3(1 - \gamma) < 0$, so the equality $2\bar{T} = 3(1 - \gamma)\overline{(\det(F))^{1-\gamma}}$ is impossible, which implies the absence of bounded for all t motions of a gaseous ellipsoid.

An important property of the dynamics of a gaseous ellipsoid under external pressure $(\sigma = -1)$ is the presence of singularities $\det(F(t_*)) = 0$. The existence of such singularities can be established with the help of estimates for the value of D. For $\sigma = -1$ the equations (1.15) take the following form:

$$\ddot{D} = 6(\gamma - 1)E + 2(5 - 3\gamma)T.$$
$$\ddot{D} = 4E + 2(5 - 3\gamma)(\det(F))^{1-\gamma}. \tag{1.19}$$

Hence for $\gamma \geqslant 5/3$ we see that

$$D \leqslant 3(\gamma - 1)Et^2 + At + B. \tag{1.20}$$

From (1.20) it follows that for $\gamma \geqslant 5/3, \sigma = -1, E < 0$ the solution exists in a finite time interval $I: t_0 \leqslant t \leqslant t_1$ (because $D = d_1^2 + d_2^2 + d_3^2 \geqslant 0$) and therefore has singularities (otherwise the solution would continue outside the interval I).

In order to analyze the singularities of the solution for all $\gamma > 1$, $\sigma = -1$ we can use the following function:

$$\left(\ln\left(\det\left(F_k^i\right)\right)\right)^{\cdot\cdot} = \sum_{i,k}^{3} \dot{F}_k^i \left(F^{-1}\right)_i^k - \sum_{i,k,\alpha,\beta}^{3} \dot{F}_k^i \left(F^{-1}\right)_i^\beta \dot{F}_\beta^\alpha \left(F^{-1}\right)_\alpha^k$$

$$= -(\gamma - 1) \sum_{i,k}^{3} \left(\left(F^{-1}\right)_i^k\right)^2 \left(\det(F)\right)^{1-\gamma} - \mathrm{Tr}\left(A^2\right),$$

(1.21)

where $A = \dot{F} \circ F^{-1}$. The matrix A is symmetric $(A = A^t)$ if $K = 0$ (see (1.12)). In this case from (1.21) by virtue of $\mathrm{Tr}(A^2) = \mathrm{Tr}(A \circ A^t) \geqslant 0$ we see that

$$\left(\ln\left(\det\left(F_k^i\right)\right)\right)^{\cdot\cdot} < 0.$$

(1.22)

Consequently for $K = 0$ the function $\det(F_k^i(t))$ is a convex up function and therefore (at least for one direction of time t) the solution has a singularity $\det(F(t_*)) = 0$.

III. Transformation of the Dynamical System. The analysis of the dynamics of the solutions (1.2) in time is reduced (as shown above) to studying the dynamics of the Lagrangian system (1.10). For the simplest classes of matrices F_j^i (scalar, axisymmetric) this system was analyzed in [156, 161] using standard methods of analysis. In a general case with non-zero integrals (1.12) the analysis of the system (1.10) requires the use of methods of the qualitative theory of differential equations.

The first step in the application of these methods is the construction of a dynamical system defined on some compact manifold S with boundary Γ, which is equivalent to the system (1.10) and has sufficiently simple critical points. Since all subsequent considerations do not depend on a particular type of potential in (1.10), we will consider a general Lagrangian system

$$\frac{d}{dt} \frac{\partial L}{\partial \dot{q}_i} = \frac{\partial L}{\partial q_i}$$

(1.23)

with the Lagrangian

$$L = \frac{1}{2} \sum_{i=1}^{n} \dot{q}_i^2 - \sigma \left(U(q_i)\right)^\alpha,$$

(1.24)

where $\alpha < 0$, $\sigma = \pm 1$ and $U(q_i)$ is a homogeneous function of order $\mu > 0$ such that the surface $U(q_i) = 0$ is of dimension $n - 1$. For the system (1.10) we have

$$\alpha = 1 - \gamma, \quad q_i = F_k^j, \quad U(q_i) = \det \| F_k^j \|, \quad n = 9, \quad \mu = 3.$$

The Lagrangian system (1.23) has the following form in phase space:

$$\dot{P}_i = -\alpha\sigma U^{\alpha-1} \frac{\partial U}{\partial q_i}, \quad \dot{q}_i = P_i, \quad i = 1, 2, \ldots, n.$$

(1.25)

We consider this system in the region $U(q_i) > 0$ (according to (1.11) the points of the boundary $U(q_i) = 0$ correspond to a physical singularity of the solution). The system (1.25) admits the group of scale transformations:

$$q_i \to \lambda q_i, \quad P_i \to \lambda^{\alpha\mu/2} P_i, \quad t \to \lambda^{1-(\alpha\mu/2)} t. \tag{1.26}$$

The transformation of energy

$$E = \frac{1}{2} \sum_{\beta=1}^{n} P_\beta^2 + \sigma U(q_i)^\alpha$$

becomes $E \to \lambda^{\alpha\mu} E$. By virtue of the presence of the scale transformations (1.26), the system (1.25) admits a reduction of order, i.e. it is equivalent to some system in $2n - 1$ variables.

In order to construct the compact manifold S of dimension $2n - 1$ we introduce two local charts W_1 and W_2. In the local chart W_1 we introduce the coordinates

$$y_i = \frac{q_i}{\left(\sum_{\beta=1}^{n} q_\beta^2 \right)^{1/2}}, \quad p_i^0 = \frac{P_i}{(U(q_i))^{\alpha/2}}.$$

In the local chart W_2 we introduce the coordinates

$$y_i, \quad p_i = \frac{P_i}{\left(\sum_{\beta=1}^{n} P_\beta^2 \right)^{1/2}}, \quad w = \frac{U^\alpha(q_i)}{\sum_{\beta=1}^{n} P_\beta^2}. \tag{1.27}$$

The coordinates y_i vary over the unit sphere

$$S^{n-1}: \sum_{i=1}^{n} y_i^2 = 1.$$

The coordinates p_i also vary over the unit sphere

$$S^{n-1}: \sum_{i=1}^{n} p_i^2 = 1,$$

whereas the coordinates p^0 vary over the entire Euclidean space E^n. The coordinates $p_i = p^0/(\sum (p_\beta^0)^2)^{1/2}$, $w = 1/(\sum (p_\beta^0)^2)^{1/2}$ compactify the Euclidean space E^n at infinity corresponding to a sphere

$$w = 0, \quad \sum_{i=1}^{n} p_i^2 = 1.$$

Thus the coordinates p_i^0, p_i and w can be thought of as varying over the unit ball

$$D^n: \sum_{i=1}^{n} x_i^2 \leqslant 1$$

(in the vicinity of the center of the ball we use the coordinates p_i^0 and in the vicinity of the boundary sphere

$$S^{n-1}\left(\sum_{i=1}^{n} x_i^2 = 1\right)$$

we use the coordinates p_i and w).

In the local charts W_1 and W_2 the manifold S is determined by the conditions

$$U(y_i) \geqslant 0, \quad w \geqslant 0.$$

The boundary Γ of the manifold S consists of two components $\Gamma = \Gamma_0 \cup \Gamma_1$. The component Γ_0 is determined by the conditions $w = 0$, $U(y_i) > 0$ and the component Γ_1 by the conditions $w > 0$, $U(y_i) = 0$. The intersection of these components $N = \Gamma_0 \cap \Gamma_1$ is of dimension $2n - 3$ and is determined by the conditions

$$U(y_i) = 0, \quad w = 0. \tag{1.28}$$

The dynamical system on the manifold S is determined by the way the system (1.25) is transformed into the coordinates W_1 and W_2 (this transformation uses the homogeneity of the function $U(q_i)$). In the coordinates W_1 and time τ_1:

$$\frac{d\tau_1}{dt} = \frac{U^{\alpha/2}(q_i)}{U(y_i)\left(\sum_{\beta=1}^{n} q_\beta^2\right)^{1/2}}$$

we obtain the following system:

$$\dot{p}_i^0 = \alpha\sigma\left(-\frac{\partial U}{\partial y_i} - \frac{\sigma}{2} p_i^0\left(\sum_{\gamma=1}^{n} \frac{\partial U}{\partial y_\gamma} p_\gamma^0\right)\right),$$

$$\dot{y}_i = U(y_j)\left(p_i^0 - y_i\left(\sum_{\gamma=1}^{n} p_\gamma^0 y_\gamma\right)\right), \tag{1.29}$$

$$\frac{dU(q_j)}{d\tau_1} = U(q_j)\left(\sum_{\gamma=1}^{n} \frac{\partial U(y)}{\partial y_\gamma} p_\gamma^0\right).$$

In the coordinates W_2 and time τ_2:

$$\frac{d\tau_2}{dt} = \frac{U^{\alpha/2}(q_i)}{U(y_i)w^{1/2}\left(\sum_{\beta=1}^{n} q_\beta^2\right)^{1/2}} \tag{1.30}$$

the system (1.8) has the following form:

$$\dot{p}_i = \sigma\alpha w\left(-\frac{\partial U}{\partial q_i} + p_i\left(\sum_{\gamma=1}^{n} \frac{\partial U}{\partial y_\gamma} p_\gamma\right)\right),$$

$$\dot{y}_i = U(y_j)(p_i - y_i\left(\sum_{\gamma=1}^{n} p_\gamma y_\gamma\right)),$$

$$\dot{w} = \alpha w(1 + 2\sigma w)\left(\sum_{\gamma=1}^{n} \frac{\partial U}{\partial y_\gamma} p_\gamma\right),$$

$$\frac{dU(q_i)}{d\tau_2} = U(q_i)\left(\sum_{\gamma=1}^{n} \frac{\partial U}{\partial y_\gamma} p_\gamma\right). \tag{1.31}$$

Obviously the boundary $\Gamma = \Gamma_0 \cup \Gamma_1$ is an invariant manifold of the systems (1.29) and (1.31). For $\sigma = -1$ the system (1.31) has another invariant manifold V: $w = 1/2$ corresponding to zero total energy

$$E = \frac{1}{2} \sum_{i=1}^{n} \dot{q}_i^2 + \sigma U^\alpha(q_i).$$

2. Oscillatory Mode of Expansion of a Rotating Gas Cloud into Vacuum

In this section we shall describe a new oscillatory mode of motion of the gas. It will be obtained by an approximation of the trajectories of the system (1.21) by sequences of critical points and their separatrices, along which these trajectories move.

I. Here is a list of the sets of critical points and their eigenvalues for the system (1.29), (1.31) when $\sigma = +1$:
1) M_ε: $w = 0$, $p_i = \varepsilon y_i$, $\varepsilon = \pm 1$. M_ε lies in the boundary Γ_0 and dim $M_\varepsilon = n - 1$. The eigenvalues of these critical points are

$$\lambda_1 = \alpha \lambda \varepsilon U(y_j) \qquad \text{(variable } w),$$

$$\lambda_2 = \ldots = \lambda_n = 0 \qquad \text{(variables } p_i),$$

$$\lambda_{n+1} = \ldots = \lambda_{2n-1} = -\varepsilon U(y_j) \quad \text{(variables } y_i).$$

The critical points in M_ε are non-degenerate, because the number of zero eigenvalues is equal to $n - 1$, which is the dimension of M_ε. Under the assumptions made earlier $(\alpha < 0, \lambda > 0$ (see (1.24))) the critical points in M_+ are attracting and the critical points in M_- are repelling.
2) N: $w = 0$, $U(y_i) = 0$. N is the corner of the boundary Γ: $N = \Gamma_0 \cap \Gamma_1$ and dim $N = 2n - 3$. At these critical points $2n - 3$ eigenvalues (corresponding to the dimension of N) are equal to zero and there are two non-zero eigenvalues:

$$\lambda_1 = \alpha\left(\sum_{\gamma=1}^{n} \frac{\partial U}{\partial y_\gamma} p_\gamma\right) \quad \text{(variable } w),$$

$$\lambda_2 = \sum_{\gamma=1}^{n} \frac{\partial U}{\partial y_\gamma} p_\gamma \qquad \text{(variables } y_i). \tag{2.1}$$

The condition

$$\sum_{\gamma=1}^{n} \frac{\partial U}{\partial y_{\gamma}} p_{\gamma} > 0$$

on N determines the set N_+ and the condition

$$\sum_{\gamma=1}^{n} \frac{\partial U}{\partial y_{\gamma}} p_{\gamma} < 0$$

determines the set N_-. Obviously these conditions mean that a point (p_{γ}, y_{γ}) contained in the set N lies in N_+ if the point (p_{γ}) on the unit sphere S^{n-1} is on the same side of the plane tangent to the surface $U(y) = 0$ at the point (y_{γ}) as the normal to this surface

$$\frac{\partial U(y)}{\partial y_{\gamma}} \bigg/ |\text{grad } U(y)|.$$

Analogously the point (p_{γ}, y_{γ}) lies in N_- if the point (p_{γ}) is on the negative side of this tangent plane.

The critical points of N_+ and N_- are non-degenerate and unstable. Each point of N_+ has a separatrix (going in Γ_1) approaching it and one separatrix (going in Γ_0) leaving it. The converse is true for N_-.

3) Degenerate critical points L, where $\partial U/\partial y_i = 0$ (critical in the geometric sense points of the surface $U(y_i) = 0$). The coordinates p_i and w are arbitrary.

II. In order to find the separatrices of critical points in M_ε and N we integrate the system (1.31) on the boundary $\Gamma = \Gamma_0 \cup \Gamma_1$. Trajectories of this system in the component Γ_0 have the following form:

$$p_i = p_i^0, \quad w = 0, \quad y_i = y_i^0 \cosh \tau_0 / \cosh \tau + p_i^0 (\sinh \tau - \sinh \tau_0) / \cosh \tau, \quad (2.2)$$

where p_i^0, y_i^0 and τ_0 are constants such that

$$\tanh \tau_0 = \sum_{i=1}^{n} p_i^0 y_i^0.$$

The time τ is related to τ_2 by the formula $d\tau = U(y_i) d\tau_2$. A trajectory in the co-ordinates y_i moves along the shortest arc of a great circle on the unit sphere S^{n-1} passing through the points (y_i^0) and (p_i^0). The endpoints of the trajectory (2.2) are contained in the sets of critical points M_+, M_- or N. In other words each trajectory is a separatrix of some critical point.

In the component of the boundary Γ_1 $(U(y_i) = 0)$ the trajectories of the system (1.31) have the following form:

$$y_i = y_i^0, \quad w = (\cosh^2 \tau_0 - \cosh^2 \tau)/(2 \cosh^2 \tau),$$
$$p_i = p_i^0 (\cosh \tau_0 / \cosh \tau) + s_i^0 (\sinh \tau - \sinh \tau_0)/\cosh \tau, \tag{2.3}$$

where

$$y_i^0, p_i^0, \tau_0, (s_i^0) = \frac{\text{grad } U(y_i^0)}{|\text{grad } U(y_i^0)|}$$

are constants such that

$$\sum_{i=1}^{n} p_i^0 s_i^0 = \tanh \tau_0 < 0, \quad \sum_{i=1}^{n} p_i s_i^0 = \tanh \tau. \tag{2.4}$$

The time τ is determined by the expression $d\tau = w|\alpha| |\text{grad } U(y_i^0)| d\tau_2$ in (2.3). The trajectory (2.3) begins at the point $(p_i^0, y_i^0, w = 0)$ for $\tau = \tau_0$ in N_-. By virtue of (2.4) the final endpoint of this trajectory $(p_i(\tau), \tau = -\tau_0, y_i^0, w = 0)$ is contained in N_+. The final point $(p_i(-\tau_0))$ is obtained from the initial point (p_i^0) by reflecting in the plane tangent to the surface $U(y_i) = 0$ at the point (y_i^0). The maximal value of w along the trajectory (2.3) is attained for $\tau = 0$ and is equal to

$$w_* = \frac{\cosh^2 \tau_0 - 1}{2} = \frac{\left(\sum\limits_{i=1}^{n} p_i^0 s_i^0 \right)^2}{2 \left(1 - \left(\sum\limits_{i=1}^{n} p_i^0 s_i^0 \right)^2 \right)}. \tag{2.5}$$

III. The form of the separatrices (2.2), (2.3) we have obtained leads to the following separatrix diagram:

$$M_- \xrightarrow{\alpha_-} N_- \underset{\beta_+}{\overset{\beta_-}{\rightleftarrows}} N_+ \xrightarrow{\alpha_+} M_+. \tag{2.6}$$

An arrow means that a critical point, which is the initial point of some separatrix leaving it, is mapped to the final point of this separatrix. The mappings thus defined are denoted $\alpha_-, \alpha_+, \beta_-, \beta_+$.

The degenerate critical points L do not appear in the separatrix diagram, because there are no separatrix steps between them and the critical points of M_ε and N. The critical points in L do not affect the modes described below.

The mappings α_-, β_+ and α_+ are realized by the separatrices (2.2) going in the component of the boundary Γ_0 ($w = 0$). Along these separatrices we have $p_y = p_y^0 = \text{const}$ and the coordinates y_y vary along the shortest arc of a great circle passing through the points (p_y^0) and (y_y^0) (where (p_y^0, y_y^0) is the initial point of the separatrix). At the repelling critical points M_- we have $p_y^0 = -y_y^0$. Through the two points $(p_y^0 = -y_y^0)$ and (y_y^0) on the unit sphere S^{n-1} passes a $(n-2)$-dimensional family of great circles. Therefore the mapping α_- is not single valued. Note that the intersection of such a circle with N_- in the case interesting to us ($\mu = 3$) is always non-empty.

The mapping β_- is realized by the separatrix (2.3) going in the component of the boundary Γ_1 ($U(y_y) = 0$). Along this separatrix we have $y_y = y_y^0 = \text{const}$. The final point $(p_y^1, y_y^0, w = 0)$ is obtained from the initial point $(p_y^0, y_y^0, w = 0)$ by way of reflection of the point (p_y^0) in the plane l tangent to the surface $U(y_y) = 0$ at the point (y_y^0). Obviously the final point belongs to N_+. Let the distance from the point (p_y^0) to the plane l equal h. For small h the shortest arc of a great circle connecting the points (p_y^1) and (y_y^0) intersects the surface $U(y_y) = 0$ at a point (y_y^1) close to the point (y_y^0) (Fig. 39). (It is assumed that the surface $U(y_y) = 0$ at the

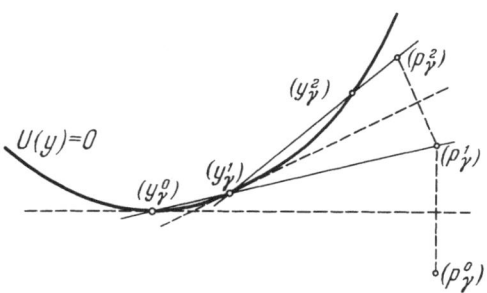

Fig. 39. Successive change of a critical point (p_γ^i, y_γ^i) during the separatrix steps (2.2) and (2.3)

point y_γ^0 is convex towards the normal. In the considered case such points (y_γ^0) do exist.)

Thus we have

$$\beta_-(p_\gamma^0, y_\gamma^0) = (p_\gamma^1, y_\gamma^0), \quad \beta_+(p_\gamma^1, y_\gamma^0) = (p_\gamma^1, y_\gamma^1). \tag{2.7}$$

For $h \to 0$ the point (p_γ^1, y_γ^1) is infinitely close to the point (p_γ^0, y_γ^0) and the repetition of mappings β_- and β_+ again leads to nearby points. Therefore for $h \to 0$ the separatrix diagram (2.6) provides for an arbitrary finite number of steps between the sets N_- and N_+:

$$M_- \xrightarrow{\alpha_-} N_- \xrightarrow{\beta_+} N_+ \xrightarrow{\beta_-} \cdots \xrightarrow{\beta_-} N_+ \xrightarrow{\alpha_+} M_+. \tag{2.8}$$

Note that under the mapping β_- the distance between the points (p_γ^i) and (y_γ^i) does not change. Under the mapping β_+ it decreases. Therefore the oscillations between the sets N_- and N_+ are always cut off when the point ends up in the set M_+, where $p_\gamma = y_\gamma$.

The motion of a point along the sequence of separatrices defined by the diagram (2.8) in the coordinates y_γ is the free motion of the point along geodesics on the unit sphere S^{n-1} in the region $U(y_\gamma) > 0$. Note that the point is reflected from the surface $U(y_\gamma) = 0$ according to the law of elastic reflection (see (2.7)). In this sense the mathematical model of the mode described below is the game of geodesic billiards on the sphere S^{n-1} with a reflecting surface $U(y_\gamma) = 0$.

IV. Now we turn our attention to the conclusions, which can be drawn from the separatrix diagram (2.6), (2.8) obtained above. The separatrix diagram (2.8) consists of separatrices going in the boundary Γ_0 and Γ_1 of the physical region S. Therefore it does not correspond to any exact physical solutions. However there exist physical trajectories $(w \neq 0, U(y_\gamma) \neq 0)$ moving arbitrarily close to the entire sequence of separatrices (2.8). A segment of such a trajectory emerging from a repelling critical point of M_- corresponds to the compression of gas inertially from an infinitely rarefied state. A segment of a trajectory approaching an attracting critical point of M_+ corresponds to the infinite free expansion of gas.

The inner part of the diagram (2.8) describes the successive contraction and expansion of gas. Indeed in the problem of a gas cloud $U(q_i) = \det \| F_{jk} \|$ and the condition

$$U(y_\gamma) = \det \| Y_{jk} \| = 0 \left(\sum_{j,k}^{3} Y_{jk}^2 = 1 \right)$$

implies that the gas is compressed along the zero eigenvector of the matrix $\| Y_{jk} \|$. Therefore the nearness of the trajectory to the unstable critical points of N_- or N_+ implies that an ellipsoid of constant gas density has a strongly compressed shape. Furthermore according to (1.31) we have

$$\frac{dU(q_1)}{d\tau_2} = U(q_i) \left(\sum_{\gamma=1}^{n} \frac{\partial U(y)}{\partial y_\gamma} p_\gamma \right).$$

Therefore in the vicinity of

$$N_- \left(\sum_{\gamma=1}^{n} \frac{\partial U}{\partial y_\gamma} p_\gamma < 0 \right)$$

$U(q_i) = \det \| F_{jk} \|$ decreases, i.e. the gas contracts. In the vicinity of

$$N_+ \left(\sum_{\gamma=1}^{n} \frac{\partial U(y)}{\partial y_\gamma} p_\gamma > 0 \right)$$

$\det \| F_{jk} \|$ increases, i.e. the gas expands. Consequently in the segment of trajectory corresponding to the step

$$N_- \xrightarrow{\beta_-} N_+$$

there exists a minimum of the gas volume and in the segment of trajectory corresponding to the step

$$N_+ \xrightarrow{\beta_+} N_-$$

there exists a maximum of the gas volume.

In order to determine the volume $(\det \| F_{jk} \|)$ at these extrema we assume that the motion of gas corresponding to the chosen trajectory of system (1.24) has total energy E. By definition we have

$$E = \frac{1}{2} \sum_{j,k}^{3} \dot{F}_{jk}^2 + (\det \| F_{jk} \|)^{1-\gamma}, \quad w = \frac{(\det \| F_{jk} \|)^{1-\gamma}}{\sum_{j,k}^{3} \dot{F}_{jk}^2}.$$

Hence

$$\det \| F_{jk} \| = \left(\frac{1 + 2w}{2Ew} \right)^{1/(\gamma-1)}. \tag{2.9}$$

Therefore the minima of volume attained on the segments of the trajectory corresponding to the steps

$$N_- \xrightarrow{\beta_-} N_+$$

are approximately equal to the following quantity:

$$\min \det \| F_{jk} \| = \left(\frac{1 + 2w_*}{2Ew_*} \right)^{1/(\gamma - 1)}, \tag{2.10}$$

where w_* is given by the formula (2.5). For a trajectory sufficiently near to the sequence of separatrices (2.8) the quantity (2.10) is attained with arbitrary precision. The maxima of volume $\det \| F_{jk} \|$ are attained on the segments of trajectory close to the separatrices in the step

$$N_+ \xrightarrow{\beta_+} N_-$$

where $w = 0$. Therefore the value of $\max \det \| F_{jk} \|$ can be arbitrarily large (see (2.9)).

Note that each trajectory of the system (1.31) in the coordinates (1.27) corresponds to a one parameter family of trajectories of the initial system (1.25) parameterized by the constant value of energy E $(0 < E < \infty)$. By virtue of scale invariance relative to the group of transformations (1.26) the time t of motion along a trajectory is related to the energy E by

$$tE^{1/2 + 1/(3(\gamma - 1))} = \text{const.} \tag{2.11}$$

Thus we have shown using the separatrix diagram (2.8) that there exist the following motions of an ideal gas:
1) initially the gas is compressed from the state of infinite rarefaction,
2) then begins an oscillatory mode: the gas contracts and expands an arbitrarily large number of times in various directions such that the amplitude of the oscillations of volume $\det \| F_{jk} \|$ can be arbitrarily large by virtue of (2.9)–(2.10) and consequently for large E the oscillations of density ρ are arbitrarily large, whereas according to (2.11) the time of oscillations can be arbitrarily small,
3) the oscillatory mode ends and is replaced by the infinite free expansion of gas.

Note that the oscillatory mode occurs even in the case when the initial distribution of the gas is close to the spherically-symmetric distribution. Furthermore from (2.10) and (2.5) it follows that for a motion with bounded energy E and with a large number of oscillations $\det \| F_{jk} \|$ is always large and therefore in the oscillatory mode the gas remains rarefied.

Special emphasis should be placed on the fact that the oscillatory mode of motion of a gaseous ellipsoid analyzed in this section is realized solely due to the forces of internal pressure in the presence of rotation of the gas (the integrals J and K are not equal to 0). In Sect. 8 we shall show that this oscillatory mode is preserved also in the presence of gravitational interaction of gas particles.

3. Analysis of a Problem in the Theory of Shallow Water

An analogue of the problem of the expansion of a gaseous ellipsoid into vacuum in two-dimensional hydrodynamics is the problem of flow of a rotating fluid ellipse in the theory of shallow water (in this case $\gamma = 2$, see part II of Sect. 6 of Chapt. VI). This problem is reduced to quadratures in [161]. However the explicit formulas obtained in [161] are so complicated that their analysis would by itself be a difficult problem. Let us show independently from the previous discourse and without analyzing the explicit formulas that in the two-dimensional problem (or in the problem of flow of an elliptic cylinder) also exists an oscillatory mode analogous to the one described in Sect. 2.

In the space of two-dimensional matrices F_{ij} we introduce coordinates d_1, d_2, Φ_1, Φ_2 such that

$$\begin{pmatrix} F_{11} & F_{12} \\ F_{21} & F_{22} \end{pmatrix} = \begin{pmatrix} \cos \varphi_1 & -\sin \varphi_1 \\ \sin \varphi_1 & \cos \varphi_1 \end{pmatrix} \begin{pmatrix} d_1 & 0 \\ 0 & d_2 \end{pmatrix} \begin{pmatrix} \cos \varphi_2 & -\sin \varphi_2 \\ \sin \varphi_2 & \cos \varphi_2 \end{pmatrix}. \quad (3.1)$$

The Lagrangian (1.24) for the two-dimensional problem has the following form:

$$L = \tfrac{1}{2}(\dot{d}_1^2 + \dot{d}_2^2 + (\dot{\varphi}_1^2 + \dot{\varphi}_2^2)(d_1^2 + d_2^2) + 4\dot{\varphi}_1\dot{\varphi}_2 d_1 d_2) - \sigma(d_1 d_2)^{-1}. \quad (3.2)$$

The coordinates φ_1 and φ_2 are cyclic and therefore the corresponding momenta

$$p_{\varphi_1} = \frac{\partial L}{\partial \dot{\varphi}_1} = J \quad \text{and} \quad p_{\varphi_2} = \frac{\partial L}{\partial \dot{\varphi}_2} = -K$$

are preserved. The integrals J and K coincide respectively with the angular momentum integral and the vorticity (see (1.12)).

In the d_1, d_2 plane we introduce polar coordinates r, φ:

$$d_1 = r \cos \varphi, \quad d_2 = r \sin \varphi. \quad (3.3)$$

The Lagrangian system with the Lagrangian (3.2) in the phase space

$$p_r = \frac{\partial L}{\partial \dot{r}}, \quad p_\varphi = \frac{\partial L}{\partial \dot{\varphi}}, \quad r, \quad \varphi$$

becomes a Hamiltonian system with the Hamiltonian

$$H = \frac{1}{2}\left(p_r^2 + \frac{1}{r^2}\left(\frac{1}{2} p_\varphi^2 + U(\varphi) \right) \right), \quad (3.4)$$

where

$$U(\varphi) = \frac{J^2 + K^2 + 2JK \sin 2\varphi}{2 \cos^2 2\varphi} + \frac{2\sigma}{\sin 2\varphi}.$$

After a change of coordinates $r = 1/x$ and a time change $d\tau/dt = 1/r^2$ this system reduces to two Hamiltonian systems with the Hamiltonians

$$H = -\tfrac{1}{2} p_r^2 - x^2 H_0, \quad H_0 = \tfrac{1}{2} p_\varphi^2 + U(\varphi). \quad (3.5)$$

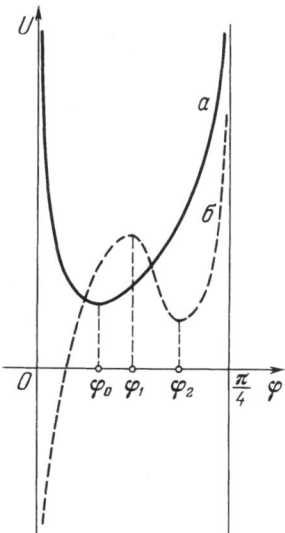

Fig. 40. The form of the potential $U(\varphi)$: a) $\sigma = +1, J \neq -K$; b) $\sigma = -1, J \approx -K \gg 1, J \neq -K$

From (3.4) it follows that (if $\sigma = +1$ and $J \neq -K$) $U(\varphi) \to \infty$ for $\varphi \to 0, \pi/4$. Thus the angle φ oscillates in a potential well determined by the potential $U(\varphi)$. Oscillations of the angle φ cause a change in the relations between the semi-axes of the ellipse d_1, d_2 and are a reflection of the general oscillatory mode of Sect. 2 (Fig. 40).

Let us evaluate the number of small oscillations of the angle φ near the state of equilibrium

$$\varphi_0 \left(\frac{dU}{d\varphi} (\varphi_0) = 0 \right)$$

over all time T of existence of the solution. In all time T the variable x varies from $0 \, (r^2 = d_1^2 + d_2^2 = \infty)$ to its maximum $x_m = |H|^{1/2} |H_0|^{-1/2}$ and again to 0. Hence $T = \pi (2H_0)^{-1/2}$.

As we know, the period T_φ of small oscillations of the angle φ is equal to $T_\varphi \cong 2\pi (U_\varphi''(\varphi_0))^{-1/2}$. Calculations show that for $J \approx K \gg 1$ the root of the equation $dU/d\varphi = 0$ is equal to $\varphi_0 \cong (2JK)^{-1/2}$. At this point we have

$$H_0 \cong U(\varphi) \cong \frac{J^2 + K^2}{2} \quad \text{and} \quad U_\varphi''(\varphi_0) = 4(2JK)^{3/2}.$$

Hence $T = \pi (J^2 + K^2)^{-1/2}$, $T_\varphi = \pi (2JK)^{-3/4}$. The number of oscillations is

$$N = \frac{T}{T_\varphi} = \frac{(2JK)^{3/4}}{(J^2 + K^2)^{1/2}}.$$

Consequently for $J \approx K \to \infty$ the number of oscillations N becomes arbitrarily large and grows like $J^{1/2}$. It is not difficult to show that small oscillations of the angle φ with the amplitude $\sim \varphi_0^{3/2} \ll \varphi_0$ cause oscillations of the quantity

$$\frac{\sin 2\varphi}{2x^2} = \det \| F_{ij} \|.$$

In the two-dimensional problem analogous to the problem of compression of a drop under pressure, where $\sigma = -1$, $U(\varphi) \to -\infty$ for $\varphi \to 0$ and $U(\varphi) \to +\infty$ for $\varphi \to \pi/4$ (see Fig. 40). For $J \approx -K \gg 1$, $J \neq -K$ and $0 < \varphi < \pi/4$ the potential $U(\varphi)$ has two extrema: a maximum φ_1 and a minimum φ_2, where $U(\varphi_2) > 0$. Solutions for which the angle φ oscillates in the vicinity of the minimum φ_2 have no physical singularities. In these solutions the compression from a rarefied state is replaced by expansion. Analogous solutions of the three-dimensional problem apparently do not exist.

Note that the infinitely tall potential barrier $U(\varphi)$ for $\varphi = \pi/4$ (which is one of the reasons for the occurence of oscillations (another reason is the barrier at $\varphi = 0$ created by pressure)) has the following purely geometric origins. In the four-dimensional space of matrices F_{ij} the set of matrices with $d_1 = d_2$ (see (3.1)) has dimension 2. Therefore for almost all trajectories of the Lagrangian system (3.2) at all times we have $d_1 \neq d_2$. Consequently almost all trajectories of the Hamiltonian system (3.4) do not intersect the surface $\varphi = \pi/4$, which is possible only in the presence of an infinite potential barrier. For the same reasons in the three-dimensional problem the relation of order among the semi-axes of an ellipsoid of constant density: $d_1 < d_2 < d_3$ is preserved for almost all motions of the gas at all times.

4. Equations of Motion of a Gravitating Gaseous Ellipsoid

Adiabatic motion of a gravitating ideal gas is determined by the equations

$$\rho \frac{du^i}{dt} = -\frac{\partial p}{\partial x_i} - \rho \frac{\partial \Phi}{\partial x_i}, \quad \frac{d\rho}{dt} = -\rho \operatorname{div} u^i, \quad \frac{d}{dt}\left(\frac{p}{\rho^\gamma}\right) = 0, \qquad (4.1)$$

where all notations are the same as in section 1 and $\Phi(x)$ is the Newtonian gravitational potential

$$\Phi(x) = G \int_V \frac{\rho(x')}{|x - x'|} dx'$$

caused at a given point x by the entire mass of gas (G is the gravitational constant). Just like the equations (1.1), in the presence of gravitational interaction the equations (4.1) have solutions with homogeneous deformation, in which the Euler coordinates of gas particles x^i are linear functions of the Lagrangian coordinates a^k:

$$x^i = F_k^i(t)a^k, \quad i, k = 1, 2, 3. \tag{4.2}$$

In order for such solutions to exist it is necessary for the gravitational potential $\Phi(x_1, x_2, x_3)$ to be a quadratic function of the coordinates x_i and this property is satisfied only for a gravitational field inside the ellipsoid filled with gas with constant density (see [172, 173]). Therefore for the solutions of type (4.2) the density of gas ρ and the pressure p are determined by the following formulas (for $a^2 = a_1^2 + a_2^2 + a_3^2 \leqslant 1$):

$$\rho = \frac{3M}{4\pi} |\det(F_k^i)|^{-1}, \quad p = \alpha \frac{3M(\gamma - 1)(1 - a^2)}{8\pi |\det(F_k^i)|^\gamma}, \tag{4.3}$$

where α and M are constants (M is the total mass of gas). For $a^2 > 1$ we assume that $\rho = 0$ and $p = 0$.

The operator F_k^i can be written in the form

$$F = Q_1 D Q_2. \tag{4.4}$$

where Q_1, Q_2 are orthogonal matrices and the matrix D is diagonal ($D_j^i = d_j \delta_j^i$). Under the mapping (4.2) the unit sphere in Lagrangian coordinates a^k becomes an ellipsoid with the semi-axes d_1, d_2, d_3 and principle directions $\bar{e}_i = Q_1 e_i$. By virtue of the conditions (4.3) the gas fills this ellipsoid with constant density. The pressure p is maximal at the center of the ellipsoid and equal to zero at its surface.

At a point (x_1, x_2, x_3) inside the ellipsoid (in the coordinates related to the principle axes) the Newtonian potential Φ has the following form (up to a constant):

$$\Phi(x_1, x_2, x_3) = \frac{3}{4} GM \int_0^\infty \left(\frac{x_1^2}{d_1^2 + s} + \frac{x_2^2}{d_2^2 + s} + \frac{x_3^2}{d_3^2 + s} \right)$$
$$\times \left((d_1^2 + s)(d_2^2 + s)(d_3^2 + s) \right)^{-1/2} ds. \tag{4.5}$$

In order to derive the equations of motion of a gravitating gaseous ellipsoid we introduce the matrix $\varphi^{ij} = F_\alpha^i F_\alpha^j$ ($\varphi = F \circ F^t$). Obviously the eigenvalues of the matrix Φ^{ij} are d_1^2, d_2^2, d_3^2. If the matrix F_k^i is diagonal ($F_k^i = D_k^i$), then the Newtonian potential (4.5) can be written in the following form:

$$\Phi = -\frac{\partial U}{\partial \varphi^{ij}} x^i x^j, \tag{4.6}$$

where

$$U = \frac{3}{2} GM \int_0^\infty \left((d_1^2 + s)(d_2^2 + s)(d_3^2 + s) \right)^{-1/2} ds. \tag{4.7}$$

Since the potential Φ is a scalar, the expression (4.6) holds in any basis. Note that the function under the integral in (4.7) can be expressed directly in terms of the components of the matrix F_k^i:

$$(d_1^2 + s)(d_2^2 + s)(d_3^2 + s)$$
$$= \det(FF^t) + \tfrac{1}{2}s[(\mathrm{Tr}(FF^t))^2 - \mathrm{Tr}(FF^tFF^t)] + s^2 \,\mathrm{Tr}\, FF^t + s^3.$$

The main equations of the dynamics (4.1) on the level of (4.3) and (4.6) turn into the following system of equations:

$$\ddot{F}_k^i = -\alpha \frac{\partial}{\partial F_k^i}(\det F)^{1-\gamma} + \left(\frac{\partial U}{\partial \varphi^{il}} + \frac{\partial U}{\partial \varphi^{li}}\right) F_k^l. \tag{4.8}$$

According to the definition $\varphi^{il} = F_\alpha^i F_\alpha^l$ we have

$$F_k^l = \frac{\partial \varphi^{il}}{\partial F_k^i} = \frac{\partial \varphi^{li}}{\partial F_k^i}.$$

Therefore the system of equations (4.8) is a Lagrangian system

$$\frac{d}{dt}\frac{\partial L}{\partial \dot{F}_k^i} = \frac{\partial L}{\partial F_k^i} \tag{4.9}$$

with the Lagrangian

$$L = \frac{1}{2}\sum_{i,k}^3 (\dot{F}_k^i)^2 - \alpha(\det F)^{1-\gamma}$$

$$+ \frac{3}{2}GM \int_0^\infty ((d_1^2 + s)(d_2^2 + s)(d_3^2 + s))^{-1/2}\, ds. \tag{4.10}$$

Thus the study of the dynamics of a gravitating gaseous ellipsoid is equivalent to the study of the dynamics of a point mass in the nine-dimensional space of matrices F_k^i in a potential field defined by (4.10). Note that the Lagrangian (4.10) depends on a characteristic parameter $\beta = \alpha/3GM$ that is not removable by a time change. Just like the system (1.10), the Lagrangian system (4.9)–(4.10) is invariant relative to the transformations $F \to Q_1 F Q_2$ (where Q_1 and Q_2 are orthogonal matrices) and has first integrals J and K (1.12).

The total energy of gas filling the ellipsoid v is

$$E = \int_v \left[\frac{\rho(u^i)^2}{2} + \frac{p}{\gamma - 1} - \rho\Phi(x)\right] dx$$

$$= \frac{M}{5}\left[\frac{1}{2}\sum_{i,k}^3 (\dot{F}_k^i)^2 + \alpha(\det F)^{1-\gamma}\right.$$

$$\left. - \frac{3}{2}GM \int_0^\infty ((d_1^2 + s)(d_2^2 + s)(d_3^2 + s))^{-1/2}\, ds\right].$$

In the important special case of spherically symmetric motion $F_k^i = F\delta_k^i$ the Lagrangian (4.10) becomes

$$L_1 = \frac{1}{2}\dot{F}^2 - \frac{\alpha}{3}F^{3(1-\gamma)} + GMF^{-1}.$$

For $\gamma > 4/3$ and a negative energy

$$E_1 = \frac{1}{2}\dot{F}^2 + \frac{\alpha}{3}F^{3(1-\gamma)} - GMF^{-1}$$

the gaseous ball oscillates in the neighborhood of the state of equilibrium

$$F_0^{3\gamma-4} = 3(\gamma - 1)\beta.$$

These oscillations of a gaseous ball were studied in the book [132] as a model of pulsations of variable stars. For $\gamma \leqslant 4/3$ and $E_1 < 0$ there is a collapse of the gaseous ball to the center of symmetry.

The dynamics of the system (4.9) for $\alpha = 0$ $(p = 0)$ (the motion of a gravitating dust ellipsoid) were analyzed for the first time in [163, 164] and have applications in the theory of formation of galaxies (see [165, 74]). The dynamics of the system (4.9)–(4.10) for diagonal matrices F_k^i with $\alpha > 0$ and $GM > 0$ were analyzed in [169, 170] by means of numerical methods.

5. Transformation of the Hamiltonian System

Turning to the analysis of the Lagrangian system (4.9) using the methods of qualitative theory of differential equations we transform this system to an equivalent system defined on some compact manifold S. To simplify notation we write the Hamiltonian corresponding to (4.10) in the following form:

$$H = \frac{1}{2}\sum_{i=1}^{n} P_i^2 + \alpha V^{1-\gamma}(q_i) - \frac{3GM}{8}U(q_i), \quad P_i = \dot{q}_i. \tag{5.1}$$

where the coordinates q_i correspond to F_{jk}, $V(q_i) = \det \| F_{jk} \|$ is a homogeneous function of order 3,

$$U(q_i) = \int_0^\infty ((d_1^2 + s)(d_2^2 + s)(d_3^2 + s))^{-1/2} ds$$

is a homogeneous function of order (-1) and $n = 9$. The Hamiltonian system corresponding to the Hamiltonian (5.1) has the following form:

$$\dot{P}_i = -\frac{\partial H}{\partial q_i} = -\alpha(1 - \gamma)V^{-\gamma}(q_i)\frac{\partial V}{\partial q_i} + \frac{3GM}{8}\frac{\partial U}{\partial q_i},$$

$$\dot{q}_i = \frac{\partial H}{\partial P_i} = P_i. \tag{5.2}$$

The system (5.2) is considered in the region S_1 of the phase space determined by the condition $V(q_i) > 0$, because according to (4.3) the points of the surface $V(q_i) = 0$ correspond to a physical singularity of the solutions. In the P_i, q_i phase space we introduce two systems of coordinates W_1 and W_2. The coordinates W_1 have the following form:

$$\bar{p}_i = \frac{P_i}{(\alpha V^{1-\gamma}(q_i) + (3GM/8)U(q_i))^{1/2}},$$

$$u = \frac{U(q_i)}{(8\alpha/3GM)V^{1-\gamma}(q_i) + U(q_i)}, \quad y_i = \frac{q_i}{\left(\sum\limits_{k=1}^{n} q_k^2\right)^{1/2}}. \tag{5.3}$$

The coordinates y_i vary over the unit sphere

$$S^{n-1}: \sum_{i=1}^{n} y_i^2 = 1,$$

the coordinates $\bar{p}_i$ vary over the entire Euclidean space E^n and the coordinate u varies over the interval $0 < u < 1$. Note that for $\gamma = 4/3$ the coordinates u and y_i becomes dependent, so below we assume that $\gamma \neq 4/3$.

The coordinates W_2 have the following form:

$$p_i = \frac{\bar{p}_i}{\left(\sum\limits_{k=1}^{n} \bar{p}_k^2\right)^{1/2}} = \frac{P_i}{\left(\sum\limits_{k=1}^{n} P_k^2\right)^{1/2}}, \quad u, y_i,$$

$$w = \frac{1}{\sum\limits_{k=1}^{n} \bar{p}_k^2} = \frac{\alpha V^{1-\gamma}(q_i) + (3GM/8)U(q_i)}{\sum\limits_{k=1}^{n} P_k^2}. \tag{5.4}$$

The coordinates p_i vary over the unit sphere

$$S^{n-1}: \sum_{k=1}^{n} p_k^2 = 1,$$

the coordinate w varies over a semiaxis $0 < w < \infty$.

In the coordinates W_1 and time τ_1

$$\frac{d\tau_1}{dt} = \frac{(\alpha V^{1-\gamma}(q_i) + (3GM/8)U(q_i))^{1/2}}{\left(\sum\limits_{k=1}^{n} q_k^2\right)^{1/2} V(y)} \tag{5.5}$$

the Hamiltonian system (5.2) takes the following form:

$$\dot{\bar{p}}_i = (1 - \gamma)(1 - u)\left(-\frac{\partial V}{\partial y_i} - \frac{1}{2}\bar{p}_i \sum_{k=1}^{n} \frac{\partial V}{\partial y_k} \bar{p}_k\right)$$

$$+ u\frac{V(y)}{U(y)}\left(\frac{\partial U}{\partial y_i} - \frac{1}{2}\bar{p}_i \sum_{k=1}^{n} \frac{\partial U}{\partial y_k} \bar{p}_k\right),$$

$$\dot{y}_i = V(y)\left(\bar{p}_i - y_i \sum_{k=1}^{n} \bar{p}_k y_k\right),$$

$$\dot{u} = u(1-u)\left(\frac{V(y)}{U(y)} \sum_{k=1}^{n} \frac{\partial U}{\partial y_k} \bar{p}_k - (1-\gamma) \sum_{k=1}^{n} \frac{\partial V}{\partial y_k} \bar{p}_k\right). \tag{5.6}$$

In the coordinates W_2 and time τ_2

$$\frac{d\tau_2}{dt} = \frac{\left(\sum_{k=1}^{n} P_k^2\right)^{1/2}}{\left(\sum_{k=1}^{n} q_k^2\right)^{1/2} V(y)} \tag{5.7}$$

the same system takes the following form:

$$\dot{p}_i = (1-\gamma)w(1-u)\left(-\frac{\partial V}{\partial y_i} + p_i \sum_{k=1}^{n} \frac{\partial V}{\partial y_k} p_k\right)$$

$$+ wu\frac{V(y)}{U(y)}\left(\frac{\partial U}{\partial y_i} - p_i \sum_{k=1}^{n} \frac{\partial U}{\partial y_k} p_k\right),$$

$$\dot{y}_i = V(y)\left(p_i - y_i \sum_{k=1}^{n} p_k y_k\right),$$

$$\dot{u} = u(1-u)\left(\frac{V(y)}{U(y)} \sum_{k=1}^{n} \frac{\partial U}{\partial y_k} p_k - (1-\gamma) \sum_{k=1}^{n} \frac{\partial V}{\partial y_k} p_k\right), \tag{5.8}$$

$$\dot{w} = w\left((1-\gamma)(1-u)(1+2w) \sum_{k=1}^{n} \frac{\partial V}{\partial y_k} p_k\right.$$

$$\left. + u(1-2w)\frac{V(y)}{U(y)} \sum_{k=1}^{n} \frac{\partial U}{\partial y_k} p_k\right).$$

Note that in accordance with the definition of the coordinates y_i and p_i the systems (5.6) and (5.8) are considered only on the manifolds

$$\sum_{k=1}^{n} y_k^2 = 1, \quad \sum_{k=1}^{n} p_k^2 = 1.$$

The region of phase space S_1, where the system (5.2) is defined, is determined by the conditions $w > 0, 0 < u < 1, V(y_i) > 0$ in the coordinates W_1 and W_2. To this region we add a boundary Γ consisting of four components, which are determined by the following conditions:

$$\Gamma_w: w = 0; \quad \Gamma_0: u = 0; \quad \Gamma_1: u = 1; \quad \Gamma_2: V(y_i) = 0.$$

Let S denote the closed manifold obtained as a result of such an addition to the boundary (on S we have $w \geq 0, 0 \leq u \leq 1, V(y_i) \geq 0$). There is a dynamical system defined on the manifold S, which coincides in the coordinates W_1 and W_2

with the systems (5.6) and (5.8) respectively. Obviously these systems can be continuously extended to the components of boundary Γ_w, Γ_0 and Γ_1.

Furthermore using the explicit form of the potential $U(q_i)$ (4.7) we can prove that for $V(y) = \det \| Y_{jk} \| \to 0$ we have $(V(y)/U(y))(\partial U/\partial y_i) \to 0$. Therefore in the component of the boundary $\Gamma_2 \, (V(y) = 0)$ we define these expressions to be zero (their limit value). As a result of such a definition the system (5.6)–(5.8) continuously extends to the component of the boundary Γ_2.

It is not difficult to verify that all components of the boundary Γ and their intersections are invariant submanifolds of the dynamical system in S, i.e. a trajectory starting in some component of the boundary Γ remains in it for all time. The system thus defined on the component of the boundary $\Gamma_0 \, (u = 0)$ is identical to the system describing the motion of a non-gravitating gaseous ellipsoid. The system defined on the component of the boundary $\Gamma_1 \, (u = 1)$ is identical to the system describing the motion of a gravitating dust ellipsoid. Thus the dynamical system on the manifold S describing the motion of a gravitating gaseous ellipsoid also contains all the information about these two limit forms of motion.

6. Oscillatory Mode of Motion with Negative Energy

I. Let us analyze the behavior of the system (5.6) for $H \leqslant 0$ and $\gamma < 4/3$. In the local coordinates W_1 the Hamiltonian H has the following form:

$$H = \frac{3GM}{8} \left(\frac{8\alpha}{3GM} \right)^{1/(4-3\gamma)} V(y)^{(1-\gamma)/(4-3\gamma)} \cdot$$
$$\cdot \left(\frac{U(y)}{u} \right)^{3(1-\gamma)/(4-3\gamma)} (1-u)^{-1/(4-3\gamma)} \left(\frac{1}{2} \sum_{k=1}^{n} \bar{p}_k^2 + 1 - 2u \right). \tag{6.1}$$

Hence the region $H \leqslant 0$ or

$$\frac{1}{2} \sum_{k=1}^{n} \bar{p}_k^2 \leqslant 2u - 1$$

lies entirely in W_1. All critical points of the dynamical system (5.6) for $H \leqslant 0$ and $\gamma < 4/3$ belong to the boundary Γ and form four sets: K_1, Φ_+, Φ_- and L.

1. The set of critical points $K_1 \, (u = 1, \, V(y) = 0)$ is the intersection of the invariant submanifolds $\Gamma_1 \, (u = 1)$ and $\Gamma_2 \, (V(y) = 0)$. These critical points are non-degenerate

$$\left(\text{for } \sum_{k=1}^{n} \frac{\partial V}{\partial y_k} \bar{p}_k \neq 0 \right)$$

and unstable. They have two non-zero eigenvalues:

$$\lambda_1 = (1 - \gamma) \sum_{k=1}^{n} \frac{\partial V}{\partial y_k} \bar{p}_k \quad \text{(variable } u\text{)},$$

$$\lambda_2 = \sum_{k=1}^{n} \frac{\partial V}{\partial y_k} \bar{p}_k \qquad \text{(variables } y_i\text{)}.$$

The remaining $2n - 2$ zero eigenvalues correspond to eigenvectors tangent to the manifold K_1. In view of the fact that $\gamma > 1$ the eigenvalues λ_1 and λ_2 have opposite signs, i.e. the points of K_1 are of saddle type. It is convenient to divide the set K_1 into two parts:

$$K_+ \left(\sum_{k=1}^{n} \frac{\partial V}{\partial y_k} \bar{p}_k \geq 0 \right) \quad \text{and} \quad K_- \left(\sum_{k=1}^{n} \frac{\partial V}{\partial y_k} \bar{p}_k \leq 0 \right).$$

Each point of K_+ has one separatrix approaching it in the manifold Γ_2 and one separatrix leaving it in the manifold Γ_1 with the opposite true for K_-.

2. The eigenvalues of the system (5.8) at the critical points of

$$\Phi_\varepsilon \left(u = 1, \, p_i = \varepsilon y_i, \, w = 1/2, \, \varepsilon = \pm 1, \, y_i = Y_{jk} = 3^{-1/2} Q_{jk} \right)$$

(where Q_{jk} is an orthogonal matrix) are

$$\lambda_1 = \frac{2}{3} \varepsilon (4 - 3\gamma) \quad \text{(variable } u\text{)},$$

$$\lambda_2 = \frac{2}{3} \varepsilon \qquad\qquad \text{(variable } w\text{)},$$

$$\lambda_3 = \lambda_4 = \frac{1}{3} \left(-\frac{\varepsilon}{2} + \left(\frac{53}{20} \right)^{1/2} \right) > 0,$$

$$\lambda_5 = \lambda_6 = \frac{1}{3} \left(-\frac{\varepsilon}{2} - \left(\frac{53}{20} \right)^{1/2} \right) < 0, \qquad\qquad (6.2)$$

$$\lambda_7 = \lambda_8 = \lambda_9 = 0,$$

$$\lambda_{13} = \lambda_{14} = \lambda_{15} = \frac{1}{3} \left(-\frac{\varepsilon}{2} + i \left(\frac{43}{20} \right)^{1/2} \right),$$

$$\lambda_{16} = \lambda_{17} = \lambda_{18} = \frac{1}{3} \left(-\frac{\varepsilon}{2} - i \left(\frac{43}{20} \right)^{1/2} \right).$$

The first six eigenvalues correspond to "diagonalizable" separatrices, i.e. solutions of the form $F(t) = Q \cdot D(t)$, where $D(t)$ is a diagonal matrix. According to (6.2) the critical points of Φ_ε are non-degenerate and unstable. Furthermore each point of the three-dimensional set Φ_- has an approaching four-dimensional separatrix, which is formed by diagonalizable solutions with spherically symmetric compression, generalizing exact spherically symmetric solutions. It also has an emerging eleven-dimensional separatrix, which lies in the boundary Γ_1 at the zero level of energy. Therefore a spherically symmetric compression is unstable (already in the class of diagonalizable solutions). The properties of the critical points in Φ_+ are identical to the properties of critical points in Φ_- with the opposite direction of time.

3. At the critical points of L: $V(y) = 0$, $\partial V/\partial y_i = 0$ ($\bar{p}_i$ and u are arbitrary) the matrix Y_{jk} is doubly degenerate. Thus for $H \leqslant 0$, $\gamma < 4/3$ the system (5.6) has no stable critical points, which is one of the reasons for the existence of the oscillatory mode.

II. As mentioned above, the separatrices of the critical points of K_+ and K_- belong to the invariant manifolds Γ_1 and Γ_2. Let us consider the system (5.6) on these manifolds.

1) On the manifold Γ_1 ($u = 1$) the system (5.6) describes the motion of a gravitating dust ellipsoid. In [164, 174] it was shown that in the process of motion of a gravitating dust ellipsoid with negative energy H the volume of the ellipsoid becomes zero twice. In other words the expansion from a compressed state is replaced by contraction. Furthermore for almost all solutions the ellipsoid is compressed to a disc in the initial and in the final state, i.e. $d_1 = 0$, $d_2 \neq 0$, $d_3 \neq 0$. In the coordinates W_1 this result implies that almost all trajectories of the system (5.6) on the manifold Γ_1 for $H < 0$ have their beginning and end in the manifold of critical points K_1 ($V(y) = 0$, $u = 1$) and that for almost all critical points in K_+ the separatrices leaving them go to some critical points of K_-.

2) On the manifold Γ_2 ($V(y) = 0$) the system (5.6) can be integrated explicitly. In the time τ determined by the expression $d\tau = 2^{1/2}(\gamma - 1)(1 - u)$ $|\text{grad } V(y_i^0)| \, d\tau_1$ the trajectories of this system are given by the following formulas

$$y_i = y_i^0, \quad \bar{p}_i = 2^{1/2} s_i \frac{\sin \tau - \sin \tau_0}{\cos \tau} + \bar{p}_i^0 \frac{\cos \tau_0}{\cos \tau},$$

$$u = \frac{\cos^2 \tau_0}{\cos^2 \tau}, \tag{6.3}$$

where

$$y_i^0, \ s_i = \frac{\text{grad } V(y_i^0)}{|\text{grad } V(y_i^0)|}, \ \tau_0, \bar{p}_i^0 \quad \text{are constants such that}$$

$$V(y_i^0) = 0, \quad \sum_{k=1}^{n} \bar{p}_k^0 s_k = 2^{1/2} \, \text{tg} \, \tau_0 < 0, \quad |\tau_0| < \pi/2, \quad \sum_{k=1}^{n} \bar{p}_k s_k = 2^{1/2} \, \text{tg} \, \tau.$$

The trajectory (6.3) is defined for $\tau_0 \leqslant \tau \leqslant -\tau_0$ and goes from a critical point ($\bar{p}_i^0$, y_i^0, $u = 1$) contained in K_- to a critical point ($\bar{p}_i^1 = \bar{p}_i(-\tau_0)$, y_i^0, $u = 1$) contained in K_+ (consequently all trajectories (6.3) are separatrices of the critical points of K_+ and K_-). It is easy to see that the final point of the trajectory (6.3) ($\bar{p}_i^1 = \bar{p}_i(-\tau_0)$) is obtained from the initial point ($\bar{p}_i$) by way of reflection in the plane tangent to the surface $V(y_i) = 0$ at the point (y_i^0).

III. The results obtained above lead to the following separatrix diagram:

$$\cdots \to K_+ \xrightarrow{\text{I}} K_- \xrightarrow{\text{II}} K_+ \xrightarrow{\text{I}} K_- \xrightarrow{\text{II}} \cdots \tag{6.4}$$

The mappings (shown by arrows) denote steps along separatrices from their initial to their final points. Mappings I and II are realized by separatrices going in the manifolds Γ_1 and Γ_2 respectively. Separatrix steps between the sets K_+, K_- and Φ_e, L are not shown in the diagram (6.4), because for almost all critical points of K_+ and K_- the whole infinite sequence of mappings (6.4) does not lead outside the limits of the sets K_+ and K_-.

The infinite sequence of separatrices defined in the diagram (6.4) is an approximation of the trajectories of the system (5.6) with large negative energies H and for $\beta = (8\alpha)/(3GM) \to 0$. Indeed the function H (see (6.1)) is bounded below everywhere on the manifold S, except on the components of the boundary Γ_1 $(u = 1)$ and Γ_2 $(V(y) = 0)$, where $H \to \infty$. Therefore the trajectories of the system (5.6) with large negative energy H always remain in close proximity with the manifolds Γ_1 and Γ_2 (the same is true for arbitrary $H < 0$, when $\beta \to 0$ (see the definition of the coordinate u (5.3))). Consequently these trajectories move along the trajectories of the system (5.6) in the manifolds Γ_1 and Γ_2, i.e. a general trajectory of the system (5.6) moves along a sequence of separatrices of the critical points of K_+ and K_-.

The obtained approximation of the trajectories of the system (5.6) by the sequence of separatrices (6.4) proves that in general the motion of a gravitating gaseous ellipsoid with large negative energy H or a small parameter β has a pulsating oscillatory nature. Indeed according to the approximation (6.4) a trajectory periodically ends up in the vicinity of the critical points of K_+ and K_-, where $\det Y_{jk} = V(y_i) = 0$, i.e. periodically the ellipsoid appears compressed to a disc. Besides, from the equation

$$\frac{dV(q_i)}{d\tau_1} = V(q_i) \sum_{k=1}^{n} \frac{\partial V(y)}{\partial y_k} \bar{p}_k \tag{6.5}$$

it follows that the volume of the ellipsoid $\det \| F_{jk} \| = V(q_i)$ reaches its maximum during the motion of a trajectory of the system (5.6) along the separatrix step I. During the motion of the trajectory along the separatrix step II (see (6.4)) $\det \| F_{jk} \|$ reaches its minimum value. Consequently the variation of density of the gas filling the ellipsoid $\rho = (3M)/(4\pi \det \| F_{jk} \|)$ also has an oscillatory nature. By virtue of the relation (see (5.5))

$$dt = d\tau_1 |H|^{-3/2} \left(\frac{8\alpha}{3GM}\right)^{1/(1-\gamma)} \left(\frac{3GM}{8}\right)^{(2-\gamma)(1-\gamma)} \frac{V(y)U(y)}{u}$$

$$\times \left| \frac{1}{2} \sum_{k=1}^{n} \bar{p}_k^2 + 1 - 2u \right|^{3/2} \tag{6.6}$$

the period of each oscillation of the ellipsoid for $H \to -\infty$ becomes arbitrarily small.

From the equation of state of an ideal gas

$$p = \rho RT$$

and (4.3) we obtain an expression for the temperature:

$$T = \alpha(\gamma - 1)R^{-1}(1 - a^2)V^{1-\gamma}(F), \tag{6.7}$$

where R is the gas constant. Obviously the oscillations of volume of the gas are accompanied by the oscillations of its temperature and other physical parameters of the gas. The constant α is defined to be the initial temperature at the center of the ellipsoid.

The described oscillatory motion of the ellipsoid for $H \to -\infty$ occurs in the state of strong compression, because the quantity

$$d_1^2 + d_2^2 + d_3^2 = \sum_{k=1}^{n} q_k^2$$

$$= \left(\frac{3GM}{8\alpha}\right)^{2/(4-3\gamma)} \left(U(y)\frac{1-u}{u}\right)^{2/(4-3\gamma)} (V(y))^{2(\gamma-1)/(4-3\gamma)} \tag{6.8}$$

tends to zero as $H \to -\infty$ (i.e. either $V(y) \to 0$ or $u \to 1$).

The separatrix approximation (6.4) implies that the asymptotic motion of the trajectories of the system (5.6) in the coordinates y_i as $H \to -\infty$ or $\beta \to 0$ proceeds as follows:

1) In the region $V(y_i) > 0$ the motion is along the trajectories corresponding to a gravitating dust ellipsoid. In general such a trajectory intersects the surface $V(y_i) = 0$ at some point y_i^0 (step I).

2) At the point of intersection the trajectory is elastically reflected from the surface $V(y_i) = 0$ (step II (see (6.3))).

3) Then the motion is again along a trajectory corresponding to a gravitating dust ellipsoid until the next intersection with the surface $V(y_i) = 0$ etc.

Thus it can be said that the model of the oscillatory mode of motion of a gravitating gaseous ellipsoid is like the game of multi-dimensional billiards in the region $\det \| Y_{jk} \| = V(y_i) \geqslant 0$ on the eight-dimensional sphere

$$S^8 \left(\sum_{j,k=1}^{3} Y_{jk}^2 = 1 \right)$$

with an elastically reflecting boundary $\det \| Y_{jk} \| = 0$. Furthermore between the collisions with the boundary a point moves along the trajectories describing the motion of a gravitating dust ellipsoid. The presence of dynamical pressure in the gas manifests itself in the property of elastic reflection of the trajectories from the boundary $\det \| Y_{jk} \| = 0$.

7. On the Impossibility of Collapse of a Gravitating Gaseous Ellipsoid in the Presence of Rotation of the Gas

In this section we shall obtain (for $\gamma < 4/3$) upper and lower estimates for the semiaxes of the ellipsoid for the motion with negative energy

$$H = \frac{1}{2} \operatorname{Tr} \dot{F}\dot{F}^t + \frac{\alpha}{(\det F)^{\gamma - 1}}$$

$$-\frac{3}{2} GM \int_0^\infty \left((d_1^2 + s)(d_2^2 + s)(d_3^2 + s) \right)^{-1/2} ds \tag{7.1}$$

and non-zero integrals J and K (1.12). First we estimate the energy integral H (7.1). We write the matrix F in the form (4.4): $F = Q_1 D Q_2$. Then $F^t = Q_2^t D Q_1^t$. Let A and B be skew-symmetric matrices such that

$$\dot{Q}_1 = -Q_1 A, \quad \dot{Q}_2 = B Q_2,$$

$$A = \begin{pmatrix} 0 & \alpha_3 & -\alpha_2 \\ -\alpha_3 & 0 & \alpha_1 \\ \alpha_2 & -\alpha_1 & 0 \end{pmatrix}, \quad B = \begin{pmatrix} 0 & \beta_3 & -\beta_2 \\ -\beta_3 & 0 & \beta_1 \\ \beta_2 & -\beta_1 & 0 \end{pmatrix}. \tag{7.2}$$

In this notation we obtain

$$\dot{F} = Q_1(-AD + \dot{D} + DB)Q_2, \quad \dot{F}^t = Q_2^t(-BD + \dot{D} + DA)Q_1^t.$$

The first integrals J and K have the following form:

$$J = F\dot{F}^t - \dot{F}F^t = Q_1 j Q_1^t, \quad K = F^t\dot{F} - \dot{F}^t F = Q_2^t k Q_2,$$

where

$$j = -2DBD + D^2A + AD^2, \quad k = -2DAD + D^2B + BD^2. \tag{7.3}$$

The three-dimensional vectors (or skew-symmetric matrices) j and k are obtained from the constant vectors J and K by means of an orthogonal rotation (Q_1^t and Q_2 respectively). Therefore

$$|j| = |J|, \quad |k| = |K|.$$

The equalities (7.3) take the following form in coordinate notation:

$$j_l = \alpha_l(d_m^2 + d_n^2) - 2d_m d_n \beta_l,$$

$$k_l = \beta_l(d_m^2 + d_n^2) - 2d_m d_n \alpha_l, \quad (m, n, l) = (1, 2, 3).$$

Hence we obtain

$$\alpha_l = \frac{j_l(d_m^2 + d_n^2) + 2k_l d_m d_n}{(d_m^2 - d_n^2)^2},$$

$$\beta_l = \frac{k_l(d_m^2 + d_n^2) + 2j_l d_m d_n}{(d_m^2 - d_n^2)^2}. \tag{7.4}$$

The kinetic energy of the gas T has the following form:

$$T = \tfrac{1}{2} \operatorname{Tr}(\dot{F}\dot{F}^t) = \tfrac{1}{2} \operatorname{Tr}(\dot{D}^2 - A^2 D^2 - B^2 D^2 + 2ADBD).$$

Using the formulas (7.2) and (7.4) we obtain

$$2T = \sum_{i=1}^{3} d_i^2 + \sum_{l \neq m \neq n}^{3} \frac{(j_l^2 + k_l^2)(d_m^2 + d_n^2) + 4j_l k_l d_m d_n}{(d_m^2 - d_n^2)^2}. \tag{7.5}$$

Obviously the kinetic energy T can be written as follows:

$$2T = \sum d_i^2 + \sum_{l \neq n \neq m} \frac{(d_m - d_n)^2(j_l^2 + k_l^2) + 2d_m d_n(j_l + k_l)^2}{(d_m - d_n)^2(d_m + d_n)^2}. \tag{7.6}$$

Let $d_1 < d_2 < d_3$. Then from (7.6) it follows that

$$T > \frac{1}{8d_3^2}(|J|^2 + |K|^2). \tag{7.7}$$

Let us estimate the integral

$$I = \int_0^\infty \frac{ds}{((d_1^2 + s)(d_2^2 + s)(d_3^2 + s))^{1/2}} = \frac{1}{d_3} \int_0^\infty \frac{ds}{((\delta_1^2 + s)(\delta_2^2 + s)(1 + s))^{1/2}},$$

where $\delta_1 = d_1/d_3$, $\delta_2 = d_2/d_3$. Obviously

$$(\delta_1^2 + s)(\delta_2^2 + s) = (\delta_1 \delta_2)^2 + (\delta_1^2 + \delta_2^2)s + s^2 > (\delta_1 \delta_2 + s)^2.$$

Therefore

$$I \leqslant \frac{1}{d_3} \int_0^\infty \frac{ds}{(\delta_1 \delta_2 + s)(1 + s)^{1/2}}$$

$$= \frac{1}{d_3} \left(\int_0^y \frac{ds}{(\delta_1 \delta_2 + s)(1 + s)^{1/2}} + \int_y^\infty \frac{ds}{(\delta_1 \delta_2 + s)(1 + s)^{1/2}} \right)$$

$$\leqslant \frac{1}{d_3} \left(\int_0^y \frac{ds}{(\delta_1 \delta_2 + s)} + \int_y^\infty \frac{ds}{s(1 + s)^{1/2}} \right)$$

$$= \frac{1}{d_3} \left(\ln \frac{\delta_1 \delta_2 + y}{\delta_1 \delta_2} + \ln \frac{\sqrt{1 + y} + 1}{\sqrt{1 + y} - 1} \right)$$

$$\leqslant \frac{1}{d_3} \left(\ln(1 + y) \frac{\sqrt{1 + y} + 1}{\sqrt{1 + y} - 1} - \ln \delta_1 \delta_2 \right).$$

The minimum value of the first summand is attained for $y = (1 + \sqrt{5})/2$. Hence we obtain the following estimate:

$$I \leqslant \frac{1}{d_3}(C_1 - \ln \delta_1 \delta_2), \quad C_1 = 5 \ln \frac{1 + \sqrt{5}}{2}. \tag{7.8}$$

Taking (7.7) and (7.8) into account we obtain an estimate for the energy integral H (7.1):

$$0 > H > \frac{|J|^2 + |K|^2}{8d_3^2} + \frac{\alpha}{d_3^{3(\gamma-1)}(\delta_1\delta_2)^{\gamma-1}} - \frac{3GM}{2}\frac{1}{d_3}(C_1 - \ln\delta_1\delta_2).$$

Multiplying this inequality by $2d_3/3GM$ we see that

$$\frac{A_0^2}{d_3} + \frac{\beta d_3^{4-3\gamma}}{x^{\gamma-1}} - (C_1 - \ln x) < 0, \tag{7.9}$$

where

$$A_0^2 = \frac{|J|^2 + |K|^2}{12GM}, \quad \beta = \frac{2\alpha}{3GM}, \quad x = \delta_1\delta_2.$$

From the inequality (7.9) (after dropping the first summand) follows the inequality

$$d_3^{4-3\gamma} < C_2\frac{\ln z}{z}, \tag{7.10}$$

where

$$C_2 = \frac{\exp(C_1(\gamma-1))}{\beta(\gamma-1)}, \quad z = \frac{\exp(C_1(\gamma-1))}{x^{\gamma-1}} > 1.$$

Since $(\ln z)/z < e^{-1}$, the inequality gives an upper estimate for d_3:

$$d_3 < \left(\frac{\exp(C_1(\gamma-1)-1)}{\beta(\gamma-1)}\right)^{1/(4-3\gamma)} = C_3. \tag{7.11}$$

In order to obtain a lower estimate for the quantity $x = \delta_1\delta_2$ we transform (7.9) (by dividing by d_3) to the following form:

$$\left(\frac{A_0}{d_3} - \frac{C_1 - \ln x}{2A_0}\right)^2 - \left(\frac{C_1 - \ln x}{2A_0}\right)^2 + \frac{\beta}{d_3^{3(\gamma-1)}x^{\gamma-1}} < 0.$$

Hence by dropping the first summand and using the estimate (7.11) we obtain the following inequality

$$D_0 < \frac{\ln z_1}{z_1}, \tag{7.12}$$

where

$$D_0 = \frac{(\gamma-1)\beta^{1/2}\exp(-C_1(\gamma-1)/2)}{C_3^{3(\gamma-1)/2}}A_0, \quad z_1 = \frac{\exp(-C_1(\gamma-1)/2)}{x^{(\gamma-1)/2}}. \tag{7.13}$$

From the inequality (7.12) it follows that $D_0 < e^{-1}$ is a bound for the values of the integrals J and K for motions with negative energy. From (7.12) we see that

$$D_0z_1 < \ln D_0z_1 - \ln D_0 < e^{-1}D_0z_1 - \ln D_0.$$

Therefore

$$z_1 < \frac{e}{e-1} \frac{|\ln D_0|}{D_0}.$$

Hence we obtain a lower estimate for x:

$$x > e^{C_1} \left(\frac{e-1}{e} \frac{D_0}{|\ln D_0|}\right)^{2/(\gamma-1)} = D_1. \tag{7.14}$$

Turning to the derivation of a lower bound for d_1, d_2 from (7.14) we show a lower bound for d_3. From the inequality (7.9) (by dropping the second summand) we see that

$$d_3 > \frac{A_0^2}{\ln[(\exp C_1)/x]}.$$

Using the lower estimate (7.14) for x we find that

$$d_3 > \frac{\gamma-1}{2} \frac{A_0^2}{\ln[(e/(e-1))((\ln D_0)/D_0)]}. \tag{7.15}$$

Now from (7.14) and (7.15) a lower estimate for the semiaxes of the ellipsoid easily follows:

$$d_2 > d_1 = \frac{d_1}{d_3} d_3 \geqslant x d_3$$

$$> e^{C_1} \left(\frac{e-1}{e} \frac{D_0}{|\ln D_0|}\right)^{2/(\gamma-1)} \frac{A_0^2}{2/(\gamma-1)|\ln[((e-1)/e)(D_0/|\ln D_0|)]|}. \tag{7.16}$$

The derived inequalities (7.11) and (7.16) provide (for $\gamma < 4/3$) both estimates for the semiaxes of the ellipsoid d_i for a motion with negative energy H and $|J|^2 + |K|^2 \neq 0$ and prove the impossibility of collapse of the ellipsoid in the presence of rotation of the gas. As shown in Sect. 6, the motion of a gaseous ellipsoid for $H \to -\infty$ occurs in an oscillatory mode in the state of strong compression. However as follows from the above, the volume of the ellipsoid (in the presence of rotation of the gas) remains bounded below.

8. Oscillatory Mode of Motion with Positive Energy

I. Consider the behavior of the dynamical system on the manifold S for $H \geqslant 0$, $\gamma > 4/3$. The critical points of the system (5.6) in the coordinates W_1 fall into the sets $K_1 = K_+ \cup K_-$, Φ_+, Φ_-, L, whose properties are the same as for $H \leqslant 0$ (see Sect. 6). We shall list the critical points which are not covered by the coordinates W_1 (i.e. which belong to the invariant manifold Γ_w ($w = 0$) with the coordinates W_2) and show the eigenvalues of the system (5.8) at these critical points.

1) $M_{1\varepsilon}$: $u = 1$, $w = 0$, $p_i = \varepsilon y_i$, $\varepsilon = \pm 1$. The critical points M_{1-} are repelling and the critical points in M_{1+} are attracting.

2) $M_{0\varepsilon}$: $u = 0$, $w = 0$, $p_i = \varepsilon y_i$, $\varepsilon = \pm 1$. These are non-degenerate and unstable critical points.

3) N_0: $u = 0$, $w = 0$, $V(y_i) = 0$. At the critical points in N_0 there are three non-zero eigenvalues:

$$\lambda_1 = -(\gamma - 1) \sum_{k=1}^{n} \frac{\partial V}{\partial y_k} p_k \quad \text{(variable } w\text{)},$$

$$\lambda_2 = (\gamma - 1) \sum_{k=1}^{n} \frac{\partial V}{\partial y_k} p_k \quad \text{(variable } u\text{)}, \tag{8.1}$$

$$\lambda_3 = \sum_{k=1}^{n} \frac{\partial V}{\partial y_k} p_k \quad \text{(variables } y_i\text{)}.$$

This implies that the critical points in N_0 are non-degenerate and unstable when

$$\sum_{k=1}^{n} \frac{\partial V}{\partial y_k} p_k \neq 0$$

4) N_1: $u = 1$, $w = 0$, $V(y_i) = 0$. These critical points are a boundary (for $w \to 0$) of the set of critical points K_1 (see Sect. 6). Just like the points of K_1 these points have two non-zero eigenvalues with opposite signs:

$$\lambda_1 = (1 - \gamma) \sum_{k=1}^{n} \frac{\partial V}{\partial y_k} p_k \quad \text{(variable } u\text{)},$$

$$\lambda_2 = \sum_{k=1}^{n} \frac{\partial V}{\partial y_k} p_k \quad \text{(variables } y_i\text{)}, \tag{8.2}$$

It is convenient to divide each of the sets N_1 and N_2 into two parts:

$$N_{0-}, N_{1-} \quad \left(\text{where } \sum_{k=1}^{n} \frac{\partial V}{\partial y_k} p_k \leq 0\right) \quad \text{and} \quad N_{0+}, N_{1+} \quad \left(\text{where } \sum_{k=1}^{n} \frac{\partial V}{\partial y_k} p_k \geq 0\right).$$

It is easy to see that the separatrices of unstable critical points of N_{0-} and N_{0+} lie in the component of the boundary Γ_w ($w = 0$) and in the corner of the boundary $\Gamma_0 \cap \Gamma_2$ ($u = 0$, $V(y_i) = 0$). The separatrices of the unstable critical points of N_{1-} and N_{1+} lie in the corners of the boundary $\Gamma_w \cap \Gamma_1$ ($w = 0$, $u = 1$) and $\Gamma_w \cap \Gamma_2$ ($w = 0$, $V(y_i) = 0$).

5) L_1: $w = 0$, $V(y) = 0$, $\sum_{k=1}^{n}(\partial V/\partial y_k)p_k = 0$, the coordinate u is arbitrary. The boundary of this set of degenerate critical points consists of the intersections $N_{0-} \cap N_{0+}$ ($u = 0$) and $N_{1-} \cap N_{1+}$ ($u = 1$).

II. Turning to the construction of a separatrix diagram for $H \geqslant 0$ we notice that the separatrices of unstable critical points of $N_{0\varepsilon}$, $N_{1\varepsilon}$, K_ε lie in the following invariant submanifolds: Γ_1, Γ_2, Γ_w, $\Gamma_0 \cap \Gamma_2$.

As noticed earlier, the system (5.6)–(5.8) on the manifold Γ_1 describes the motion of a gravitating dust ellipsoid. This type of motion was studied in [163–165, 174]. The system (5.6) on the manifold Γ_2 $(V(y) = 0)$ for $u \neq 0$ is integrated explicitly (see (6.3)). The system (5.8) on the manifolds Γ_w and $\Gamma_0 \cap \Gamma_2$ can also be integrated explicitly.

1) The trajectories of the system (5.8) on the manifold Γ_w $(w = 0)$ have the following form:

$$p_i = p_i^0, \quad w = 0, \quad y_i = y_i^0 \frac{\cosh \tau_0}{\cosh \tau} + p_i^0 \frac{\sinh \tau - \sinh \tau_0}{\cosh \tau},$$

$$u = \frac{C(\cosh \tau)^{3\gamma-4} U(y_i)(V(y_i))^{\gamma-1}}{1 + C(\cosh \tau)^{3\gamma-4} U(y_i)(V(y_i))^{\gamma-1}}. \tag{8.3}$$

where p_i^0, y_i^0, τ_0 and C are constants such that $\tanh \tau_0 = \sum_{k=1}^{n} p_k^0 y_k^0$. The time τ is related to τ_2 by the formula $d\tau = V(y_i) d\tau_2$. In the coordinates y_i for $\tau > \tau_0$ the trajectory (8.3) moves along an arc of a great circle on the unit sphere S^{n-1} passing through the points y_i^0 and p_i^0. The endpoints of the trajectory (8.3) belong to the sets of critical points M or N, i.e. each trajectory is a separatrix of some critical point. The corner of the boundary $\Gamma_w \cap \Gamma_2$ $(w = 0, V(y) = 0)$ contains trajectories, along which only the coordinate u varies (from 0 to 1).

2) The trajectories of the system (5.8) on the manifold $\Gamma_0 \cap \Gamma_2$ $(u = 0, V(y) = 0)$ have the following form:

$$y_i = y_i^0, \quad p_i = p_i^0 \frac{\cosh \tau_0}{\cosh \tau} + s_i^0 \frac{\sinh \tau - \sinh \tau_0}{\cosh \tau},$$

$$w = \frac{\cosh^2 \tau_0 - \cosh^2 \tau}{2 \cosh^2 \tau}. \tag{8.4}$$

where y_i^0, p_i^0 and $s_i^0 = \operatorname{grad} V(y_i^0)/|\operatorname{grad} V(y_i^0)|$ are constants such that

$$\sum_{k=1}^{n} p_k^0 s_k^0 = \tanh \tau_0 < 0, \quad \sum_{k=1}^{n} p_k s_k^0 = \tanh \tau. \tag{8.5}$$

The time τ in (8.4) is determined by the expression $d\tau = d\tau_2 w(\gamma - 1)|\operatorname{grad} V(y_i^0)|$. The trajectory (8.4) has its initial point $(p_i^0, y_i^0, w = 0, u = 0)$ for $\tau = \tau_0$ in N_{0-}. By virtue of (8.5) the final point of this trajectory $(p_i(\tau), \tau = -\tau_0, y_i^0, w = 0, u = 0)$ is contained in N_{0+}. The final point $p_i(-\tau_0)$ is obtained from the initial point (p_i^0) by way of reflection in the plane tangent to the surface $V(y_i) = 0$ at the point (y_i^0). The maximal value of w along the trajectory (8.4) is attained when $\tau = 0$ and is equal to

$$w_* = \frac{\cosh^2 \tau_0 - 1}{2} = \frac{\left(\sum_{k=1}^{n} p_k^0 s_k^0\right)^2}{2\left(1 - \left(\sum_{k=1}^{n} p_k^0 s_k^0\right)^2\right)}. \tag{8.6}$$

Table 7. Separatrix diagram of the dynamical system (5.8) for $H > 0, \gamma > 4/3$

	8 M_{1-}	8 M_{1+}	16 K_-	16 K_+	15 N_{1-}	15 N_{1+}	15 N_{0-}	15 N_{0+}	8 M_{0-}	8 M_{0+}	S_1
M_{1-}											
M_{1+}	$18a_2^1$			$17a_2^4$		$16a_2^6$		$17a_2^8$		$9a_2^{10}$	a_2^{11}
K_-	$17a_3^1$			$17a_3^4$							
K_+			$17a_4^3$								
N_{1-}	$16a_5^1$					$16a_5^6$					
N_{1+}								$16a_6^8$			
N_{0-}	$17a_7^1$				$16a_7^5$			$17a_7^8$	$16a_7^9$		
N_{0+}							$16a_8^7$				
M_{0-}	$9a_9^1$										
M_{0+}								$16a_{10}^8$	$17a_{10}^9$		
S_1	a_{11}^1										

The results of integration of the separatrices are assembled in the separatrix diagram (Table 7), where we use the following notation:

a) In a square with an entry the symbol a_i^k denotes a separatrix going from the set of critical points in the top row to the set of critical point in the leftmost column. The number before the symbol a_i^k is the total dimension of this separatrix. An empty square represents the absence of a separatrix. The numbers above the letters in the top row are the dimensions of the sets of critical points (recall that the dimension of the manifold S is 18).

b) The letter S_1 denotes invariant submanifolds of the physical region of S, onto which (generally speaking) some separatrices of the critical points of $M_{1\varepsilon}$ can wind.

The critical points Φ_ε, L and L_1 are not included in the separatrix diagram, because their separatrices have measure zero in the space of all separatrices and for almost all other critical points an unlimited iteration of mappings determined by the diagram (Table 7) does not lead outside the frame of this diagram.

For the case $H \geqslant 0, \gamma < 4/3$ the separatrix diagram differs from Table 7 by a certain change in the separatrices of the critical points of $M_{0\varepsilon}$, $M_{1\varepsilon}$ and the inclusion of unstable states of equilibrium P contained in the physical region S_1. The oscillatory mode of motion of the ellipsoid described below pertains equally well to the case $\gamma \leqslant 4/3, H \geqslant 0$.

III. According to the separatrix diagram (Table 7) there are the following sequences of separatrices on the boundary Γ:

$$M_{1-} \xrightarrow{a_7^1} N_{0-} \xrightarrow{a_8^7} N_{0+} \xrightarrow{a_7^8} \cdots \xrightarrow{a_8^7} N_{0+} \xrightarrow{a_2^8} M_{1+}. \tag{8.7}$$

Let us show that there are sequences in (8.7) in which the number of steps between the sets N_{0-} and N_{0+} is arbitrarily large. Let (y_i^0) be a point on the surface $V(y_i) = 0$. Let n^i be the normal and l the tangent plane to the surface $V(y_i) = 0$ at the point (y_i^0). Let l' be a two-dimensional plane passing through the normal n^i. Choose the direction of the normal in such a way that the curve carved out in the plane l' by the surface $V(y_i) = 0$ is convex (towards the normal n^i). Let (p_i^0) be a point contained in the plane l' to the negative side of the normal at a small distance h from the plane l. Then the point $P_0 = (p_i^0, y_i^0, w = 0, u = 0)$ belongs to N_{0-}. The choice of the point P_0 made here assumes the presence of rotation of the ellipsoid. According to the definition of the mapping a_8^7 (see (8.4)) the point $P_1 = a_8^7(P_0)$ has coordinates $(p_i^1, y_i^0, w = 0, u = 0)$, where the point (p_i^1) is obtained from (p_i^0) by way of reflection in the plane l. Obviously the point P_1 belongs to N_{0+}. By virtue of definition of the mapping a_7^8 (see (8.3)) the point $P_2 = a_7^8(P_1)$ has coordinates $(p_i^1, y_i^1, w = 0, u = 0)$, where the point (y_i^1) is the point of intersection of the shortest arc of a great circle on S^{n-1} passing through the points (p_i^1) and (y_i^0) with the surface $V(y_i) = 0$. Obviously for small h the point $P_2 = a_7^8 \circ a_8^7(P_0)$ is arbitrarily close to the initial point P_0 (an analogous situation was elucidated in Fig. 39 (see Sect. 2)). Consequently as $h \to 0$ any finite number of iterations of the mapping $a_7^8 \circ a_8^7$ leads to nearby points and therefore the corresponding sequence of separatrices (8.7) contains an arbitrary finite number of steps between the sets N_{0-} and N_{0+}.

In general the steps between the sets N_{0-} and N_{0+} are terminated by the entry of the point into the set $M_{1\varepsilon}$, where $p_i = \varepsilon y_i$. The reason for this is that under the mapping $a_7^8 \circ a_8^7$ the distance between the points $(p_i^{(k)})$ and $(y_i^{(k)})$ decreases. Note that along with the cyclic steps between the sets N_{0-} and N_{0+} the separatrix diagram (Table 7) also contains cyclic steps between the sets K_-, K_+ and N_{0-}, N_{0+}, N_{1+}, N_{1-}. However as $w \to 0$ the corresponding sequences of separatrices tend to the sequences in (8.7).

Consider a trajectory of the system (5.8) emerging from a critical point in M_{1-} near the component of the boundary $\Gamma_w (w = 0)$. Such trajectories always remain near the sequence of separatrices (8.7) and also end up in the set of attracting critical points M_{1+}. The presence of a trajectory of the system (5.8) in the vicinity of the critical points of M_{1-} and M_{1+} corresponds respectively to the inertial compression of the gaseous ellipsoid from an infinitely rarefied state and to the inertial infinite expansion of the ellipsoid. The presence of a trajectory in the vicinity of critical points of N_{0-} and N_{0+} (where $\det \| Y_{jk} \| = V(y_i) = 0 \|$ means that the ellipsoid is compressed to a disc along an eigenvector of the matrix Y_{jk}. Therefore the steps of a trajectory between the sets N_{0-} and N_{0+} correspond to some oscillatory mode of motion of the ellipsoid. Using the definitions of H and w

it is not difficult to see that

$$\det \| F_{jk} \| = V(q_i) = \left(\alpha \frac{1 + 2(1 - 2u)w}{2H(1 - u)w} \right)^{1/(\gamma - 1)}. \tag{8.8}$$

From the equation (6.5) it follows that during the motion of a trajectory of the system (5.8) along the separatrix a_8^7 the volume of the ellipsoid $\det \| F_{jk} \|$ reaches its minimum. By virtue of (8.4) and (8.8) this minimum is equal to

$$\min \det \| F_{jk} \| = \left(\alpha \frac{1 + 2w_*}{2Hw_*} \right)^{1/(\gamma - 1)}, \tag{8.9}$$

where w_* is defined by (8.6). During the motion of a trajectory along the separatrix a_7^8 $\det \| F_{jk} \|$ reaches its maximum, which according to (8.3) and (8.8) can be arbitrarily large. Note that for an arbitrary value of the energy $H > 0$ there exist trajectories of the system (5.8), which are sufficiently near the sequence of separatrices (8.7). Hence it follows that for large H there are motions of gas, for which $\det \| F_{jk} \|$ oscillates from an arbitrarily small minimum (see (8.9)) to an arbitrarily large maximum such that the oscillations of gas density

$$\rho = \frac{3M}{4\pi \det \| F_{jk} \|}$$

are arbitrarily large. The period of oscillations depends on the way a trajectory approaches the sequence of separatrices (8.7).

Thus we have just shown (using the separatrix approximation (8.7)) that in the presence of rotation and for $w \to 0$ (the kinetic energy is much greater than the potential energy) there exist the following motions of a gravitating gaseous ellipsoid:

1) initially the gas contracts from the state of infinite rarefaction;

2) then the oscillatory mode sets in: the gas contracts and expands an arbitrarily large number of times over varying directions in such a way that for large energies H the amplitude of oscillations of density is arbitrarily large;

3) the oscillatory mode ends and is replaced by the infinite expansion of gas.

Note that the motion of a point in the manifold S along the sequence of separatrices (8.7) in the coordinates y_i is a free motion of the point along the geodesics on the eight-dimensional sphere S^8 in the region $\det \| Y_{jk} \| \geqslant 0$ such that this point is reflected from the boundary $\det \| Y_{jk} \| = 0$ according to the law of elastic reflection. In this sense the mathematical model of the oscillatory mode of motion of a gas in the presence of rotation is like the game of geodesic billiards on the sphere S^8 with an elastically reflecting surface $\det \| Y_{jk} \| = 0$. In the absence of rotation of the gas (the matrix Y_{jk} is diagonal) the surface $\det \| Y_{jk} \| = 0$ degenerates into three coordinate planes. Therefore only three oscillations are realized. These are the three consecutive contractions and expansions of the ellipsoid along the orthogonal axes.

The oscillatory mode described above can serve as a model for the motion of a cloud of expanding and rotating gas formed as a result of an explosion of a rotating supernova.

9. Concluding Remarks

In this section we shall give a short description of the oscillatory modes of motion of a gravitating gaseous ellipsoid that have been found earlier in this chapter. As shown in Sect. 4, the analysis of motion of a gravitating gaseous ellipsoid is equivalent to the analysis of the dynamics of the following Lagrangian system defined in the space of three-dimensional matrices:

$$\frac{d^2 F_k^i(t)}{dt^2} = -\alpha \frac{\partial V^{1-\gamma}(F)}{\partial F_k^i} + \frac{3}{2} GM \frac{\partial U(F)}{\partial F_k^i},$$

$$V(F) = \det F, \quad U(F) = \int_0^\infty \left((d_1^2 + s)(d_2^2 + s)(d_3^2 + s) \right)^{-1/2} ds. \tag{9.1}$$

If the semiaxes of the ellipsoid d_1, d_2 and d_3 are comparable to each other $(d_i \sim d)$, then

$$\frac{3}{2} GMU(F) \sim \frac{3}{2} GMd^{-1}, \quad \frac{3}{2} GM \frac{\partial U(F)}{\partial F_k^i} \sim d^{-2},$$

$$\alpha V^{1-\gamma} \sim \alpha d^{3(1-\gamma)}, \quad \alpha \frac{\partial V^{(1-\gamma)}}{\partial F_k^i} \sim \alpha d^{3(1-\gamma)-1}. \tag{9.2}$$

These simple estimates will be used below.

I. Oscillatory Mode of Expansion into Vacuum of a Rotating Gas Cloud. Consider the motion of a strongly expanded gaseous ellipsoid with total energy $E > 0$ and $\gamma > 1$, where

$$E = T + \alpha V^{1-\gamma}(F) - \frac{3}{2} GMU(F),$$

$$T = \frac{1}{2} \sum_{i,k}^3 (\dot{F}_k^i)^2. \tag{9.3}$$

Here T is the kinetic energy of gas. Let

$$d_1, d_2, d_3 \sim d \gg \frac{GM}{T} + \left(\frac{\alpha}{T} \right)^{1/3(\gamma-1)}. \tag{9.4}$$

Then the change in velocity $\Delta |dF_k^i/dt|$ during the motion in a segment of a trajectory of length $\sim d$ admits (after substituting (9.4) into (9.2) and (9.1)) the following estimate:

$$\varDelta \left| \frac{dF_k^i}{dt} \right| \sim \left| \frac{d^2 F_k^i}{dt^2} \right| \frac{d}{\sqrt{T}} \ll \sqrt{T} \sim \left| \frac{dF_k^i}{dt} \right|.$$

Therefore under the conditions (9.4) the coefficients $F_k^i(t)$ in first approximation vary over the straight lines

$$F_k^i(t) = A_k^i t + B_k^i. \tag{9.5}$$

For an appropriate choice of the constants A_k^i and B_k^i the straight line (9.5) for some $t = t_0$ intersects the surface L: $V(F) = 0$. In other words as $t \to t_0$ the ellipsoid is compressed to a disc along some direction. However the pressure hindering contraction grows without bound and the components of velocity of the gas and the gravitational forces remain finite. Therefore compression is replaced by expansion. Such a change is represented by the elastic reflection of the straight line (9.5) from the surface L at the point of intersection for $t = t_0$. Then the coefficients $F_k^i(t)$ again vary over a straight line of the type (9.5) with new constants $(A_k^i)^1$ and $(B_k^i)^1$. This straight line can again intersect the surface L, which means a new compression of the ellipsoid etc. Thus the change of coefficients $F_k^i(t)$ on the entire time axis in the first approximation occurs along broken lines, which are elastically reflected from the surface L at the points of intersection.

Using the fact that L is a strongly curved surface we can direct the initial straight line (9.5) in such a way that the broken line constructed on it has an arbitrarily large (but finite) number of intersections with this surface.

During the change of $F_k^i(t)$ along any segment of the broken line the volume of the ellipsoid $(V(F))$ reaches a maximum and then decreases, i.e. the gas is in an oscillatory mode. The amplitude of oscillations of the volume of the ellipsoid and the gas density

$$\rho(t) = \frac{3M}{4\pi \det(F_k^i)}$$

can be arbitrarily large. Obviously the above reasoning is also applicable for $G = 0$, so the oscillatory mode also occurs in the absence of gravitational interaction among the gas particles.

The oscillatory mode terminates when a segment of the broken line does not intersect the surface L even with unbounded continuation. In this case we have an unbounded free expansion of the gas.

In order to have an oscillatory mode the presence of rotation of the gas is necessary because in the absence of rotation (the matrix $\| F_{ik}(t) \|$ is diagonal) the surface L degenerates into three coordinate planes and the broken line has only three reflections, which correspond to three consecutive contractions and expansions of the ellipsoid along the orthogonal axes.

II. Oscillatory Mode of Motion with Negative Energy. Consider the motion of a gravitating gaseous ellipsoid for $E < 0$, $\gamma < 4/3$ in the state of strong compression. Suppose that $d_i \sim d \ll (GM/\alpha)^{1/(4-3\gamma)}$. Then for $\gamma < 4/3$ by virtue of (9.2) we see that

$$\left| \frac{3}{2} GM \frac{\partial U(F)}{\partial F_k^i} \right| \gg \alpha \left| \frac{\partial V^{1-\gamma}(F)}{\partial F_k^i} \right|.$$

Therefore the motion of the ellipsoid is determined by the gravitational forces and is approximated by the motion of a gravitating dust ellipsoid. According to [164], in general the dust ellipsoid is compressed to a disc under the action of gravitational forces (i.e. $d_1 \to 0$, $d_2 \to C_2 > 0$, $d_3 \to C_3 > 0$ and $V(F) = d_1 d_2 d_3 \to 0$). However in the presence of pressure an unbounded compression of an ellipsoid to a disc is impossible, because the components of velocity and the forces of gravity remain finite during such a compression, but the pressure opposing contraction grows without bound. Therefore the compression to a disc is replaced by expansion (such a change is represented by the elastic reflection of the velocity vector dF_k^i/dt from the surface L) as a result of which the semiaxes d_1, d_2 and d_3 again become comparable to each other and the motion is again determined by the forces of gravity and is approximated by the motion of a dust ellipsoid. This leads to a new compression of the ellipsoid to a disc (possibly along another direction), which after an elastic reflection of the velocity vector from the surface L will again be replaced by expansion and so on to infinity.

In order for the described approximation of the oscillatory mode to be valid a strong compression of the ellipsoid is not required. It is sufficient that $\beta = 8\alpha/(3GM) \ll 1$ (this inequality is satisfied when the initial temperature of the gas is low). The above approximation of the oscillatory mode is precise for a strong compression of the ellipsoid $(E \to -\infty)$ or for $\beta \to 0$.

For these values of the parameters the only solutions of the system (9.1) without oscillations are solutions with a spherically symmetric mode of compression [156]. However such a mode of compression is unstable. Therefore if in some interval of time the motion of the ellipsoid is close to the spherically symmetric motion, then during subsequent motion the ellipsoid departs from this mode and oscillations start again. The oscillations of volume of the ellipsoid are accompanied by the oscillations of temperature and other physical parameters of the gas.

Chapter VIII
The Dynamics of Perturbations of the Periodic Toda Lattice

1. Hamiltonian Perturbations of the Toda Lattice

In 1970 M. Toda in the course of numerical analysis of various models of interaction of atoms in a crystalline lattice discovered the absence of stochastization in a system of unit mass particles on a straight line, whose interaction is determined by the following potential:

$$V = \sum_i \exp(q_i - q_{i+1}),$$

where q_i is the deviation of the i-th particle from its state of equilibrium [175]. In further research of this problem several first integrals were found [176]. With the help of a choice of an appropriate $L - A$ pair[1] the total integrability of the Toda lattice was proved in [177, 178]. In the periodic case $q_i \equiv q_{i+n+1}$ the Toda lattice has the following Hamiltonian:

$$H = \frac{1}{2} \sum_{i=1}^{n+1} p_i^2 + \sum_{i=1}^{n} \exp(q_i - q_{i+1}) + \exp(q_{n+1} - q_1). \tag{1.1}$$

The periodic Toda lattice was also studied in [179] with the help of algebraic-geometric methods.

Obviously in a real physical situation it is more plausible to have some general perturbation of the Hamiltonian (1.1). This happens (for example) due to the inclusion of pairwise interaction among all particles and not just among the nearest neighbors and due to the differences in mass of the interacting particles. In view of this it seems important to analyze the most general mode of the dynamics of perturbations of the periodic Toda lattice and determine the character of symmetry, which distinguishes the completely integrable Toda lattice from its general non-integrable perturbations.

In [180] (independently from the work of the author [20]) numerical methods were used to analyze the dynamics of a system of two particles with arbitrary masses and the Toda lattice potential. The results of [180] show the presence of stochastization of the trajectories for $H \gg 1$, which agrees with the results of [20] obtained for a wider class of perturbations of the Toda lattice with an arbitrary number of particles.

---

[1] *Translator's note.* This is usually called a "Lax pair".

In this chapter the above questions are resolved for arbitrary perturbations of the Toda lattice in the class of Hamiltonian systems of the type

$$\dot{p}_i = -\frac{\partial H}{\partial q_i}, \quad \dot{q}_i = \frac{\partial H}{\partial p_i},$$

$$H = \frac{1}{2}\sum_{i,j}^{n} a_{ij}p_ip_j + \sum_{k,l}^{n+1} b_{kl}\exp(\{\alpha_k, q\} + \{\alpha_l, q\}). \tag{1.2}$$

where $\alpha_1, \ldots, \alpha_{n+1}$ are vectors in the n-dimensional space $\mathbb{R}^n$ with coordinates $\alpha_k = (d_{k1}, \ldots, d_{kn})$ and $q = (q_1, \ldots, q_n)$. In $\mathbb{R}^n$ there are two scalar products defined as follows:

$$(x, y) = \sum_{i,j}^{n} a_{ij}x_iy_j, \quad \{x, y\} = \sum_{i=1}^{n} x_iy_i. \tag{1.3}$$

The vectors α_k and quadratic forms a_{ij}, b_{kl} satisfy the following conditions A and B:

A. For each vector p in $\mathbb{R}^n$ we have $\max_k(\alpha_k, p) > 0$.

B. For all k we have $(\alpha_k, \alpha_k)b_{kk} > 0$.

The Hamiltonian (1.1) takes the form (1.2) after changing to the "center of mass" system (i.e. $p_1 + p_2 + \cdots + p_{n+1} = 0$). Note that Hamiltonian systems of type (1.2) also occur in the theory of homogeneous cosmological models. For example the homogeneous model of type IX in vacuum (at the level $H = 0$) is described by a Hamiltonian system with the Hamiltonian

$$H = 2\sum_{i<j}^{3} p_ip_j - \sum_{i=1}^{3} p_i^2 + 2\sum_{i<j}^{3} \exp(q_i + q_j) - \sum_{i=1}^{3} \exp(2q_i). \tag{1.4}$$

As a consequence of the obvious resemblance of the Hamiltonians (1.2) and (1.4), the oscillatory modes in these systems display common properties and admit a unique conclusion shown below (in a more general case) in Sect. 4.

2. Separatrix Approximation of the Oscillatory Mode

In order to study the Hamiltonian system (1.2) using the methods of qualitative theory of differential equations, we change over to the coordinates

$$r_k = \frac{Q_k}{G}, \quad s_i = \frac{p_i}{P}, \quad w = \frac{G^2}{P^2}, \tag{2.1}$$

where

$$Q_k = \exp(\{\alpha_k, q\}), \quad k = 1, \ldots, n+1, \quad Q_k > 0,$$

$$P = (p_1^2 + \cdots + p_n^2)^{1/2}, \quad G = (Q_1^2 + \cdots + Q_{n+1}^2)^{1/2},$$

and make a time change $d\tau_1 = P\,dt$. In the coordinates (2.1) and time τ_1 the system

(1.2) takes the following form:

$$\dot{r}_k = r_k\left(\sum_{i,j} d_{ki}a_{ij}s_j - \sum_{i,j,l} r_l^2 d_{li}a_{ij}s_j\right),$$

$$\dot{s}_i = w\left(-\sum_{k,l} b_{kl}(d_{ki} + d_{li})r_k r_l + s_i\sum_{k,l,m} s_l b_{km}(d_{kl} + d_{ml})r_k r_m\right), \qquad (2.2)$$

$$\dot{w} = 2w\left(\sum_{i,j,k} r_k^2(d_{ki}a_{ij}s_j) + w\sum_{i,k,l} s_i b_{kl}(d_{ki} + d_{li})r_k r_l\right).$$

This sytem is defined on the $2n$-dimensional invariant manifold

$$r_1^2 + \cdots + r_{n+1}^2 = 1, \quad s_1^2 + \cdots + s_n^2 = 1, \quad w \geqslant 0, \quad r_k \geqslant 0.$$

Obviously the system (2.2) can be continued smoothly to the boundary Γ of the manifold V, where $w = 0$, $r_k = 0$. The components of the boundary Γ are invariant submanifolds of the system (2.2).

Later we will need the following transformations of the system (1.2) into the coordinates r_i, p_i, w:

$$\dot{r}_k = r_k\left(\sum_{i,j} d_{ki}a_{ij}p_j - \sum_{i,j,l} r_l^2 d_{li}a_{ij}p_j\right),$$

$$\dot{p}_i = -\sum_{k,l} b_{kl}(d_{ki} + d_{li})r_k r_l P^2 w, \qquad (2.3)$$

$$\dot{w} = 2w\left(\sum_{i,j,k} r_k^2(d_{ki}a_{ij}p_j) + w\sum_{i,k,l} p_i b_{kl}(d_{ki} + d_{li})r_k r_l\right).$$

The system (1.2) is equivalent to the system (2.3) considered on the invariant $2n$-dimensional manifold

$$(wP^2)^{(c_1 + \cdots + c_{n+1})/2} r_1^{c_1} \cdots r_{n+1}^{c_{n+1}} = 1, \quad r_1^2 + \cdots + r_{n+1}^2 = 1,$$

where the choice of numbers c_i is determined by the conditions

$$c_1\alpha_1 + \cdots + c_{n+1}\alpha_{n+1} = 0, \quad c_1 + c_2 + \cdots + c_{n+1} = c,$$

where c can be either zero or one.

The system (2.2) has $n + 1$ sets $M_i (i = 1, \ldots, m)$ of critical points. Each set M_i is an $(n-1)$-dimensional sphere and has coordinates

$$r_k = \delta_{ki}, \quad w = 0, \quad s_1^2 + \cdots + s_n^2 = 1.$$

A point in the set M_k is denoted by a pair $\{s, k\}$.

The eigenvalues of the system (2.2) on the manifold V at a critical point $\{s, k\}$ are (the corresponding eigenvectors are shown in parentheses):

$$\lambda_l = (\alpha_l, s) - (\alpha_k, s) \quad \text{(variables } r_l; l = 1, \ldots, n + 1, l \neq k),$$

$$\lambda_n = 2(\alpha_k, s) \qquad\qquad \text{(variable } w), \qquad (2.4)$$

$$\lambda_{n+1} = \cdots = \lambda_{2n} = 0 \quad \text{(variables } s_i).$$

From the condition A (sec. 1) and (2.1) it follows that all critical points in M_k, except for the submanifolds of smaller dimension, are non-degenerate and unstable (a critical point is called non-degenerate if the number of its zero eigenvalues is equal to the dimension of the set M_k).

In each set M_k we distinguish two subsets V_k and W_k:

$$V_k(\alpha_k, s) < 0,$$

$$W_k(\alpha_k, s) = \max_l (\alpha_l, s) > 0.$$

The separatrices emerging (for $\tau_1 \to -\infty$) from a point $\{s^0, j\}$ contained in the set V_j have the following form:

$$r_m(\tau_1) = C_m \exp((\alpha_m, s^0)\tau_1)\left(\sum_l C_l^2 \exp(2(\alpha_l, s^0)\tau_1)\right)^{-1/2},$$
$$(2.5)$$

$$w \equiv 0, \quad s_i \equiv s_i^0, \quad C_m \geqslant 0.$$

where m and l vary over those numbers i for which $\lambda_i > 0$ (see (2.1)). If however $\lambda_i < 0$, then $r_i \equiv 0$. Let $(\alpha_k, s^0) = \max_l(\alpha_l, s^0)$. Obviously all separatrices (2.2) for which $C_k > 0$ approach the critical point $\{s^0, k\}$ in the set W_k as $\tau_1 \to +\infty$.

According to (2.4), from the critical point $\{s^0, k\}$ contained in the set W_k emerges a unique separatrix $\mathscr{L}$ lying in the invariant manifold $r_i = \delta_{ik}$. In order to integrate this separatrix it is convenient to refer to the system (2.3), because all trajectories of the system (2.2) (in the manifold $s_1^2 + \cdots + s_n^2 = 1$) are obtained from the trajectories of the system (2.3) under the mapping

$$s_i = p_i/P. \tag{2.6}$$

After a time change $d\tau/dt = P^2 w$ the trajectories of the system (2.3) in the manifold $r_i = \delta_{ik}$ are easily integrated:

$$p_i = s_i^0 - \tau b_{kk} d_{ki}, \quad w(\tau) = \tau \frac{(\alpha_k, s^0) - \dfrac{\tau}{2} b_{kk}(\alpha_k \alpha_k)}{\{s^0 - \tau b_{kk}\alpha_k, s^0 - \tau b_{kk}\alpha_k\}} \tag{2.7}$$

(note that instead of integrating the equation (2.3) for w it is convenient to use the integral H (1.2)).

For $\tau = \tau_* = 2(\alpha_k, s^0)/b_{kk}(\alpha_k, \alpha_k)$ ($\tau_* > 0$ according to the definition of s^0 and the condition B in Sect. 1) the trajectory (2.7) approaches the point

$$p^1 = \tau_k(s^0) = s^0 - \frac{2(\alpha_k, s^0)}{(\alpha_k, \alpha_k)}\alpha_k, \quad w(\tau_*) = 0. \tag{2.8}$$

Obviously the mapping τ_k is a reflection in a plane orthogonal (in the metric a_{ij}) to the vector α_k and $(p^1, \alpha_k) = -(s^0, \alpha_k) < 0$. The separatrix $\mathscr{L}$ obtained from the trajectory (2.7) under the mapping (2.6) for $\tau = \tau_1$ approaches a critical point contained in the set V_k (because $(p^1, \alpha_k) < 0$).

The above integration of the separatrices (2.5) and (2.7) shows that the separatrices emerging from the sets of critical points V_k and W_i again approach these same sets. Therefore there exist infinite sequences formed by separatrices going between these critical sets:

$$\cdots \to V_j \overset{\text{I}}{\to} W_k \overset{\text{II}}{\to} V_k \overset{\text{I}}{\to} W_i \overset{\text{II}}{\to} \cdots \tag{2.9}$$

A trajectory of the system (2.2) starting in a sufficiently small neighborhood of one of the critical sets V_j, W_k will move for an arbitrarily long time along the sequence of separatrices (2.9). The corresponding trajectory in the initial coordinates q_i moves in the following way: during the step I (2.5) the motion is with approximately constant direction of momentum p_i. This motion ends, when some $Q_k \gg Q_1$ for all l. Then during the step II (2.7) there is an effective "reflection" of momentum described by the mapping (2.8). Then again we have motion with constant direction of momentum and so on.

In the case of a positive definite metric a_{ij} this oscillatory mode is realized (with asymptotic precision) for $H \gg 1$. Indeed from the condition A (Sect. 1) it follows that at least one $Q_i > 1$. Therefore

$$w = \frac{G^2}{P^2} > \frac{1}{P^2}.$$

Furthermore the energy integral H in the coordinates (2.1) has the following form:

$$H = \frac{1}{2} P^2 \left(\sum_{i,j}^{n} a_{ij} s_i s_j + w_i \sum_{k,l}^{n+1} b_{kl} r_k r_l \right). \tag{2.10}$$

Since during the separatrix steps I we have $w \ll 1$, for a positive definite metric a_{ij} (without a loss of generality it can be considered Euclidean) we have $H \gg 1$. Note that from (2.10) it also follows that during the transitions I not only the direction but also the length of the momentum vector p_i does not change to the order of magnitude. In the case of an indefinite metric a_{ij} (as example (1.4) shows) the oscillatory mode is realized even for $H = 0$.

According to (2.8) the approximately constant vectors giving the direction of the momentum s_N under the successive transitions along the separatrices I (2.5) are obtained from one another by the action of the mapping T:

$$s_{N+1} = T(s_N) = \frac{\tau_k(s_N)}{\{\tau_k(s_N), \tau_k(s_N)\}}, \tag{2.11}$$

where k is determined by the condition

$$(\alpha_k, s_N) = \max_l (\alpha_l, s_N), \tag{2.12}$$

and the mapping τ_k is a reflection in a plane orthogonal to the vector α_k and is defined in (2.8).

The mapping T for the Toda lattice (1.1) is periodic. Here the vectors α_i are roots of a simple Lie algebra of the type $A_n(SL(n+1))$. For the Hamiltonian system (1.2) the mapping T is periodic if the Coxeter group G generated by the reflections (2.8) is finite (if d is the order of the group G, then $T^{d!} \cdot T^d = T^d$, but the mapping T can be not invertible). All finite Coxeter groups G are known [91] and except for three exceptional cases are Weyl groups of simple Lie algebras (under the condition that G cannot be written as a product of two other groups). The exceptions are two Coxeter groups in three-dimensional and four-dimensional spaces and an infinite series of dihedral groups (groups of symmetries of right polygons) in two-dimensional space.

The property of periodicity of the mapping T reflects a deep algebraic symmetry of the Hamiltonian (1.1) and distinguishes the periodic Toda lattice from its general perturbations (1.2). In the general case (1.2) the Coxeter group is infinite and its closure (for a positive definite metric a_{ij}) coincides with the orthogonal group $O(n)$ and for a general vector p the set $T^k(p)$ is everywhere dense on the sphere $(p, p) = \text{const}$. The general Hamiltonian system (1.2) apparently has no first integrals apart from the energy H.

3. Hamiltonian Systems Connected with Simple Lie Algebras

In this section we exhibit new examples of Hamiltonian systems of the type (1.2), for which, just like for the Toda lattice (1.1), the mapping T is periodic and which admit a representation in the form of an $L - A$ pair (and therefore have a large set of first integrals). The construction of these systems uses the theory of simple Lie algebras $\mathfrak{G}$.

Let us summarize the prerequisite facts about the Cartan-Weyl basis e_{α_i}, h_k in $\mathfrak{G}$ [91]. In the Cartan subalgebra H (H is the maximal commutative subalgebra of $\mathfrak{G}$) there is a set of vectors $\alpha_1, \dots, \alpha_s$ called roots and some basis $h_1, \dots, h_n$ ($n = \dim H$ is the rank of the algebra $\mathfrak{G}$) is chosen. The vectors e_{α_i}, h_k form a basis for the algebra $\mathfrak{G}$ and satisfy the following commutation relations:

$$[e_{\alpha_i}, e_{\alpha_j}] = N_{\alpha_i \alpha_j} e_{\alpha_i + \alpha_j}, \quad [e_{\alpha_i}, e_{-\alpha_i}] = \alpha_i,$$
$$[h_k, e_{\alpha_i}] = (h_k, \alpha_i) e_{\alpha_i}, \quad [h_k, h_j] = 0. \tag{3.1}$$

The scalar product (x, y) is determined by the Killing-Cartan form

$$(x, y) = \text{Tr}(\text{ad } x \circ \text{ad } y), \quad \text{ad } x(z) = [x, z].$$

A set of roots $\alpha_1, \dots, \alpha_N$ is called admissible if for all $i, j \leqslant N$ the vector $\alpha_i - \alpha_j$ is not a root. In this case

$$[e_{\alpha_i}, e_{-\alpha_j}] = 0.$$

In each simple Lie algebra $\mathfrak{G}$ there is one important admissible set of roots

$$\omega_1, \omega_2, \dots, \omega_n, -\Omega, \tag{3.2}$$

where ω_i are simple roots (all roots α_i are linear combinations of ω_k over the integers) and $\Omega = k_1 \omega_1 + \cdots + k_n \omega_n$ is the so-called maximal root $(\Omega + l_1 \omega_1 + \cdots + l_n \omega_n$ is not a root for all $l_i \geqslant 0)$. All subsets of this set of roots are also admissible.

Theorem 1. *Suppose that $\mathfrak{G}$ is a simple Lie algebra of rank n and $h_1, \ldots, h_n$ is a basis for its Cartan subalgebra. Suppose that $\alpha_1, \ldots, \alpha_N$ is an admissible set of roots, $\alpha_i = d_{i1} h_1 + \cdots + d_{in} h_n$, b_j are arbitrary real constants and the scalar product (x, y) in $\mathfrak{G}$ is determined by the Killing-Cartan form. Then the Hamiltonian system*

$$\dot{p}_i = -\frac{\partial H}{\partial \dot{q}}, \quad \dot{q}_i = \frac{\partial H}{\partial p_i},$$

$$H = \frac{1}{2} \sum_{k,l}^{n} (h_k, h_l) p_k p_l + \sum_{j=1}^{N} b_j \exp\left(2 \sum_{k=1}^{n} d_{jk} q_k\right) \tag{3.3}$$

in the 2n-dimensional phase space p_i, q_i admits a representation in the form of an $L - A$ pair.

Proof. Under the mapping

$$l_j = b_j^{1/2} \exp\left(\sum_{k=1}^{n} d_{jk} q_k\right), \quad j = 1, \ldots, N, \tag{3.4}$$

all solutions of the system (3.3) become solutions of the system

$$\dot{p}_k = -2 \sum_{j=1}^{N} l_j^2(t) d_{jk},$$

$$\dot{l}_j = l_j \sum_{k,l}^{n} d_{jl} p_k(h_k, h_l). \tag{3.5}$$

In the Lie algebra $\mathfrak{G}$ consider the following equation [181]:

$$\dot{i} = [l, A(l)], \tag{3.6}$$

where the vectors $l(t)$ and $A(l(t))$ have the following form:

$$l(t) = \sum_{j=1}^{N} l_j(t)(e_{\alpha_j} + e_{-\alpha_j}) + \sum_{k=1}^{n} p_k h_k,$$

$$A(l(t)) = \sum_{j=1}^{N} l_j(t)(e_{\alpha_j} - e_{-\alpha_j}). \tag{3.7}$$

Using the commutation relations (3.1) and the definition of an addmisible set of roots $\alpha_1, \ldots, \alpha_N$ it is easy to verify that the system (3.5) is equivalent to the equation (3.6) (under the conditions (3.7)). For any linear representation T of the algebra $\mathfrak{G}$ the equation (3.6) determines an $L - A$ pair:

$$T(l)^{\cdot} = [T(l), T(A(l))]. \tag{3.8}$$

Thus under the mappings (3.4) and (3.7) all solutions of the system (3.3) become solutions of the equation (3.6) and therefore solutions of the equation (3.8). Let T be an exact representation of the algebra $\mathfrak{G}$ with minimal dimension. Then for $l(t)$ given by (3.7) and (3.4) the equation (3.8) is equivalent to the system (3.3), which proves the theorem.

From the representation in the form of an $L - A$ pair, it follows that the system (3.3) has first integrals

$$I_k = \text{Tr}\big(T^k(l(t))\big).$$

The equation (3.6) always has an integral (l, l) (I_2), which under the condition (3.7) has the following form:

$$(l, l) = \sum_{k,l}^{n}(h_k, h_l)p_k p_l + 2\sum_{j=1}^{N} l_j l_j = 2H.$$

Here are some examples of systems (3.3). We will use the classification of simple Lie algebras and the standard form of the roots of the algebra $\mathfrak{G}$ in an orthonormal basis $e_1, \ldots, e_n$ (see [91]) (for algebras of types A_n, E_6, E_7 and G_2 it is convenient to expand the Cartan subalgebra by adding an element, which commutes with the entire algebra (in this expansion the basis is $e_1, \ldots, e_n, e_{n+1}$)). Let $h_i = e_i$ and let the admissible set of roots be the set (3.2). The corresponding Hamiltonian (3.3) has the following form:

$$H = \frac{1}{2}\sum_{i=1}^{m} p_i^2 + V_{\mathfrak{G}}(q_i). \tag{3.9}$$

where $m = n + 1$ for the algebras of types A_n, E_6, E_7, G_2 and $m = n$ for the other types (n is the rank of the algebra). Another class of completely integrable Hamiltonian systems related to simple Lie algebras is given in [182], where the Moser-Calogero construction of the $L - A$ pair is used [183, 184]. Let

$$V_k = \sum_{i=1}^{k} \exp(q_i - q_{i+1}).$$

Depending on the type of $\mathfrak{G}$ the explicit form of the potentials $V_{\mathfrak{G}}(q_i)$ is

$$V_{A_n} = V_n + \exp(q_{n+1} - q_1), n \geqslant 2,$$

$$V_{B_n} = V_{n-1} + \exp(q_n) + \exp(-q_1 + q_2)), n \geqslant 2,$$

$$V_{C_n} = V_{n-1} + \exp(2q_n) + \exp(-2q_1), n \geqslant 3,$$

$$V_{D_n} = V_{n-1} + \exp(q_{n-1} + q_n) + \exp(-q_1 - q_2), n \geqslant 4,$$

$$V_{E_6} = V_5 + \exp\left(\frac{1}{2}(-q_1 - q_2 - q_3 + q_4 + q_5 + q_6)\right.$$

$$\left. + \frac{1}{\sqrt{2}}q_7\right) + \exp(-\sqrt{2}q_7),$$

$$V_{E_7} = V_5 + \exp\left(\frac{1}{2}(-q_1 + q_2 + \cdots + q_7 - q_8)\right)$$
$$+ \exp(-q_1 - q_2) + \exp(-q_7 + q_8),$$

$$V_{E_8} = V_6 + \exp\left(\frac{1}{2}(-q_1 + q_2 + \cdots + q_7 - q_8)\right)$$
$$+ \exp(-q_1 - q_2) + \exp(q_7 + q_8),$$

$$V_{F_4} = \exp(q_1 - q_2) + \exp(q_2 - q_3) + \exp(q_3)$$
$$+ \exp\left(\frac{1}{2}(-q_1 - q_2 - q_3 + q_4)\right) + \exp(-q_1 - q_4),$$

$$V_{G_2} = \exp(q_1 - q_2) + \exp(-2q_1 + q_2 + q_3)$$
$$+ \exp(q_1 + q_2 - 2q_3). \tag{3.10}$$

Using standard linear representations of simple Lie algebras it can be shown that Hamiltonian systems with the Hamiltonians (3.9) have exactly m integrals. In some cases it can be shown that they are involutive. In recent works [185, 186] it was shown that the Hamiltonian systems (3.9) for all simple Lie algebras are completely integrable. Furthermore in [185] there is an explicit integration of the systems (3.9), for which the maximal root is dropped from the potential (3.10) and therefore the particles fly away to infinity. Most recently the connection between the integrability of Hamiltonian systems (3.9) and the theory of representations of semi-simple Lie algebras was studied in [187].

For a Lie algebra of the type $A_n(SL(n + 1))$ the Hamiltonian (3.9) determines a periodic Toda lattice. For other types we obtain new lattices of particles with a large number of integrals of motion (note however that the system (3.9) for the type C_n can be included in a Toda lattice with $2n$ particles). In all these systems stochastization is impossible.

For a system of two particles we obtain (from (3.10)) in addition to the Toda lattice two other integrable Hamiltonian systems with potentials
$$V_{B_2} = \exp(q_1 - q_2) + \exp(q_2) + \exp(-q_1 - q_2),$$
$$V_{G_2} = \exp(q_1) + \exp(\sqrt{3}q_2) + \exp\left(-\frac{3}{2}q_1 - \frac{\sqrt{3}}{2}q_2\right).$$

Note that both these systems differ from the system (considered in [179]) with the potential
$$V_T = \exp(q_1 - q_2) + \exp(q_2) + \exp(-q_1)$$
describing a Toda lattice with one fixed particle.

Using Theorem 1 we can exhibit many other examples of Hamiltonian systems admitting a representation in the form of an $L - A$ pair. For example such systems can be obtained from (3.9) by dropping several summands in the potential $V_{\mathfrak{G}}$.

4. Non-Linear Oscillatory Modes in Systems of Hydrodynamical Type

The concept of a system of hydrodynamical type was introduced in [188]. Such systems appear in finite-dimensional approximation of the equations of fluid dynamics using the Galerkin method. They have the following form:

$$\dot{u}^i = \Gamma^i_{jk} u^j u^k \qquad (4.1)$$

for a constant Γ^i_{jk}. According to the definition in [188], a system of hydrodynamical type has a quadratic (in u^i) energy integral. The flux determined by the system (4.1) preserves the phase volume (div $\dot{u}^i = 0$). Hamiltonian systems (1.2) for $\alpha_1 + \cdots + \alpha_{n+1} = 0$ under the mapping $Q_k = \exp(\{\alpha_k, q\})$ become systems of hydrodynamical type.

Let us distinguish a general class of dynamical systems containing the systems (1.2), (1.4) and some systems of hydrodynamical type (4.1) for which we will prove the existence of complex non-linear oscillatory modes admitting a separatrix approximation similar to the one described in Sect. 2.

Consider a system of differential equations of the following type (the variables are divided into two groups: $Q^\alpha \in R^m$, $p^i \in R^n$):

$$
\begin{aligned}
\dot{Q}^\alpha &= a^\alpha_{\beta\gamma} Q^\beta Q^\gamma + Q^\alpha (b^\alpha_i p^i + \lambda^\alpha), \\
\dot{p}^i &= c^i_{\beta\gamma} Q^\beta Q^\gamma + d^i_{\beta j} Q^\beta p^j + \lambda^i_\beta Q^\beta
\end{aligned}
\qquad (4.2)
$$

where there is implicit summation over indices repeating in one part of the equations.

Proposition 1. *The following conditions A, B and C are sufficient for the existence of a non-linear oscillatory mode in the system* (4.2), *which admits a separatrix approximation:*

A. *For almost all $p^i \in \mathbb{R}^n$*

$$\max_\alpha (b^\alpha_i p^i + \lambda^\alpha) = b^\gamma_i p^i + \lambda^\gamma > 0, \quad \gamma = \gamma(p). \qquad (4.3)$$

For each α the vector $(b^\alpha_i) \neq 0$.

B. *For each α the $(n + 1) \times (n + 1)$ matrices*

$$
S_\alpha = \begin{vmatrix} a^\alpha_{\alpha\alpha} & b^\alpha_j \\ c^i_{\alpha\alpha} & d^i_{\alpha j} \end{vmatrix}
$$

have a diagonal Jordan canonical form and all their eigenvalues are purely imaginary (Re $\lambda_i = 0$).

C. *For $\beta \neq \alpha a^\alpha_{\beta\beta} = 0$.*

In order to analyze the oscillatory mode we change over to the coordinates

$$q^\alpha = Q^\alpha/\Omega, \quad \Omega = ((Q^1)^2 + \cdots + (Q^m)^2)^{1/2}. \qquad (4.4)$$

In the coordinates q^α, p^i, Ω the system (4.2) has the following form:

$$\dot{q}^\alpha = a^\alpha_{\beta\gamma}q^\beta q^\gamma \Omega + q^\alpha(b^\alpha_i p^i + \lambda^\alpha)$$
$$- q^\alpha(a^\delta_{\beta\gamma}q^\delta q^\beta q^\gamma \Omega + (q^\delta)^2(b^\delta_i p^i + \lambda^\delta)),$$
$$\dot{\Omega} = \Omega(a^\delta_{\beta\gamma}q^\delta q^\beta q^\gamma \Omega + (q^\delta)^2(b^\delta_i p^i + \lambda^\delta)),$$
$$\dot{p}^i = \Omega(c^i_{\beta\gamma}q^\beta q^\gamma \Omega + d^i_{\beta j}q^\beta p^j + \lambda^i_\beta q^\beta).$$

$$(4.5)$$

By virtue of the change (4.4) the system (4.5) is considered on the invariant manifold $(q^1)^2 + \cdots + (q^m)^2 = 1$, $\Omega > 0$ and is continuously extended to the boundary $\Omega = 0$, which is also an invariant manifold. This system has $2m$ sets of critical points $M^\varepsilon_\alpha = \mathbb{R}^n$ ($\alpha = 1, \ldots, m$; $\varepsilon = \pm 1$) with coordinates

$$q^\beta = \varepsilon\delta^\beta_\alpha, \quad \Omega = 0, \quad p^i \in R^n.$$

The points contained in M^ε_α are denoted by triples (p, α, ε).

The eigenvalues of the system (4.5) at the critical points (p, α, ε) are (the corresponding eigenvectors are shown in parentheses)

$$\mu_\beta = (b^\beta_i p^i + \lambda^\beta) - (b^\alpha_i p^i + \lambda^\alpha) \quad \text{(variables } q^\beta; \beta = 1, \ldots, m, \beta \neq \alpha),$$
$$\mu_m = b^\alpha_i p^i + \lambda^\alpha \quad \text{(variable } \Omega),$$
$$\mu_{m+1} = \cdots = \mu_{m+n} = 0 \quad \text{(variables } p^i).$$

From the condition A it follows that almost all critical points in M^ε_α are non-degenerate (i.e. the number of zero eigenvalues is equal to the dimension of M^ε_α) and unstable.

Let us distinguish two subsets V^ε_α and W^ε_α in each M^ε_α: $V^\varepsilon_\alpha(b^\alpha_i p^i + \lambda^\alpha) < 0$ and $W^\varepsilon_\alpha \alpha = \gamma(p)$ (see (4.3)). Separatrices emerging (for $t \to -\infty$) from a point $(p_0, \alpha, \varepsilon) \in V^\varepsilon_\alpha$ have the following form:

$$q^\beta = C^\beta \exp((b^\beta_i p^i_0 + \lambda^\beta)t)\left(\sum_\delta (C^\delta)^2 \exp(2t(b^\delta_i p^i_0 + \lambda^\delta))\right)^{-1/2},$$
$$\Omega(t) \equiv 0, \quad p^i(t) = p^i_0, \quad C^\alpha = \varepsilon, \quad -\infty < t < +\infty$$

$$(4.6)$$

where β varies over those indices, for which $\mu_\beta > 0$. For other indices $\beta q^\beta \equiv 0$. Let $\gamma = \gamma(p_0)$ (see (4.3)). If $C^\gamma > 0$, then all separatrices (4.6) for $t \to +\infty$ approach a critical point $(p_0, \gamma(p_0), +1) \in W^{+1}_\gamma$. If $C^\gamma < 0$, then they approach a critical point $(p_0, \gamma(p_0), -1) \in W^{-1}_\gamma$. The exceptional separatrices (for which $C^\gamma = 0$) are unstable. Below in the definition of the mapping T_1 they will not be considered.

Separatrices emerging from the critical points $(p_0, \alpha, \varepsilon) \equiv W^\varepsilon_\alpha$ lie in the invariant (by virtue of the condition C) manifold $q^\beta = \varepsilon\delta^\beta_\alpha$. After a time change $d\tau/dt = \Omega > 0$ on this manifold the system (4.5) is transformed into a linear system

$$\dot{X} = \varepsilon S_\alpha \cdot X + \varepsilon \cdot Y,$$

$$(4.7)$$

where the vectors X and Y have coordinates $X = (\varepsilon\Omega, p^i)$, $Y = (\lambda^\alpha, \lambda^i_\alpha)$. By virtue of the condition B each trajectory of the system (4.7) beginning at the plane $\Omega = 0$

for $b_i^\alpha p^i + \lambda^\alpha > 0$ (a separatrix emerging from the critical points of W_α^ε) again intersects this plane (for $b_i^\alpha p^i + \lambda^\alpha < 0$). Let T_α^ε denote the sequential function thus defined.

Thus we have defined a mapping T_1 on the sets of critical points V_α^ε and W_α^ε, which takes an initial critical point to the final point of the separatrix emerging from it. The mapping T_1 has the following form:

$$(p, \alpha, \varepsilon) \in W_\alpha^\varepsilon: \quad T_1(p, \alpha, \varepsilon) = (T_\alpha^\varepsilon(p), \alpha, \varepsilon) \in V_\alpha^\varepsilon,$$

$$(p, \alpha, \varepsilon) \in V_\alpha^\varepsilon: \quad T_1(p, \alpha, \varepsilon) = \begin{cases} (p, \gamma(p), +1) \in W_\gamma^{+1}, \\ (p, \gamma(p), -1) \in W_\gamma^{-1}. \end{cases} \qquad (4.8)$$

In the latter case the mapping T_1 is double-valued.

A trajectory of the system (4.5) which begins in a sufficiently small neighborhood of the critical sets V_α^ε and W_α^ε, will move for an arbitrarily long time along some sequence of separatrices in the following diagram:

$$
\begin{array}{c}
\nearrow \quad W_\alpha^{+1} \to V_\alpha^{+1} \quad \nearrow \cdots \\
\cdots \to V_\alpha^\varepsilon \qquad\qquad \searrow \cdots \\
\searrow \quad W_\alpha^{-1} \to V_\alpha^{-1} \quad \nearrow \cdots \\
\qquad\qquad\qquad\qquad\qquad \searrow \cdots
\end{array}
\qquad (4.9)
$$

(an arrow denotes a separatrix going between critical sets). In other words, the oscillatory mode in the system (4.2)–(4.5) admits a separatrix approximation. Below we shall exhibit a class of systems (4.2)–(4.5), for which the Q^α, p^i space is divided into invariant regions, in each of which all transitions in the diagram (4.9) are single-valued. In the initial coordinates Q^α, p^i the oscillatory mode has the following dynamics. A trajectory beginning for sufficiently small values of the coordinates Q^α periodically appears in the vicinity of points in the plane P ($Q^1 = \cdots = Q^m = 0$) obtained one from another by the successive application of a mapping T (double-valued) defined as follows. The plane P is divided into subsets P_α such that in P_α we have $\gamma(p) = \alpha$ (see (4.3)). On each P_α the mapping $T = T_\alpha^\varepsilon$ (see (4.8)), i.e. $T(p) = T_{\gamma(p)}^\varepsilon$, $\varepsilon = \pm 1$. During the transition of a trajectory between two consecutive points of the plane P the coordinate $Q^{\gamma(p)}$ varies strongly and sign $Q^{\gamma(p)} = \varepsilon$, whereas for the other coordinates $|Q^\beta| \ll |Q^{\gamma(p)}|$. The stochastic properties of the oscillatory mode are determined by the properties of the mapping T.

The mapping T (and the separatrix diagram (4.9)) is reduced to single-valued mappings in an important special case of the systems (4.2), when the coefficients

$$a_{\beta\gamma}^\alpha = 0 \quad \text{for } \beta, \gamma \neq \alpha. \qquad (4.10)$$

In this case all planes $Q^\alpha = 0$ are invariant manifolds of the system and therefore each of the 2^m regions σ, where sign $Q^\alpha = \varepsilon_\alpha$ is also invariant. If a trajectory in the region σ moves along some separatrix (4.6), then we must have sign $C^\gamma = \varepsilon_\gamma$ (see (4.6)). Therefore all transitions in the diagram (4.9) for the trajectories in the region σ are single-valued. The corresponding mapping $T = T_\sigma$ on the set P_α has the form $T_\sigma = T_\sigma^{\varepsilon_*}$ and is also single-valued. All mappings T_σ and the corresponding oscillatory modes are (generally speaking) different.

If a system of the type (4.2) (when the condition (4.10) is satisfied) is considered only in some one region σ (sign $Q^\alpha = \varepsilon_\alpha$), then the sufficient conditions for the existence of an oscillatory mode can be weakened. Instead of the condition B it is sufficient to demand that Re $\lambda_i \geqslant 0$, where λ_i are the eigenvalues of matrices $\varepsilon_\alpha \cdot S_\alpha$.

Examples of systems satisfying the conditions of Proposition 1 are the Hamiltonian systems (1.2) and (1.4). Under the mapping $Q_k = \exp\{\alpha_k, q\}$ the systems (1.2) take the form (4.2) and satisfy the conditions A, B, C (4.10) (by virtue of the conditions A and B in Sect. 1). The above oscillatory mode for these systems becomes the oscillatory mode found in Sect. 2. Under the mappings $Q_i = \exp q_i$ the Hamiltonian system (1.4) takes the form (4.2) and satisfies the condition A at the critical points $H = 0, Q_i = 0$ and the conditions B and C (4.10). In this case the oscillatory mode is isomorphic to the oscillatory mode of relativistic cosmology (see Sect. 6 of Chapt. II).

The systems of hydrodynamical type (4.1), which admit a representation (4.2) (with $\lambda^\alpha = \lambda_\beta^i = 0$) and have an energy integral of the form $E = (u^1)^2 + \cdots + (u^n)^2$, automatically satisfy the conditions A (by virtue of the preservation of volume ($\operatorname{div} \dot{u}^i = 0$)) and B (by virtue of the existence of the energy integral). In general the satisfaction of the condition C must be demanded additionally.

The simplest example of a system of hydrodynamical type are Euler's equations of motion of a rigid body, i.e. a triple [189]:

$$\dot{v}_1 = l(v_2^2 - v_3^2), \quad \dot{v}_2 = -lv_1v_2, \quad \dot{v}_3 = lv_1v_3,$$

obviously satisfying the conditions A, B, C (4.10). Here the role of the coordinates p^i is played by the unstable mode v_1. The coordinates $Q^\alpha - v_2$ and v_3 are linear combinations of stable modes.

There is a special class of systems of hydrodynamical type which automatically satisfy the conditions A, B, C. Such systems can be obtained by way of superposition [189] of triples according to the following rule: if two triples are connected in some mode, then in both triples this mode should be either stable or unstable. Furthermore double connections of triples are possible and each triple can be connected with any number of others. The systems of hydrodynamical type thus obtained have an additional integral $F = Q^1 \cdot \ldots \cdot Q^m$ so that in the oscillatory mode the dimensionless integral $F \cdot E^{-m/2} \ll 1$.

For such systems the matrices S_α have rank 2 and each reflection $T_\alpha^{+1} = T_\alpha^{-1}$ is a reflection τ_α in a plane orthogonal to the vector (b_i^α). Thus, just as in the perturbations of the Toda lattice, the mapping T corresponds to a Coxeter group G generated by the reflections τ_α. As is well known, for a general set of vectors (b_i^α) the closure of the group G coincides with the orthogonal group $O(n)$ (for $m > n$). Apparently in the general case the mapping T is ergodic on spheres of constant radius (the invariant measure for the invertible mapping T is the usual Euclidean measure). However the proof of this fact is difficult already in the two-dimensional case, where the problem is reduced to permuting segments in a circle.

Bibliography

1. Poincaré, H.: Les Methodes Nouvelles de la Mechanique Celeste. Paris 1899
2. Bendixon, I.: Acta Math., v. 24, p. 1 (1901)
3. Andronov, A.A.: Collection of Works. Akad. Nauk SSSR Press, Moscow 1956
4. Andronov, A.A., Vitt, A.A., Khaikin, C.E.: Theory of Oscillations. 2nd ed. Fizmatgiz, Moscow 1959
5. Andronov, A.A., Leontovich, E.A., Gordon, I.I., Mayer, A.G.: Bifurcation Theory of Dynamical Systems in a Plane. Nauka, Moscow 1967 (trans: Israel Program of Scientific Translations, Jerusalem 1971)
6. Andronov, A.A., Leontovich, E.A., Gordon, I.I., Mayer, A.G.: Theory of Bifurcations of Dynamical Systems in a Plane. Nauka, Moscow 1967
7. Sedov, L.I.: Similarity Methods and Dimensional Analysis in Mechanics. 3rd ed. Nauka, Moscow 1954; 8th ed. Nauka, Moscow 1977
8. Birkhoff, G.D.: Dynamical Systems. AMS Colloquim Publications IX, New York 1927
9. Bruno, A.D.: Local Method of Non-linear Analysis of Differential Equations. Nauka, Moscow 1979; In: Trudy Mosk. Mat. Ob. (Proc. Mosc. Math. Soc.) MGU Press, v. 25, p. 119 (1975); v. 26, p. 199 (1976)
10. Anosov, D.V.: Geodesic Flows on Closed Riemannian Manifolds with Negative Curvature. In: Trudy Mat. Inst. im. V.A. Steklova (Proc. Steklov Math. Inst.), Nauka, Moscow 1967
11. Nitecki, Z.: Differential Dynamics; an introduction to the orbit structure of diffeomorphisms. MIT Press 1971
12. Bogoyavlensky, O.I., Novikov, S.P.: Zhurn. Eksp. i Teor. Fiz. (J. Exp. & Theor. Phys.), v. 64, No. 5, p. 1475 (1973)
13. Bogoyavlensky, O.I.: Usp. Mat. Nauk (Succ. Math. Sc.), v. 28, No. 5, p. 1973 (1973)
14. Bogoyavlensky, O.I.: Pis. v Astron. Zhurn. (Lett. J. Astron.), v. 1, No. 9, p. 22 (1975)
15. Bogoyavlensky, O.I., Novikov S.P.: Trudy Sem. im. I.G. Petrovskogo (Proc. Petrovsky Sem.), MGU Press, Moscow, v. 1, p. 7 (1975)
16. Bogoyavlensky, O.I.: Trudy Sem. im. I.G. Petrovskogo (Proc. Petrovsky Sem.), MGU Press, Moscow, v. 2, p. 67 (1976)
17. Bogoyavlensky, O.I.: Prikl. Mat. i Mekh. (Appl. Math. & Mech.), v. 40, No. 2, p. 270 (1976)
18. Bogoyavlensky, O.I.: Zhurn. Eksp. i Teor. Fiz. (J. Exp. & Theor. Phys.), v. 70, No. 2, p. 361 (1976)
19. Bogoyavlensky, O.I.: Teor. Mat. Fiz. (Theor. Math. Phys.), v. 27, No. 2, p. 184 (1976)
20. Bogoyavlensky, O.I.: Comm. Math. Phys., v. 51, No. 3, p. 201 (1976)
21. Bogoyavlensky, O.I., Novikov, S.P.: Usp. Mat. Nauk (Succ. Math. Sc.), v. 31, No. 5, p. 33 (1976)
22. Bogoyavlensky, O.I.: Dokl. Akad. Nauk SSR (Rep. Acad. Sc. USSR), v. 232, No. 6, p. 33 (1977)
23. Bogoyavlensky, O.I.: Pis. v Zhurn. Eksp. i Teor. Fiz. (Lett. J. Exp. & Theor. Phys.), v. 26, No. 2, p. 63 (1977)
24. Bogoyavlensky, O.I.: Phys. Lett., v. 60A, No. 3, p. 163 (1977)
25. Bogoyavlensky, O.I.: Zhurn. Eksp. i Teor. Fiz. (J. Exp. & Theor. Phys.), v. 73, No. 4, p. 1201 (1977)
26. Bogoyavlensky, O.I.: Pis. v Zhurn. Eksp. i Teor. Fiz. (Lett. J. Exp. & Theor. Phys.), v. 27, No. 2, p. 91 (1978)
27. Bogoyavlensky, O.I.: Astrofizika (Astrophysics), v. 14, No. 3, p. 501 (1978)
28. Bogoyavlensky, O.I.: Pis. v Zhurn. Astron. (Lett. J. Astron.), v. 4, No. 9, p. 397 (1978)

29. Bogoyavlensky, O.I.: J. Geoph. Astroph. Fl. Dyn., v. 12, No. 1/2, p. 117 (1979)
30. Bogoyavlensky, O.I.: Pis. v Zhurn. Eksp. i Teor. Fiz. (Lett. J. Exp. & Theor. Phys.), v. 29, No. 10, p. 622 (1979)
31. Bogoyavlensky, O.I., In: Proc. Intern. Congr. of Math., Helsinki, v. 1, p. 395 (1980)
32. Bogoyavlensky, O.I.: Trudy Sem. im. I.G. Petrovskogo (Proc. Petrovsky Sem.), MGU Press, Moscow, v. 6 (1980)
33. Pontryagin, L.S.: Ordinary Differential Equations. 4th ed. Nauka, Moscow 1975
34. Bautin, N.N., Leontovich, E.A.: Methods and Techniques of Qualitative Analysis of Dynamical Systems in a Plane. Nauka, Moscow 1976
35. Petrovsky, I.G.: Lectures on the Theory of Ordinary Differential Equations. 6th rev. ed. Nauka, Moscow 1970
36. Coddington, E.A., Levinson, N.: Theory of Ordinary Differential Equations. McGraw-Hill, New York 1955
37. Bogolyubov, N.N.: Selected Works. Naukova Dumka, Kiev, v. I (1969)
38. Bogolyubov, N.N., Mitropolsky, Yu.A.: Asymptotic Methods in the Theory of Non-linear Oscillations. Fizmatgiz, Moscow 1974
39. Dorodnitsyn, A.A.: Prikl. Mat. i Mekh. (App. Math. & Mech.), v. 11, No. 3, p. 313 (1947)
40. Tikhonov, A.N.: Mat. Sborn. (Math. Coll.), v. 22, No. 2, p. 193 (1948)
41. Mischenko, E.F., Rozov, N.Kh.: Differential Equations with a Small Parameter and Relaxatory Oscillations. Nauka, Moscow 1975
42. Hartman, P.: Ordinary Differential Equations. John Wiley and Sons, New York 1964; 2nd ed., Birkhäuser, Boston 1982
43. McGehee, R.: Invent. Math., v. 27, p. 191 (1974)
44. Mather, J., McGehee, R., In: Lecture Notes in Physics. ed. J. Moser. New York, v. 38, p. 673 (1975)
45. Friedmann, A.A.: Z. Phys., v. 10, p. 377 (1922)
46. Friedmann, A.A.: Z. Phys., v. 21, p. 326 (1924)
47. Taub, A.H.: Ann. Math., v. 53, p. 472 (1951)
48. Lifshitz, E.M., Khalatnikov, I.M.: Usp. Fiz. Nauk (Succ. Phys. Sc.), v. 80, p. 391 (1963)
49. Belinsky, V.A., Khalatnikov, I.M.: Zhurn. Eksp. i Teor. Fiz. (J. Exp. & Theor. Phys.), v. 56, p. 1700 (1969)
50. Belinsky, V.A., Lifshitz, E.M., Khalatnikov, I.M.: Usp. Fiz. Nauk. (Succ. Phys. Sc.), v. 102, p. 463 (1970)
51. Belinsky, V.A., Lifshitz, E.M., Khalatnikov, I.M.: Zhurn. Eksp. i Teor. Fiz. (J. Exp. & Theor. Phys.), v. 60, p. 1969 (1971)
52. Belinsky, V.A., Khalatnikov, I.M.: Zhurn. Eksp. i Teor. Fiz. (J. Exp. & Theor. Phys.), v. 63, p. 1121 (1972)
53. Penrose, R.: Phys. Rev. Lett., v. 14, p. 57 (1965)
54. Hawking, S.W.: Phys. Rev. Lett., v. 15, p. 689 (1965)
55. Geroch, R.P.: Phys. Rev. Lett., v. 17, p. 445 (1966)
56. Penrose, R.: Structure of Space-Time., In: Battelle Rencontres, 1967; Lec. in Math. and Phys., ed: C.M. DeWitt and J.A. Wheeler, W.A. Benjamin Co. 1968
57. Arnowitt, R., Deser, S., Misner, C.W.: Phys. Rev., v. 118, p. 1100 (1960)
58. Misner, C.W.: Phys. Rev. Lett., v. 22, p. 1071 (1969)
59. Hechman, O., Schücking, E., In: Gravitation: An Introduction to Current Research (ed: L. Witten) Wiley, New York, p. 438 (1962)
60. Newman, E., Tamburino, L. Unti, T.: J. Math. Phys., v. 4, p. 915 (1963)
61. Grischuk, L.P., Doroshkevich, A.G., Novikov, I.D.: Zhurn. Eksp. i Teor. Fiz. (J. Exp. & Theor. Phys.), v. 55, p. 2281 (1968)
62. Ellis, G.F.R., MacCallum, M.A.H.: Comm. Math. Phys., v. 12, p. 108 (1969); Comm. Math. Phys., v. 19, p. 31 (1970)
63. Matzner, R.A., Shepley, L.C., Warren, I.C.: Ann. Phys., v. 57, p. 401 (1970)
64. Faddeev, L.D.: Teor. Mat. Fiz. (Theor. Math. Phys.), v. 1, p. 3 (1969)

65. Faddeev, L.D., Popov, V.N.: Usp. Fiz. Nauk (Succ. Phys. Sc.), v. 111, p. 427 (1973)
66. MacCallum, M.A.H.: Comm. Math. Phys., v. 20, p. 57 (1971)
67. Doroshkevich, A.G., Lukash, V.N., Novikov, I.D.: Zhurn. Eksp. i Teor. Fiz. (J. Exp & Theor. Phys.), v. 60, p. 1201 (1971)
68. Grischuk, L.P., Doroshkevich, A.G., Lukash, V.N.: Zhurn. Eksp. i Teor. Fiz. (J. Exp. & Theor. Phys.), v. 61, p. 3 (1971)
69. Novikov, S.P.: Zhurn. Eksp. i Teor. Fiz. (J. Exp. & Theor. Phys.), v. 62, p. 1977 (1972)
70. Doroshkevich, A.G., Lukash, V.N., Novikov, I.D.: Zhurn. Eksp. i Teor. Fiz. (J. Exp. & Theor. Phys.), v. 64, p. 1475 (1973)
71. Lukash, V.N., Starobinsky, A.A.: Zhurn. Eksp. i Teor. Fiz. (J. Exp. & Theor. Phys.), v. 66, p. 1515 (1974)
72. Peebles, P.J.E.: Physical Cosmology. Princeton University Press, Princeton, N.J. 1971
73. Zeldovich, Ya.B., Novikov, I.D.: Theory of Gravity and Evolution of Stars. Nauka, Moscow 1971
74. Zeldovich, Ya. B., Novikov, I.D.: Structure and Evolution of the Universe. Nauka, Moscow 1975
75. Ryan, M.P., Shepley, L.C.: Homogeneous Relativistic Cosmologies. Princeton University Press, N.J. 1975
76. Weinberg, S.: Gravitation and Cosmology: Principles and Applications of the General Theory of Relativity. Wiley, New York 1972
77. Shikin, I.S.: Dokl. Akad. Nauk SSSR (Rep. Acad. Sc. USSR), v. 176, p. 1048 (1967)
78. Collins, C.B.: Comm. Math. Phys., v. 23, p. 137 (1971)
79. Shikin, I.S.: Zhurn. Eksp. i Teor. Fiz. (J. Exp. & Theor. Phys.), v. 63, p. 1529 (1972); Zhurn. Eksp. i Teor. Fiz., v. 68, p. 1583 (1975)
80. Collins, C.B.: Comm. Math. Phys., v. 39, No. 2, p. 131 (1974)
81. Belinsky, V.A., Khalatnikov, I.M.: Pis. v Zhurn. Eksp. i Teor. Fiz. (Lett. J. Exp. & Theor. Phys.), v. 21, p. 223 (1975); Zhurn. Eksp. i Teor. Fiz. (J. Exp. & Theor. Phys.), v. 69. p. 401 (1975); Zhurn. Eksp. i Teor. Fiz., v. 72, p. 3 (1977)
82. Grigoryan, S.D.: Izv. Akad. Nauk Arm. SSR (News Acad. Sc. of Armenia), v. 11, No. 5, p. 468 (1976); Matem. Zametki (Math. Notic.), v. 26, No. 2, p. 235 (1979)
83. Peresetsky, A.A.: Usp. Mat. Nauk (Succ. Math. Sc.), v. 31, No. 5, p. 251 (1976): Matem. Zametki (Math. Notic.), v. 21, No. 1, p. 71 (1977)
84. Landau, L.D., Lifshitz, E.M.: Field Theory. Nauka, Moscow 1973
85. Misner, C.W., Thorne, K.S., Wheeler, J.A.: Gravitation. W.H. Freeman and Co., San Francisco 1973
86. Dubrovin, B.A., Novikov, S.P., Fomenko, A.T.: Modern Geometry. Nauka, Moscow 1979
87. Petrov, A.Z.: Einstein Spaces. Fizmatgiz, Moscow 1961
88. Kasner, E.: Am. J. Math., v. 43, p. 217 (1921)
89. Kruskal, M.: Phys. Rev., v. 119, p. 1743 (1960)
90. Bianchi, L.: Mem. Soc. It. della Sc., v. 11, p. 267 (1897)
91. Bourbaki, N.: Groupes et Algèberes de Lie. Hermann, Paris 1968
92. Lukash, V.N.: Pis. v Zhurn. Eksp. i Teor. Fiz. (Lett. J. Exp. & Theor. Phys.), v. 19, p. 499 (1974)
93. Brill, D.R.: Phys. Rev., v. B133, p. 845 (1964)
94. Doroshkevich, A.G.: Astrofizika (Astrophysics), v. 1, p. 255 (1965)
95. Shikin, I.S.: Dokl. Akad. Nauk SSSR (Rep. Acad. Sc. USSR), v. 171, p. 73 (1966)
96. Khalatnikov, I.M.: Pis. v Zhurn. Eksp. i Teor. Fiz. (Lett. J. Exp. & Theor. Phys.), v. 5, p. 595 (1967)
97. Hughston, L.P., Jacobs, K.C.: Astroph. J., v. 160, p. 147 (1970)
98. Collins, C.B.: Comm. Math. Phys., v. 27, p. 37 (1972)
99. Belinsky, V.A., Khalatnikov, I.M., In: Rand. Sem. Mat. Univ. Pol. Torino. Turin, v. 35, p. 159 (1977)
100. Arnold, V.I.: Mathematical Methods in Classical Mechanics. Nauka, Moscow 1974 (trans: Springer-Verlag 1978)

101. Skripkin, V.A.: Zhurn. Prikl. Mekh. Tekhn. Fiz. (J. Appl. Mech. Techn. Phys.), v. 4, p. 3 (1960); Astronom. Zhurn. (Astron. J.), v. 38, p. 192 (1961)
102. Stanyukovich, K.P.: Dokl. Akad. Nauk SSSR (Rep. Acad. Sc. USSR), v. 150, p. 77 (1966); Zhurn. Eksp. i Teor. Fiz. (J. Exp. & Theor. Phys.), v. 66, p. 826 (1974)
103. Gurovich, B.Ts., Stanyukovich, K.P., Sharshekeev, O.Sh.: Dokl. Akad. Nauk SSSR (Rep. Acad. Sc. USSR), v. 165, p. 510 (1965)
104. Cahill, M.E., Taub, A.H.: Comm. Math. Phys., v. 21, p. 1 (1971)
105. Taub, A.H.: Comm. Math. Phys., v. 29, p. 79 (1973)
106. Eardley, D.M.: Comm. Math. Phys., v. 37. p. 287 (1974)
107. Ibragimov, N.Kh.: Group Properties of some Differential Equations. Nauka, Novosibirsk 1967
108. Carr, B.J., Hawking, S.W.: Mon. Not. R. Astr. Soc., v. 168, p. 399 (1974)
109. Sibgatullin, N.R., Dinariev, O.Yu.: Zhurn. Eksp. i Teor. Fiz. (J. Exp. & Theor. Phys.), v. 73, p. 1599 (1977)
110. Landau, L.D., Lifshitz, E.M.: Mechanics of Continuous Media. Gostekhizdat, Moscow-Leningrad, 1954 (trans: Addison-Wesley, Reading, Mass.)
111. Oppenheimer, J.R., Volkoff, G.: Phys. Rev., v. 55, p. 374 (1939)
112. Harrison, B.K., Thorne, K.S., Wakano, M., Wheeler, J.A.: Gravitation Theory and Gravitational Collapse. Univ. of Chicago Press, Chicago 1965
113. Dmitriev, N.A., Kholin, S.A.: Questions of Cosmogony. Akad. Nauk SSSR Press, Moscow, v. 9, p. 254 (1963)
114. Saakyan, G.S., Vartanyan, Yu.L.: Soobsch. Byurakanskoy Obs. (Comm. Byurakan Obs.), v. 33, p. 55 (1963); Astronom. Zhurn. (Astron. J.), v. 41, p. 193 (1964)
115. Zeldovich, Ya. B., Rayzer, Yu.p.: Physics of Shock Waves and High-temperature Hydrodynamic Phenomena. Fizmatgiz, Moscow 1963
116. Carrus, P.A., Fox, A.A., Haas, S., Kopal, Z.: Astrophys. J., v. 113, No. 3, p. 496 (1951)
117. Kopal, Z.: Astrophys. J., v. 120, No. 1, p. 159 (1954)
118. Yavorskaya, I.M.: Dokl. Akad. Nauk SSSR (Rep. Acad. Sc. USSR), v. 111, No. 4, p. 783 (1956)
119. Sakurai, A.: J. Fluid Mech., v. 1, No. 1, p. 436 (1956)
120. Rogers, M.N.: Astroph. J., v. 125, p. No. 2, p. 478 (1957)
121. Korobeinikov, V.P., Melnikova, N.S., Ryazanov, E.V.: Theory of a Point Explosion. Fizmatgiz, Moscow 1961
122. Kurth, R.: Dimensional Analysis and Group Theory in Astrophysics. Pergamon Press, Oxford-New York 1972
123. Dibay, E.A., Kaplan, S.A.: Dimensions and Similarity of Astrophysical Quantities. Nauka, Moscow 1976
124. Carrus, P.A., Fox, A.A., Haas, F., Kopal, Z.: Astrophys. J., v. 113, No. 1, p. 193 (1951)
125. Lidov, M.L.: Astronom. Zhurn. (Astron. J.), v. 34, No. 4, p. 603 (1957)
126. Kazhdan, Ya.M., Lutsky, A.E.: Astrofizika (Astrophysics), v. 13, No. 3, p. 535 (1977)
127. Cheng, A.F.: Astrophys. J., v. 213, No. 2, p. 537 (1977)
128. Taylor, G.: Proc. Roy. Soc. London, v. A201, p. 192 (1950)
129. Anisimov, S.I., Zeldovich, Ya. B.: Pis. v Zhurn. Teor. Fiz. (Lett. J. Theor. Phys.), v. 3, No. 20, p. 1081 (1977)
130. Bautin, N.N.: Prikl. Mat. i Mekh. (Appl. Math. & Mech.), v. 18, No. 1, p. 36 (1954)
131. Chandrasekhar, S.: An Introduction to the Study of Stellar Structure. Dover Publications, New York 1957
132. Rosseland, S.: The Pulsation Theory of Variable Stars. Oxford University Press 1949
133. Sedov, L.I.: Mechanics of a Dense Medium. 3rd rev. ed. Nauka, Moscow, v. 1 & 2 (1976)
134. Zmitrienko, N.V., Imshennik, V.S., Khlopov, M.Yu., Chechetkin, V.M.: Zhurn. Eksp. i Teor. Fiz. (J. Exp. & Theor. Phys.), v. 75, No. 10, p. 1169 (1979)
135. Nadezhin, D.K.: Astronom. Zhurn. (Astron. J.), v. 45, p. 1166 (1968)
136. Guderley, G.: Luftfahrtforschung, v. 19, p. 302 (1942)
137. Stanyukovich, K.P.: Unestablished Motions of a Dense Medium. Gostekhizdat, Moscow 1955
138. Ovsyannikov, L.V.: Group Analysis of Differential Equations. Nauka, Moscow 1978

300 Bibliography

139. Oseen, C.W.: Ark. Mat. Astr. Fys., No. 7 (1911)
140. Bellamy-Knights, P.G.: J. Fluid Mech., v. 41, No.3, p. 673 (1970)
141. Pukhnachev, V.V.: Dokl. Akad. Nauk SSSR (Rep. Acad. Sc. USSR), v. 202, No. 2, p. 302 (1972)
142. Byteev, V.O.: Zhurn. Prikl. Mekh. Tekhn. Fiz. (J. Appl. Mech. Techn. Phys.), No. 6, p. 56 (1972)
143. Kochin, N.E., Kibel, I.A., Roze, N.B.: Theoretical Hydrodynamics. 6th rev. ed. Fizmatgiz, Moscow, pts. 1 & 2 (1963)
144. Birkhoff, G.: Hydrodynamics; a study in losic, fact and similitude. Dover Publications, New York 1955
145. Birkhoff, G., Zarantonello, E.H.: Jets, Wakes and Cavities. Academic Press, New York 1957
146. Sobolev, S.L.: Izv. Akad. Nauk SSSR: Ser. Mat. (News Acad. Sc. USSR: Math. Ser.), v. 18, No. 1, p. 3 (1954)
147. Whitham, G.B.: Linear and Nonlinear Waves. Wiley, New York 1974
148. Kapitansky, L.V.: Zap. Nauchn. Sem. LOMI (Notes Scient. Sem. LOMI), v. 84, p. 89 (1979)
149. Kulikovsky, A.G., Lyubimov, G.A.: Magnetic Hydrodynamics. Nauka, Moscow 1963
150. Kadomtsev, B.B.: Collective Phenomena in Plasma. Nauka, Moscow 1976
151. Stoker, J.J.: Water Waves; the mathematical theory with applications. Interscience Publishers, New York 1957
152. Shklovsky, I.S.: Supernovae and related Problems. Nauka, Moscow 1976
153. Dirichlet, G.L.: J. und angew. Math., v. 58, No. 1, p. 181 (1860)
154. Lamb, H.: Hydrodynamics. Dover Publications, New York 1945
155. Chandrasekhar, S.: Ellipsoidal Figures of Equilibrium. Yale University Press, New Haven 1969
156. Sedov, L.I.: Dokl. Akad. Nauk SSSR (Rep. Acad. Sc. USSR), v. 90, No. 5, p. 735 (1953)
157. Lidov, M.L.: Dokl. Akad. Nauk SSSR (Rep. Acad. Sc. USSR), v. 97, No. 3, p. 409 (1954)
158. Ovsyannikov, L.V.: Dokl. Akad. Nauk SSSR (Rep. Acad. Sc. USSR), v. 111, No. 1, p. 47 (1957)
159. Dyson, J.F.: J. Math. Mech., v. 18, No. 1, p. 91 (1968)
160. Nemchinov, I.V.: Prikl. Mat. i Mekh. (Appl. Math & Mech.), v. 29, No. 1 (1965)
161. Anisimov, S.I., Lysikov, Yu.I.: Prikl. Mat. i Mekh. (Appl. Math. & Mech.), v. 34, No. 5, p. 926 (1970)
162. Anisimov, S.I., Inogamov, N.A.: Pis. v Zhurn. Eksp. i Teor. Fiz. (Lett. J. Exp. & Theor. Phys.), v. 20, No. 3, p. 174 (1974)
163. Lynden-Bell, D.: Proc. Cambridge Phys. Soc., v. 58, No. 4, p. 709 (1962)
164. Zeldovich, Ya. B.: Astronom. Zhurn. (Astron. J.), v. 41. No. 5, p. 873 (1964)
165. Zeldovich, Ya. B.: Astrofizika (Astrophysics), v. 6, No. 2, p. 319 (1970)
166. Nevzglyadov, V.G.: Theory of a Body with Homogeneous Deformation and its Application to Atomic Nuclei. DGU Press, Vladivostok 1970
167. Andreev, V.K., Pukhnachev, V.V.: Zhurn. Prikl. Mech. Techn. Fiz. (J. Appl. Mech. Techn. Phys.), No. 2, p. 25 (1979)
168. Kulikovsky, A.G.: Dokl. Akad. Nauk SSSR (Rep. Acad. Sc. USSR), v. 114, No. 5, p. 984 (1957)
169. Fujimoto, M.: Astroph. J., v. 152, No. 2, p. 523 (1968)
170. Hara, T., Matsuda, T., Nakasawa, K.: Progr. Theor. Phys., v. 49, No. 2, p. 460 (1973)
171. Landau, L.D., Lifshitz, E.M.: Mechanics. Nauka, Moscow 1965 (trans: Addison-Wesley)
172. Vladimirov, V.S.: Equations of Mathematical Physics. Nauka, Moscow 1971 (trans: M. Deckker Co.)
173. Sretensky, L.N.: Theory of Newtonian Potential. OGIZ, Moscow-Leningrad 1946
174. Novikov, I.D.: Astronom. Zhurn. (Astron. J.), v. 52, No. 5, p. 1038 (1975)
175. Toda, M.: Progr. Theor. Phys. Suppl., v. 45, p. 174 (1970)
176. Henon, M.: Phys. Rev., v. B9, p. 1921 (1974)
177. Flascka, H.: Phys. Rev., v. B9, p. 1924 (1974)
178. Manakov, S.V.: Zhurn. Eksp. i Teor. Fiz. (J. Exp. & Theor. Phys.), v. 67, No. 2, p. 543 (1974)
179. Dubrovin, B.A., Matveev, V.B., Novikov, S.P.: Usp. Mat. Nauk (Succ. Math. Sc.), v. 31, No. 1, p. 55 (1976)

180. Cassati, G., Ford, J.: Phys. Rev., v. A12, p. 1702 (1975)
181. Lax, P.D.: Comm. Pure Appl. Math., v. 21, p. 467 (1968)
182. Olshanetsky, M.A., Perelomov, A.M.: Inv. Math., v. 37, p. 93 (1976)
183. Moser, J.: Adv. Math., v. 16, p. 197 (1965)
184. Calogero, F., Ragnisco, O., Matchioro, C.: Lett. Nuovo Cimento, v. 13, p. 383 (1975)
185. Olshanetsky, M.A., Perelomov, A.M.: Preprint ITEPh-157, Moscow 1978
186. Reiman, A.G., Semenov-Tyan-Shansky, M.A., Frenkel, I.B.: Dokl. Akad. Nauk SSSR (Rep. Acad. Sc. USSR), v. 247, No. 4, p. 802 (1979)
187. Kostant, B.: The solution to a Generalized Toda Lattice and Representation Theory. Advance in Math., v. 34, pp. 195–338 (1979)
188. Obukhov, A.M.: Dokl. Akad. Nauk SSSR (Rep. Acad. Sc. USSR), v. 184, No. 2, p. 309 (1969)
189. Dolzhansky, F.V., Klyatskin, V.I., Obukhov, A.M., Chusov, M.A.: Non-linear Systems of Hydrodynamical Type. Nauka, Moscow 1974

C. Godbillon

Dynamical Systems on Surfaces

Translated from the French by
H. G. Helfenstein

Universitext

1983. 70 figures. VII, 201 pages
ISBN 3-540-11645-1

Contents: Vector Fields on Manifolds. – The
Local Behaviour of Vector Fields. – Planar
Vector Fields. – Direction Fields on the Torus
and Homeomorphisms of the Circle. – Vector
Fields on Surfaces. – Bibliography.

This volume presents results both old and
new on the qualitative study of ordinary differential equations – mostly in two dimensions –
and concentrates in particular on local
behavior in a neighborhood of a singular point
and of a periodic orbit, Poincaré-Bendixson
theory on surfaces, direction fields on the
torus, and diffeomorphisms of the circle.
These questions have consistently been of
interest ever since Poincaré's work in the
1880's, and this interest has been renewed by
the recent remarkable development of the
theory of dynamical systems. The approach in
this book is very much inspired by the new
geometric methods developed in the theory of
foliated manifolds and the book will thus be
valuable as an introduction not only to
dynamical systems but also to foliations.

Springer-Verlag
Berlin
Heidelberg
New York
Tokyo

V. I. Arnold

Catastrophe Theory

Translated from the Russian by R. K. Thomas

1984. 65 figures. IX, 79 pages
ISBN 3-540-12859-X

Contents: Singularities, Bifurcations, and Catastrophe Theories. – Whitney's Singularity Theory. – Applications of Whitney's Theory. – A Catastrophe Machine. – Bifurcations of Equilibrium States. – Loss of Stability of Equilibrium and the Generation of Auto-Oscillations. – Singularities of Stability Boundaries and the Principle of the Fragility of Good Things. – Caustics, Wave Fronts and Their Metamorphoses. – Large Scale Distribution of Matter in the Universe. – Singularities in Optimization Problems, the Maxima Function. – Singularities of Accessibility Boundaries. – Smooth Surfaces and Their Projections. – Problems of By-Passing Obstacles. – Symplectic and Contact Geometries. – The Mystics of the Catastrophe Theory. – References.

Catastrophe Theory is a new field; and its value has become an issue of heated controversy, not only among specialists but also in the popular press. It has been called a "revolution in mathematics" comparable with Newton's invention of the differential and integral calculus. While Newtonian theory only considers smooth continuous processes, **Catastrophe Theory** provides a universal method for the study of jump transitions, discontinuities, and sudden quantitative changes. This little book, translated by R. K. Thomas, from the Russian original, clearly explains what **Catastrophe Theory** is and why it has aroused considerable fervor among defenders and detractors. The book also contains uncontroversial results from the mathematical theories of singularities and bifurcation. Among the aspects described are studies of the shapes of caustics and wavefronts and of their metamorphoses, of the universal large-scale structure, of optimal control problems and of the calculus of variation singularities, of the singularities of visual contours, and of symplectic and contact geometries.

The author, a leading Soviet mathematician, illustrates fundamental results and their far-reaching applications in a style which will make sense to readers with a minimal background in mathematics. The book includes a series of remarkable figures, which form an integral part of the text.

Springer-Verlag
Berlin
Heidelberg
New York
Tokyo